Roadwork

Theory and Practice

Fifth Edition

Peter Kendrick
Malcolm Copson
Steve Beresford
Paul McCormick

ELSEVIER
BUTTERWORTH
HEINEMANN

AMSTERDAM • BOSTON • HEIDELBERG • LONDON • NEW YORK • OXFORD
PARIS • SAN DIEGO • SAN FRANCISCO • SINGAPORE • SYDNEY • TOKYO

Elsevier Butterworth-Heinemann
Linacre House, Jordan Hill, Oxford OX2 8DP
30 Corporate Drive, Burlington, MA 01803

First published 1981
Reprinted 1982
Second edition 1988
Reprinted 1989
Third edition 1991
Reprinted 1993, 1995, 1997
Fourth edition 1999
Reprinted 2001, 2002
Fifth edition 2004

British Library Cataloguing in Publication Data
A catalogue record for this book is available from the British Library

Library of Congress Cataloguing in Publication Data
A catalogue record for this book is available from the Library of Congress

ISBN 0 7506 6470 3

For information on all Elsevier Butterworth-Heinemann
publications visit our website at http://books.elsevier.com

Printed and bound in Great Britain by Biddles Ltd, Kings Lynn, Norfolk

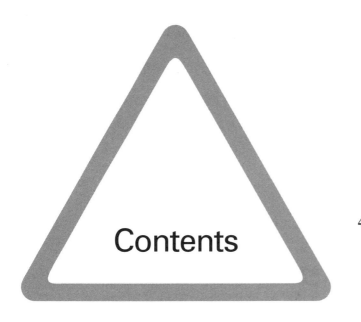

Contents

Preface

The aim of this book is to provide an explanation of the basic theory for work in road construction and related activities. In addition, it provides examples of the practical application of the basic theory.

It is intended to be of particular value to persons preparing for National Vocational Qualifications (NVQ), Business and Technician Education Council (BTEC) Courses and also undergraduates undertaking further studies in civil engineering. The content will also benefit students and engineers seeking Chartered, Incorporated or Technician Engineer status. Those seeking Construction Skills Certification through the NVQ route will also find this book to be of enormous value.

Roadwork: Theory and Practice fills the gap between design and construction techniques and provides an interesting and informative introduction to the great variety of construction and maintenance work that the term 'roadwork' embraces. The fifth edition has been updated to include new European standards for definition of terms and changes in specifications and technology.

The authors wish to express their thanks and gratitude to all those engineers, colleagues and friends who have so willingly contributed their time and knowledge in suggesting improvements to the text and especially, for the forebearance of our wives.

Particular thanks go to David Fanthorpe for updates to the materials and testing section, together with Andy Hurrock, Senior Engineer, Carillion WSP for the Temporary Traffic Management drawings produced using 'Cone 7.0'.

Peter Kendrick
Malcolm Copson
Steve Beresford
Paul McCormick

1
Highway development

1.1 Introduction: Early Road Systems

The need for roads stems from the invention of the wheel some 5000 years ago, probably originating in Samaria (about 3000 BC). In Britain, the earliest wheels date back to the Bronze Age. Some brief notes on early roads in various parts of the world are as follows:

(a) Chinese Civilization

One of the earliest and best known roads was the Chinese Silk Route which dates back to 2600 BC. The Chinese discovered the secret of silk weaving and sent this precious material by road to India and returned with ivory tusks.

(b) Persian Empire

This was a great trading organization. Silk imported from China was re-exported to Europe along the roads they had built. They also sold Chinese porcelain and precious woodware.

(c) Britain 2500 BC

A log-raft type of road has been discovered; this crosses the Somerset peat bogs to Glastonbury dating back to 2500 BC. The Berkshire Ridgeway was used to bring flint axes and weapons from Grimes Graves in Norfolk over the Chiltern and Berkshire Downs and Salisbury Plain to Stonehenge.

(d) Europe

In the low countries log roads similar to the Somerset ones have been found and there is evidence of the same type of road existing in the Swiss Lakeside Villages and across the Pangola Swamps in Hungary.

(e) India

Their early civilization was centred around the Indus Valley where archaeologists in 1922 discovered roads constructed of bricks and proper piped surface water drainage systems.

(f) Mesopotamia and Egypt

Moving to the Middle East and forward in time to about 1100 BC, Syrian troops constructed a new road through the mountains of northern Mesopotamia. Streets paved in asphalt and brick have been found in the Cities of Nineveh and Babylon. The Egyptians built roads to cart the stone required to construct the pyramids.

(g) Great Britain

Even before Christ, the Iberians and Celts were quite active constructing trackways and a good example is the Wyche cutting which was part of the salt route from Droitwich to Wales. This crossed the Malvern Hills.

(h) Roman Roads

The Roman era was undoubtedly the greatest road building age not only in Britain but throughout Europe. Five thousand miles of their superb highways stretched from Cadiz on the west coast of Spain through France, Germany, Italy, the Adriatic coast to Turkey, through Syria at the eastern end of the Mediterranean, back along the north coast of Africa via Alexandria, Carthage and so on to Tangier to complete the loop.

Their roads were renowned for their straightness but they were only straight in most cases between one hill top and another, i.e. as far as the eye could see. There is less chance of ambush on a straight road and the use of four-wheeled wagons posed no problems. (They had not learnt to pivot the front axle.)

Roman roads were generally constructed well above the ground level, being in some cases on embankments up to 2 m high. The first operation was to cut deep ditches or fosses (hence Fosseway) and then build up an embankment with layers of chalk, flint, sand and gravel topped off with huge stone slabs. Any marauder would have to cross the ditches and scramble up the embankment first.

Three classes of road structure were used by the Romans, these were:

1 Levelled earth
2 Gravelled surface
3 Paved (*see* Figure 1.1).

This conforms roughly with current road structure (i.e. four layers). The carriageway width seldom exceeded

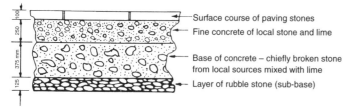

Figure 1.1 Roman road structure

4.25 m. The carriageway had drainage ditches on each side. After the withdrawal of the Romans from Britain at the start of the fifth century AD (AD 407) their road system fell into decay and disuse.

1.2 Terms Used and Their Derivations

1.2.1 Road

This is a recent term only used since the Civil Wars. It is derived from 'Ride' then 'Rode'.

1.2.2 Street

The Romans called their roads VIA STRAETA meaning of course a route or way that had been built up in layers. In time the *via* was dropped and *straeta* became STREET. It is town roads that tend to be called streets because in mediaeval times, and indeed up until the sixteenth century, only the roads in towns were paved.

1.2.3 Pavement

The top layer of a Roman road consisted of large stone slabs and these were referred to as the PAVIMENTUM. A lot of stone slabs were used for FOOTWAYS in towns in Victorian times and now, of course, concrete flags are used, hence the term PAVEMENT, although strictly, the pavement is the actual road structure.

1.2.4 Way

This word, originating from the Dark Ages, is a most useful one as it can be prefixed by almost anything and a thumbnail sketch of the History of Roads can be written around it.

In Neolithic times the tracks usually ran along the top of escarpments and were known as RIDGEWAYS of which the Berkshire Ridgeway was the most famous. One had to get down off the top of the hill some time and in so doing the side of the hill became eroded and a hollow formed, hence the HOLLOWAYS – such as Allesborough Hill, Pershore, Kempsey Common, Green Street (The Holloway).

In mediaeval times the only all-weather all-year-round routes were established on high ground to avoid drainage problems, and the principal routes were, therefore, known as HIGHWAYS. During the summer months one could cut across marshes or river meadows and these routes which were of less importance, for obvious reasons were the BYWAYS.

By the eighteenth century wheeled vehicles such as the stage coaches and carriages were in much more frequent use and it soon became necessary to segregate them from the pedestrian traffic and so the highway, particularly in towns, was divided up to give a central CARRIAGEWAY with FOOTWAYS on either side. Routes that were inadequate for wheeled traffic but quite suitable for anyone on horseback were naturally BRIDLEWAYS.

The early nineteenth century saw the first canals in this country and even today they are still officially classed as WATERWAYS. The Americans were very far sighted and started landscaping roads as early as in 1923 and called them PARKWAYS. In the early 1950s congestion of trunk roads was becoming a serious problem and so the Ministry of Transport forbade stopping on certain lengths designated as CLEARWAYS. This did not do the trick and special roads later to become MOTORWAYS had to be built. The Americans called these roads EXPRESSWAYS.

1.3 The Transition Period

When the Romans left, the Britons settled back into self-sufficient communities – they had no desire to travel and hence no maintenance was carried out and woods began to encompass roads. A few Roman roads survived and King Harold, on hearing of the landing of the Conqueror, was able to ride from York to London in four days.

On the Continent the picture was much the same. Numerous small and often warring states were formed and they had neither the will nor the technology to keep their Roman roads repaired. Throughout history it has been shown that a powerful central authority seems essential for a good road network.

In the early tenth century conditions in Europe became more settled, trade developed and, therefore, overland travel began to expand. Routes from Italy to the towns of Flanders were busy carrying considerable quantities of merchandise.

1.3.1 Britain

At home mediaeval man became short of stone and plundered the Roman roads for building material. Roads declined into packhorse tracks, disappearing almost completely in boggy winter conditions.

In Henry II's time, however, the ancient shire towns such as Oxford, Winchester, Northampton and so forth became important administrative centres – courts were held there and the justices travelled about the country.

Politicians started making whistle-stop tours, London had begun to benefit from the upsurge in European trade and traffic was thus increasing with a consequent increase in the need for adequate roads.

By the thirteenth century Englishmen began to improve their communications but their efforts were directed largely to bridge building. A new stone bridge, London Bridge, was constructed to replace the earlier timber one and King John spanned the Avon with a new bridge at Tewkesbury. Road maintenance was still carried out, however, by the Lords of the Manor and the Monasteries. The Church used its influence and urged its members to maintain their roads as a pious duty like visiting the sick. Piers Ploughman, the poet and philosopher, exhorted the charitable to repair their wicked ways: the ways in this instance means 'roads'. Chapels on Old London Bridge were erected and there is still one at Bradford-on-Avon in Wiltshire.

Mediaeval Englishmen had three primary civic duties:

1 Defence of the Realm
2 Repair of Bridges
3 Maintenance of the Road.

1.3.2 The Tudor Government

The Government became very concerned about the condition of the country roads on which virtually no maintenance was done following the dissolution of the monasteries. They put an Act through Parliament in 1555 under which every Parish had to elect a surveyor annually (the forerunner of the Divisional Surveyor). His job was to nominate four days between Easter and Whitsuntide when every landowner had to provide two men with teams of horses and tools to MEND THEIR WAYS. He stopped wagons drawn by more than the statutory number of horses. This system of forced labour was not very satisfactory with everyone busy on farms at this time of year. Pepys, Defoe and Bunyan all wrote complaining bitterly about road conditions in their days. Every journey was an adventure, often with mud up to the bellies of the animals.

1.3.3 Europe

Henry IV of France appointed a Grand Vizier with responsibility for all main roads and Louis XIV set up perhaps the first Roads and Bridges Department. He included roads in his General Economic Policy.

Germany had three hundred separate states or independent towns, therefore no progress was made, but Switzerland, Austria, and Italy had, by the seventeenth century, good roads.

There was not much commerce in the early middle ages, but under the rule of the Stuarts throughout most of the seventeenth century there was a large increase in trade. The roads were inadequate to cope with it. This, therefore, brings us to the second important landmark since Roman times; this was the Turnpike road system.

1.3.4 Turnpikes

The 1663 Act provided for the first Turnpike to be constructed in Hertfordshire. These roads were provided on the principle that those who used the roads paid for them rather than the cost being borne by local people.

The first Turnpike was established by an Act of Parliament in 1706. These Turnpike roads eventually extended to some 22 000 miles of road which were controlled by 1100 Trusts. As examples of such Turnpikes, in 1779 Bromyard Trust had a working capital of £450 and the same year Broadway Trust reported 900 four-wheeled vehicles and 1700 two-wheeled vehicles in the year. The Droitwich and Bromsgrove Trust fined their Surveyor 20 shillings (£1) for allowing streets in Droitwich to become dangerous.

At this time the Scots remained quite content to take coal, fish, salt, and oatmeal by pack horses, showing no desire to use wheeled vehicles. Whilst numerous Turnpike Trusts were being set up in England, General Wade (1723), who was in charge of the English Army in Scotland after the first Jacobite rising, started building new roads to open up the Highlands, using military labour as the early Syrians and Romans had done. Wade used gunpowder to blast his way through the rocks.

There is an inscription on a stone outside Fort William which reads:

'Had you seen these roads before they were made,
You would hold up your hands and bless General Wade.'

1.3.5 France

Jean Radolphe Perronet, an eighteenth century Frenchman, discovered a Roman pavement in his back garden and applied the same construction principle to the Paris streets and elsewhere in France. Many had metalled, straight, well-drained roads before the Revolution.

1.4 In Later Years

During the eighteenth century, Britain's modern road builders began their work. The best known and most important of these were as follows.

1.4.1 Robert Phillips

Phillips was the real pioneer of road design. In 1736 he presented a paper to the Royal Society entitled 'A dissertation concerning the Present State of the High Roads of England'. This contained the suggestion that on the clay and gravel roads then existing, a layer of gravel, if resting on a well-drained base, would be beaten by traffic into a solid road surface.

1.4.2 John Metcalf

He was known as 'Blind Jack of Knaresborough'. Born in 1717, he was blinded by smallpox at the age of six. After

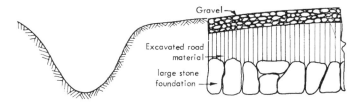

Figure 1.2 Metcalf's road structure

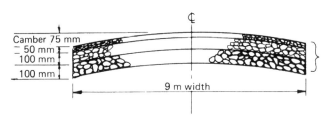

Figure 1.4 Macadam's road structure

a varied life as musician, soldier, waggoner, and horse trader, he became a road maker at the age of forty.

In 1765 he was given a contract to construct 5 km of turnpike between Harrogate and Boroughbridge. This was so successful a road that he eventually built 290 km of roads in Yorkshire, including all bridges, culverts, retaining walls, etc.

He insisted on good drainage and a good foundation (now referred to as 'sub-base'). The carriageway was arched to assist surface water drainage. (*See* Figure 1.2.)

For crossing boggy/soft ground, Metcalf introduced a sub-base raft of bundled heather.

1.4.3 Thomas Telford

Born in 1757, Telford began his career as an apprentice stonemason, eventually becoming one of the greatest civil engineers of all time.

He designed 1600 km of new roads in the Scottish Highlands, also the Caledonian Canal, the Menai Suspension Bridge and many other major civil engineering works in Britain and other parts of Northern Europe.

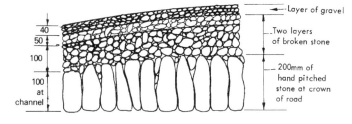

Figure 1.3 Telford's road structure

Telford's form of road construction is shown in Figure 1.3. N.B.: The formation is horizontal, the sub-base formed to camber, and other layers of even thickness.

1.4.4 John Macadam

Born in 1756, Macadam was the first true highway engineering specialist. He went to America, returning in 1783 to his native Scotland. Although interested in roads

and their improvement, he did not take an active part in their construction until 1816 when he became Surveyor to the Bristol Turnpike Trust. By 1826 he had become Advisor to the Government on Road Construction.

Macadam's construction method differed from Telford's in that the formation was shaped to the road camber and was a cheaper form of construction (*see* Figure 1.4).

1.5 Road Development

Due to the efforts of Macadam and Telford, there was a vast improvement in communications, roads being provided to suit the needs of the traffic – for the first time in many years. This traffic included the first experimental steam carriages.

The development of the railways at this time, however, led to a reduction in the number of people travelling by stagecoach. This reduction caused a loss of revenue to the Turnpike Trusts, which consequently became insolvent and could no longer carry out their duties of repairing the roads. Due to this, the liability for road maintenance again fell upon the parish.

Early in the second half of the nineteenth century Turnpike Trusts were wound up and roads passed to Parish Councils and Road Boards and later to County Councils, Urban District Councils and Rural District Councils, etc.

1.6 Highway Legislation

1.6.1 Highways Act 1835

This Act retained the parish as the administrative unit for all roads, other than turnpikes, within its boundaries. It also authorized the formation of groups of parishes into Joint Highway Boards. Salaried officers were appointed and 'statute' labour was replaced by hired labour, but this was only operated on a very small scale.

1.6.2 Highways Act 1862

This gave Local Justices the power to combine parishes into 'Highway Districts' in an effort to improve road conditions by enlargement of the impossibly small administrative area of a single parish.

1.6.3 Red Flag Act 1865

This allowed maximum vehicle speeds of 2 mph in towns and 4 mph in the country.

1.6.4 Government grant of £200 000 per annum

This grant was made from 1876 to ease the ratepayers' burden caused by their being required to maintain the former turnpike roads.

1.6.5 The Highways and Locomotives Act of 1878

This Act adjusted the burden between different parishes by requiring half of the maintenance costs for the former turnpike roads to be met from county funds.

1.6.6 1880

The first motor vehicles appeared on roads.

1.6.7 Local Government Act 1888

This created the County Councils which were required to pay the full maintenance costs for all main roads in the country, except for those within the County Boroughs.

1.6.8 1890

The first steam rollers – which improved road surfaces and made travel by pneumatic-tyred vehicle possible.

1.6.9 Local Government Act 1894

This Act abolished both Highway Parishes and Highway Districts and constituted the Urban and Rural District Councils in an attempt to give wider areas of administration for the up-keep of minor roads.

1.6.10 Repeal of the 'Man and Red Flag Act' 1896

The lifting of the speed restrictions began the modern era of the motor vehicle.

1.6.11 Development and Road Improvement Act 1909

This Act created the Road Board, the first national road authority in the country, which had at its disposal the income from motor vehicle licences and petrol tax.

1.6.12 1919

The Ministry of Transport was created.

1.6.13 1920

Government Grants to Local Authorities for road work began. Tar spraying of roads started to reduce the problem of dust created by pneumatic-tyred vehicles.

1.6.14 1930s

Unemployment relief schemes resulted in the construction of the Kingston Bypass, the Liverpool, Manchester, East Lancs Road, and Mersey Tunnel.

1.7 Mid-twentieth Century Roads, Prior to 1940

The outstanding development at this time was the construction of dual carriageway roads, with controlled access at infrequent intervals.

In America, these are called 'superhighways' or 'throughways'. In Germany (in the 1930s) they were called 'autobahnen', and in Britain they are known as 'motorways'.

Included in the concept of this highway design are the following features:

(a) grade-separated junctions (roads crossing at different levels);
(b) elimination of steep gradients;
(c) elimination of sharp curves;
(d) reasonably direct routes between large centres of population with short link roads into these centres;
(e) blending of the highway into the natural features of the countryside through which it is routed.

The main advantages of motorways, or express highways, are:

(a) greater safety than ordinary roads;
(b) greater speed and higher average journey speeds;
(c) more comfort and convenience for drivers or passengers;
(d) lower vehicle operating costs.

In Italy, construction of 'autostrada' began in 1924. Although these were not up to the later standards required for express highways, they did include limited access and grade separated junctions. The autostrada were built and owned by private companies and paid for by tolls and advertising.

The first true express highways were the 'autobahnen' built in Germany in the 1930s. The ideas and plans for the system of autobahnen on a national network basis were formulated during 1930–32. By 1942 the system, known as 'reichsautobahnen' extended to about 2100 km. The military benefit of the system was probably the main reason behind this rapid construction.

Holland was the only other European country to undertake motorway-type construction prior to the 1939–45 war.

1.8 Road Construction After 1940, in Great Britain

The Special Roads Act of 1949 provided for a network of about 1100 km of motorway. By the mid-1960s nearly 420 km of motorway had been built, the most important being the M1 (London to Birmingham) motorway which opened in 1959. The two lane dual carriageway, the M5 near Worcester, was one of the first motorways to be built, but this fell short of the usual requirement of three lanes per carriageway. Financial limitations imposed by the Government were responsible for the limit of two lanes, the motorway having been designed as a three lane dual carriageway.

By 1986 the planned motorway system was virtually complete. The emphasis is now on improving existing trunk roads by converting them into dual carriageways, by constructing bypasses around main urban areas and by traffic management schemes.

In the 1970s and 1980s, severe cuts in Government and Local Authority expenditure on roads slowed down the rate of development and also cut back on the essential maintenance of the existing road network. So severe were the cuts that the maintenance of roads to their existing standard was also seriously jeopardized.

In the late 1990s, multi-modal studies were commissioned. The outcome proposed additional motorways, the widening of existing motorways and the use of intelligent telematics to achieve maximum use of road capacity.

There are signs in the twenty-first Century that this underfunding is being reversed. However, the maintenance backlog is further compounded by a skills shortage in workforces.

1.9 Definition of Highway

In simple terms, a highway is defined as 'a way over which the public have a right to pass and repass as of right and not by sufferance or by licence'. It is an interesting point to note that a waterway can be a 'highway'.

Highways are classified according to the use made of them by the public and also according to the body, such as a County Council, which is responsible for their maintenance.

As from January 2002, changes to terminology for Highway construction layers were made which are now used in European Standards.

    'Surface course' – replaces – 'wearing course'
    'binder course' – replaces – 'base course'
    'base (roadbase)' – replaces – 'roadbase'
    'materials to BS 4887' – replaces – 'macadam'

For further details on material specifications and construction layers, refer to Section 15 NRSWA.

1.9.1 Creation of Highways

The most common methods are:

(1) by dedication, i.e. the owner of the land allows the public at large to pass;
(2) by statute, under the various provisions in the Highways Act 1980;
(3) by agreement or declaration as part of a development plan under a Town and Country Planning Act.

1.10 Classification of Highways by Use

Classification by use depends upon whether the public right to pass and repass may be exercised on foot only, on horseback or with vehicles, etc.

It is important to understand the difference between a footpath and a footway because of the special rules for the creation, diversion, etc. of footpaths which do not apply to footways.

A *footpath* is 'a highway over which the public have a right of way on foot only, not being a footway'. This means that it is across fields, through woods, etc. but not alongside a carriageway.

A *footway* is 'a way within a highway which also comprises a carriageway and is a way over which the public has a right on foot only'. In this case it is either directly alongside the carriageway, or generally parallel to it but separated from it by a verge.

Bridleways are highways over which the public has a right of way on foot or on horseback or leading a horse, but no other rights of way.

Cycle tracks are ways constituting or comprised in a highway over which the public has a right of way on pedal cycles with or without a right of way on foot.

Carriageway means a 'way constituting or comprised in a highway, being a way (other than a cycle track) over which the public has a right of way for the passage of vehicles'.

This type of highway, which may or may not include footways and/or cycle tracks, permits the widest number of rights of use by the public at large.

Motorways are special roads reserved for particular classes of 'traffic'. 'Traffic', as defined in Section 329 of the Highways Act 1980, includes pedestrians and animals. As these are generally excluded from special roads (motorways), this precludes them from being highways.

1.11 Highway Management

The legal aspects of management of the highway have become very complex and consist of a mixture of common law, local bylaws and the Highways Act 1980, The 1980

Highways Act, the main constituent of this mixture, comprises:

Part I   Highway Authorities and Agreements between Authorities
Part II   Trunk, Classified, Metropolitan and Special Roads
Part III   Creation of Highways
Part IV   Maintenance of Highways
Part V   Improvement of Highways
Part VI   Construction of Bridges/Tunnels/Diversion of Watercourses
Part VII   Provision of Special Facilities for Highways
Part VIII Stopping Tip, Diversion of Highways and Means of Access
Part IX   Lawful and Unlawful Interference with Highways and Streets
Part X   New Streets
Part XI   Making up of Private Streets
Part XII   Acquisition, Vesting and Transfer of Land etc.
Part XIII Financial Provisions
Part XIV   Miscellaneous and Supplementary Provisions Schedules

1.11.1 Highway Authorities

The majority of the non-strategic highway network in the UK is still managed by local Highway Authorities, which range from County Councils, Metropolitan Boroughs, District Councils or Boroughs.

The Highways Agency, an executive agency of the Department for Transport (DfT), is the Highway Authority for trunk and special roads, most of the latter being motorways outside London.

County Councils or Unitary Authorities used to have responsibility for maintaining and improving these roads within their area, which was delegated to them under an agency agreement until these arrangements were revised in 1997. The DfT than re-defined these areas across county boundaries and also invited consultancies to fulfil this role.

Within London, the Transport for London (TfL) is the Highway Authority for the existing trunk road network and primary strategic routes.

Revision Questions

1   What were the three classes of road structure used by the Romans?
2   Thomas Telford's method of constructing roads used which construction layer to form the camber?
3   What is the definition of a 'FOOTPATH'?
4   The Highways Agency is an executive agency of which Government department?

Note: See page 305 for answers to these revision questions

2 Maps, drawings & surveying

2.1 Types of Drawings

There are some five types of drawings commonly used in highway work. These are ordnance survey maps, plans, longitudinal sections, transverse sections and detail drawings.

2.1.1 Maps

Since the Système International (SI), now commonly termed the Metric System, has replaced the Imperial System of measurement, a restructuring of the map scales has taken place. The scales now used are:

1 : 50 000   (i.e. 1 mm to 50 m, or if preferred 1 m to 50 km)
1 : 25 000
1 : 10 000
1 : 2500
1 : 1250

Ordnance survey maps are available in a range of scales. Some of the earlier imperial scale maps are still in use:

Scale	Referred to as
1 : 63 360	1 inch to 1 mile
1 : 25 000	2½ inches to 1 mile (approximately)
1 : 10 560	6 inches to 1 mile
1 : 2 500	25 inches to 1 mile (approximately)
1 : 1 250	50 inches to 1 mile (approximately)

2.1.2 Plans

These are drawn to a larger scale than ordnance survey maps and are produced or drawn from a survey made of the site. The SI range of scales for sites and sections is:

1 : 500
1 : 200
1 : 100
1 : 50

The usual scale for a highway plan is 1:500 which means that 1 m on the plan represents 500 m on the site. *See* Figure 2.1 for an example of a highway plan.

2.1.3 Longitudinal Section

The purpose of a longitudinal section is to show the height or the elevation of the ground along a line which is the centre line of proposed highway (*see* Figure 2.2).

The horizontal scale is 1:500 which is the same as the plan and so its length on the drawing corresponds to the plan. Both the plan and the associated longitudinal section are often shown on the same drawing sheet. However, the vertical scale of the longitudinal section is usually exaggerated and drawn to 1:100 or 1:50, which allows the differences in height to be increased five or ten times and thus seen more clearly.

2.1.4 Transverse Section (see *Figure 2.3*)

This is a cross-section of the highway and is usually drawn to an exaggerated scale, the vertical scale being 10 × the horizontal scale. If shown as 1/100 natural it means both vertical and horizontal scales are the same. Transverse sections are numbered for easy identification and are usually grouped together on a separate drawing sheet.

2.1.5 Detail Drawing

This type of drawing is made when it is necessary to see all the details clearly. An example could be the detail of a culvert or perhaps the method of bedding a pipe in concrete at a certain point (*see* Figure 2.4).

The SI range of scales for detailed drawings is:

1 : 20
1 : 10
1 : 5
1 : 1 (actual size)

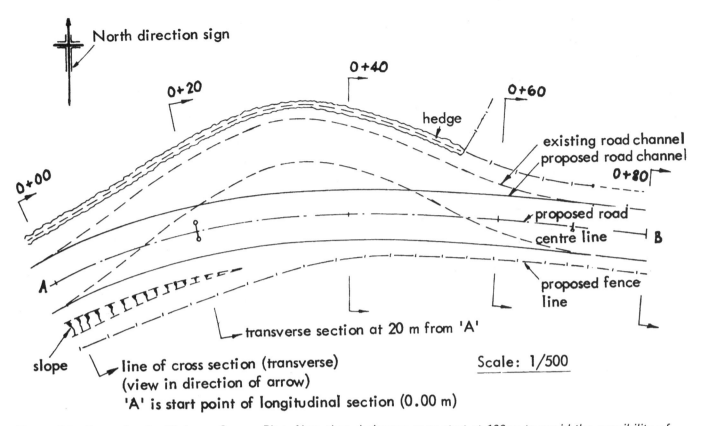

Figure 2.1 *Example of a Highway Survey Plan. Note that chainages may start at 100 m to avoid the possibility of negative chainage values. Also, in order to be computer compatible, chainages may be shown as 000, 020, rather than 0+00, 0+20, etc.*

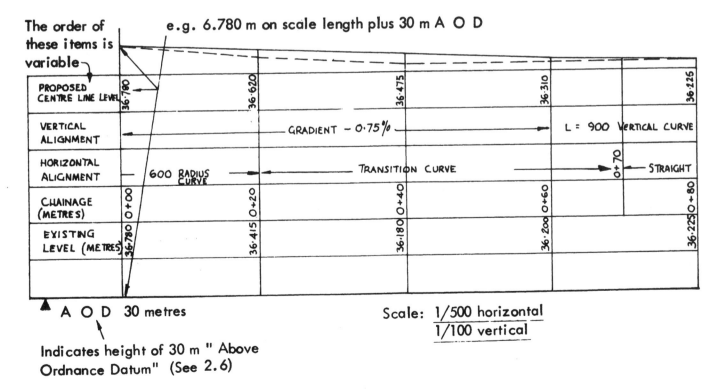

Figure 2.2 *Longitudinal section along line AB in Figure 2.1 above*

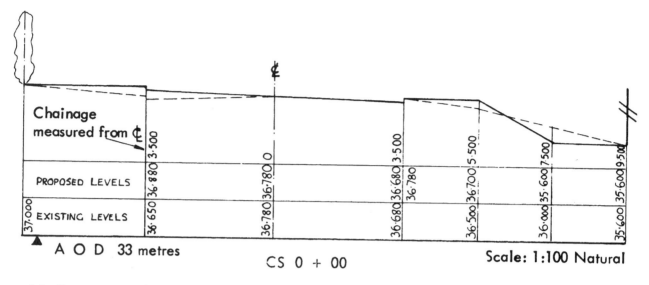

Figure 2.3 Transverse section

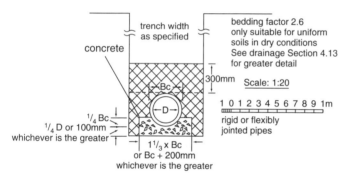

Figure 2.4 Detail of pipe bedding (reduced from true scale)

2.1.6 Some Signs and Symbols Used on a Highway Survey (Figure 2.5)

The increasing use of Autocad and Microstation has enabled the computer aided designer to quickly standardize details and produce varying additional sections/drawings.

2.1.7 Some Difficult Symbols and Terms Explained

Designers have varying drawing styles. Figures 2.6 and 2.7 may make clearer some of these differences.

The terms 'super elevation', 'balanced sections', etc., are used in designs illustrated in Figures 2.8(*a*) and (*b*). Alternatively, a technique known as the 'moving crown, is used (Figures 2.9(*a*) and (*b*)).

2.1.8 An Example of Part of a Typical Improvement Scheme

Figures 2.10(*a*), (*b*) and (*c*) are examples of the plan, longitudinal section and transverse section of the same

short length of road. These diagrams are not drawn to the scale shown.

2.1.9 Example of a Longitudinal Section

Consider the solution to a typical problem. The following table gives the existing ground levels and the proposed centre line levels for a section of new carriageway. The levels have been taken at 20 m intervals.

Ground levels (AOD)   20.500 22.750 23.000 20.000 18.500 21.250
Proposed ₵ levels (AOD) 21.500 21.200 20.900 20.600 20.300 20.000

(*a*) To a horizontal scale of 1:500 and a vertical scale of 1:100 draw the longitudinal section, showing the ground and centre line levels.

(*b*) State the depth of excavation or fill at each station.

(*c*) State the overall gradient of the section of proposed carriageway.

Answer

(*a*) (not to scale): *see* Figure 2.11.

(*b*) Excavation depths
     —   1.550 m 2.100 m   —   — 1.250 m
     Fill depths
     1.000 m   —   — 0.600 m 1.800 m   —

(*c*)

$$\text{Overall gradient} = \frac{\text{length}}{\text{height change}} = \frac{100}{1.500} = 66.67$$

say 1 in 66.7

$$\text{or expressed as a \%} = \frac{1.5}{100} \times 100 = \underline{-1.5\%}$$

(N.B: / = + ve, \ = −ve)

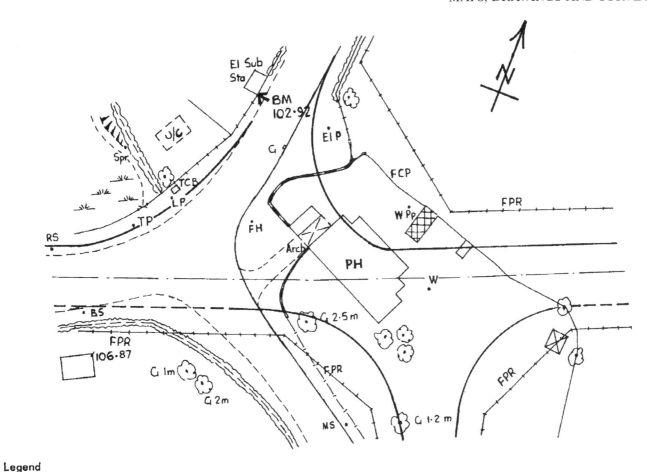

Legend

Buildings		**Walls, Fences, etc.**		**Street Furniture**	

Buildings

Glasshouse

Ruins

under construction

Roads

Kerbed

Unkerbed

Footpath   FP

Railways

Signal Box   SB

Signal Post   SP·

Level Crossing   LC

Water Features

Culvert   Cul

Sluice   Sl

Spring   Spr

Well   W·

Walls, Fences, etc.

Walls under 0.2 m

over 0,2 m

Retaining Wall   RW

Sloping Masonry

Walls, Fences, etc.

Chestnut Paling   FCP

Close Boarding   FCB

Post and Rail   FPR

Interwoven   F Int

Hedge

Miscellaneous Features

Chimney   Chy

Electricity Pole   EP·

Electricity Power Pole   El P·

Electricity Pylon   El Pylon

Electricity Sub Station El Sub Sta

Footbridge   FB

Gate

Marsh

Petrol Pump   P Pp·

Post   P·

Post Office   PO

Public House PH

Stile

Telephone Pole   TP·

Telephone Box   TCB

Tree and Girth   G 2m

Water Pump   W Pp·

Street Furniture

Belisha Beacon   BB·

Bollard   Boll·

Box (Electric)   Box Elec

Bus Stop   BS·

Fire Hydrant   FH·

Gas Stop Valve   Gsv·

GPO or POTele Cover   POT

Gully   G▫

Inspection Cover   IC

Lamp Post   LP·

Letter Box   LB

Manhole   MH

Mile Stone   MS·

Name Plate   NP

Pedestrian Crossing   PC

Police Call Box   PCB

Reflector Post   RP·

Road Sign   RS·

Telephone Call Box   TCB▫

Traffic Lights   TL

T.L. Contact Pad TLCP

Water Stop Valve Wsv·

Survey Information

Triangulation Station   △ TS 4

Traverse Station   ⊖ ts 2

O.S. Bench Mark   ⬆ BM 101·12

Spot Height   ⁻ 110·27

Figure 2.5   Highway plan 1:500

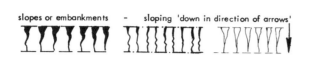

Figure 2.6   Variations in representation of symbols

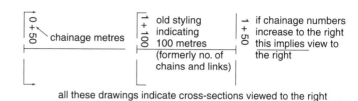

all these drawings indicate cross-sections viewed to the right

Figure 2.7   Cross-sections

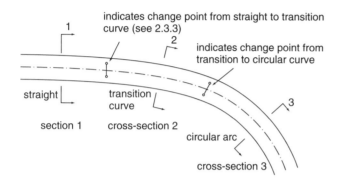

Figure 2.8(a)   Transition curve

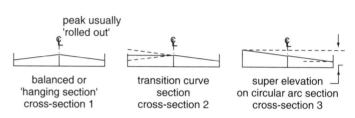

Figure 2.8(b)   Cross-sections of Figure 2.8(a)

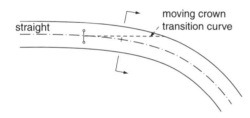

Figure 2.9(a)   Moving crown

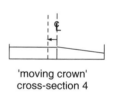

Figure 2.9(b)   Moving crown cross-section

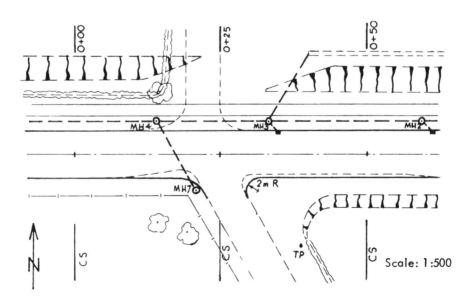

Figure 2.10(a)   Typical plan

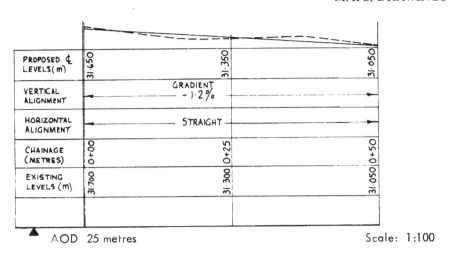

Figure 2.10(b) Longitudinal section

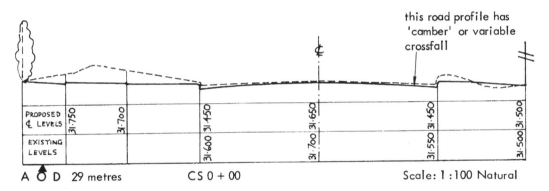

Figure 2.10(c) Transverse section

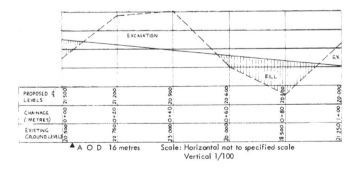

Figure 2.11 Longitudinal section

2.2 Simple Setting Out

The ability to set out a straight line and to take off, from a point on this line, a further line at right-angles to the first, is most useful knowledge for any road worker.

Setting out curves is more difficult, but with practice can also be undertaken.

2.2.1 Straight Line Setting Out

In setting out a straight line between two points, either ranging rods or steel pins may be used. Ranging rods are round wooden poles having a pointed steel shoe or tip to enable them to be driven into the ground. They vary from 2 to 4 m in length and are painted in red and white bands. The steel pins are from 900 mm to 1 m in length and are usually 20 mm in diameter.

(a) Use of Ranging Rods

Ranging rods are more clearly visible than pins and are used to establish a temporary line or to locate the position of a steel pin or wooden peg when more permanent marking is needed (*see* Figure 2.12). When using ranging rods, place the first and the last and eye-in the intermediates. If the rod

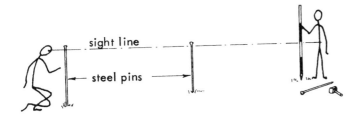

Figure 2.12   Sighting a line with ranging rods

is held loosely between the finger and the thumb near the top, and just clear of the ground, then its weight will cause it to hang vertically before it is driven in.

(b)   Ranging Over a Hill

When the first and last points are not visible to each other, a method of repeated alignment is used (*see* Figures 2.13(*a*) and (*b*)).

Figure 2.13(a)   Ranging over a hill

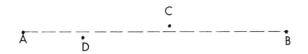

Figure 2.13(b)

1 Set out C and D as nearly as can be judged on line AB and so that C is just visible from A and D is visible from B.

2 Line in D, between A and C (Figure 2.14).

3 Line in C, between D, and B (Figure 2.15).

4 Continue the process until no further correction is visibly required – then A, B, C and D are correctly lined up.

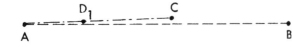

Figure 2.14

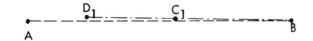

Figure 2.15

(c)   Use of Steel Pins (often known as Kerbing Pins)

Pins are used when more permanent marking is necessary and a nylon line is to be erected. After the first and last pins are placed, further pins are driven in the ground at both ends to act as anchor pins. The line is made taut on the anchor pins and tied off on the first, last and intermediate pins (*see* Figure 2.16). It is important that intermediate pins are accurately placed and are vertical. The method of placing and of adjustment is shown in Figure 2.17.

2.2.2   Right-Angled Off-Set from a Point on a Line

(*a*) By using the properties of a particular right-angled triangle (Pythagorean) which has sides in the ratio of 3, 4 and 5, an off-set line at 90° may be set out. Figure 2.18 shows the way in which a tape or line marked off at 3 m, 4 m and 5 m may be used. One man holds the 12 m mark to the end of the tape and places a pin at position C. The off-set is then at right-angles from the point D on the original line AB.

(*b*) Alternatively, mark off an equal distance each side of the point at which the off-set is required – say 3 m – then securing or holding the tape at A and B, pull a length of tape or line out and place pin at the centre of this length (*see* Figure 2.19). One man holds the tape at the 4 m mark and places a pin at point C.

(*c*) When an obstruction is in the line of sight the above method may be varied slightly (*see* Figure 2.20). The operation is carried out twice to position pins C and D. The line through the pins will give an off-set at right-angles to the original line.

2.2.3   Right-Angled Off-Set from a Point to the Line

(*a*) The method shown in Figure 2.18 can also be used to establish a right-angled off-set from a given point to the line AB.

(*b*) Alternatively, the tape is swung with its end at the given point and the minimum reading at which it crosses the line AB is noted. This happens when the tape is perpendicular or at right-angles to the line AB, but the method can only be used on smooth ground where a free swing of the tape is possible.

(*c*) Where the above method is not possible, with the end of the tape at the point C (*see* Figure 2.21) strike an arc to cut the base line at points A and B. The right-angled off-set is then found by joining C to the point D, which is the mid-point between A and B.

(*d*) Another method is to swing the tape with the end at the given point C, to cut the base line at A (preferably measuring an even number of metres) – see Figure 22.2(*a*). Then taking the mid-point D of the line CA, swing an arc having a radius of length $\frac{CA}{2}$ as shown in Figure 2.22(*b*).

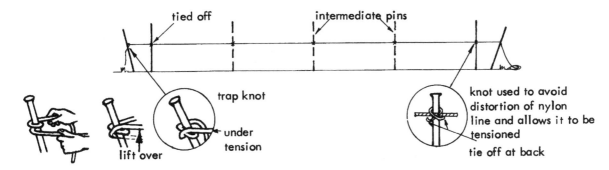

Figure 2.16   Setting a line using steel pins

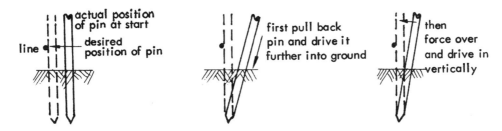

Figure 2.17   Placing intermediate pins to the line

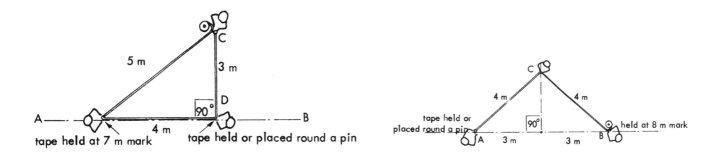

Figure 2.18

Figure 2.19

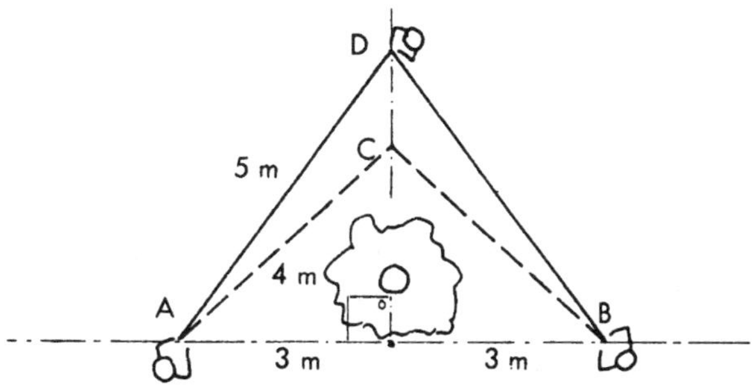

Figure 2.20

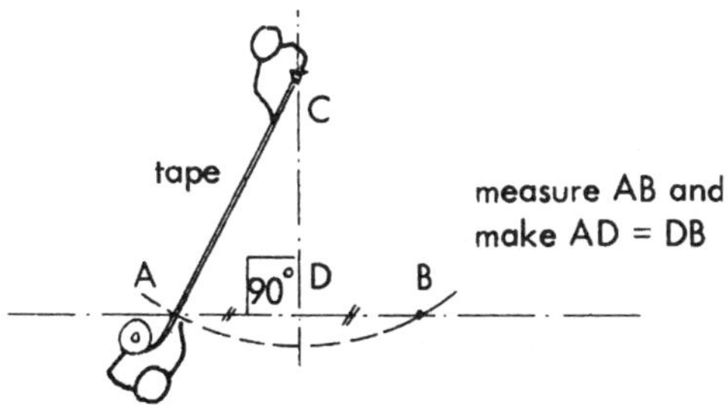

Figure 2.21

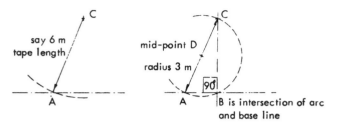

Figure 2.22(a)   Figure 2.22(b)

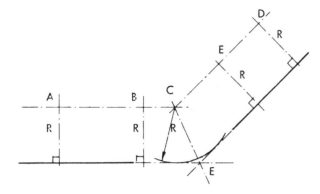

Figure 2.24 Setting out a corner radius for any angle

The right-angled off-set is then given by a line sighted between pins B and C (*see* Figure 2.22(*b*)).

2.2.4 Setting Out a Radiused Corner

Skilled road workers and supervisors should be able to swing a radius linking two straight lines and also put in a circular curve between two points, given the off-set distance.

(a) When Corner is at Right-Angles

If a 3 m radius is required, first mark out 3 m in both directions, then pull 6 m of tape by its middle from A to B.

The centre of the radius C will be found at the 3 m mark on the tape (*see* Figure 2.23).

(c) Alternative Method

Measure back from intersection point E the same distance to A and B. Then find point C by swinging tapes of the same length from A and B (*see* Figure 2.25). Sight through line CE and swing the desired radius R with its centre on line CE. The correct position to swing the radius will be found by trial and error.

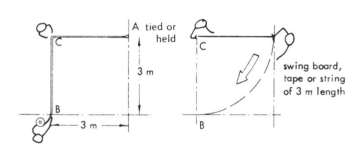

Figure 2.23 Setting out a right-angle corner radius

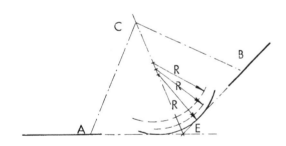

Figure 2.25 Setting out a corner radius for any angle

See Chapter 7, Kerbing, for alternative method where C is obstructed (see page 122).

(b) When the Two Lines are at any Angle
  (see *Figure 2.24*)

1 Erect perpendiculars to lines;
2 Measure desired radius R along perpendiculars;
3 Sight through A and B then D and E to fix point C;
4 Swing radius R from point C.

2.2.5 Curved Circular Arc – Given an Off-Set between Two Points (Halving and Quartering)

This technique is used for longer circular curves than illustrated in the previous examples on corners. The method known as 'halving and quartering' will quickly produce an approximately correct circular curve using a tape, line, and pins.

More accurate methods of curve ranging are shown in Section 2.3 and should be used for work other than just smoothing out, say, an existing line of kerbs.

Taking an example where two points 20 m apart have a central off-set of 1 m, the procedure is as shown in Figure 2.26(*a*)–(*f*):

(a) Set out the off-set of 1 m

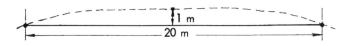

(b) Place the centre pin halfway and stretch the line over the pin

(c) Halve the 10 m lengths and quarter the off-set i.e. 1/4 m = 250 mm

(d) Stretch line over pins

(e) Halve the 5 m lengths and quarter the off-sets, i.e. 250 mm/4 = 62.5 mm

(f) Stretch line over pins and 'eye-in'

Finished line

Figure 2.26(a)–(f)

2.3 More Difficult Setting Out

This section is concerned with the more accurate techniques when setting out right-angles and curves. These methods will involve more difficult calculations.

2.3.1 Measuring Right-angles

Right-angles are usually set out using a theodolite. Except for the cross staff, the following instruments are still in use, but a laser version of the optical square is often found in the building trades.

(a) The Cross Staff (Out-moded Device)

Two types are shown in Figure 2.27, both being set up by sighting along the base line through one pair of slits. The line of sight at right-angles to the base line is then found by sighting through the second pair of slits whose axes are perpendicular to the first pair.

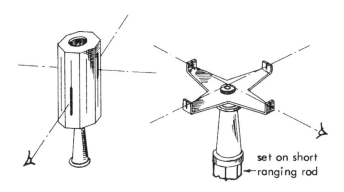

Figure 2.27 Cross staff

(b) The Site Square (see Figure 2.28)

This is basically two telescopes set with their axes at right-angles and mounted on a tripod. This instrument allows right-angles to be set out very quickly, and with an accuracy of 5 mm in 25 m.

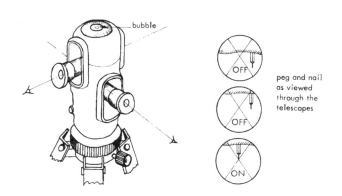

Figure 2.28 Site square

(c) The Optical Square

There are two types of optical square, one using two mirrors and the other a prism. These are more accurate than the cross staff and a more compact alternative to the site square.

The mirror type makes use of the fact that a ray of light reflected from two mirrors set apart 45° is turned through twice the angle between the mirrors, giving a right-angle. See Figure 2.29(a).

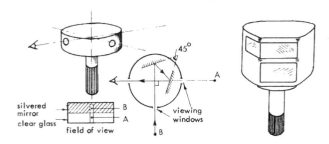

Figure 2.29(a) and (b)   Optical square

The surveyor stands at a point on the base line facing rod A on the line and signals his assistant to move rod B. When the image of B is in line with the bottom half of A, seen directly through the plain glass, then A and B form a right-angle with the instrument.

The prismatic type of optical square has a prism cut so that two faces contain an angle of 45°. It is used in the same way as the mirror square but is even more accurate. *See* Figure 2.29(*b*).

2.3.2   Curve Ranging

The setting out or ranging of curves varies from the approximate methods already given in Section 2.2, such as the 'halving and quartering' techniques for a circular arc (Method I), to the more accurate methods II and III described here. The very accurate methods using theodolites are beyond the scope of this section.

Method I

To establish a circular curve given the tangent points and the off-set distance: an example of this method is given as follows.

(*a*) Describe the setting out of a circular horizontal curve where the chord distance between tangent points is 140 m and the mid-chord off-set is 20 m.
(*b*) Calculate the size of the radius of the curve.

Answer (a)

1 The main chord is set out with a central offset (i.e. 70 m from each tangent point: TP) of 20 m. This off-set may be set out by the 3:4:5 triangle method.

2 The half chord is then set out between each tangent point and A (shown by—·—·—). Having halved the chord, the off-set must then be quartered (i.e. off-set = 20/4 = 5 m).

3 This off-set is set out from the mid point of A–TP.

4 Again halve the chord and quarter the off-set (shown by– – –), i.e. this off-set = 1.25 m (5/4).

5 Continue this reducing of the chords until an adequate number of points is obtained to set the curve accurately (*see* Figure 2.30(*a*)).

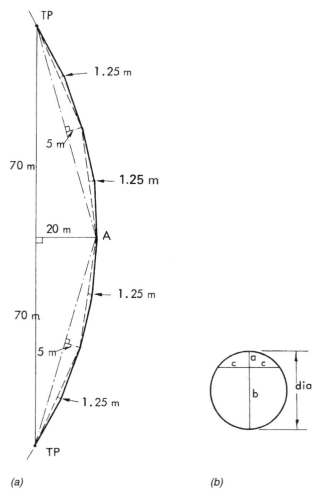

Figure 2.30(a) and (b)

(*b*) Using intersecting chord formula (*see* Figure 2.30(*b*)),

$$a \times b = c \times c$$

$$b = \frac{c^2}{a}$$

$$= \frac{70 \times 70}{20}$$

$$= 245 \text{ m}$$

So radius $= \dfrac{a+b}{2} = \dfrac{20 + 245}{2}$

$$= 132.5 \text{ m}$$

Method II

To establish a circular curve given the tangent points and the radius: this may be done by calculating the off-set distance and then setting out the curve as shown in Method I.

Consider the case shown where the tangent points are given and it is known that the required radius is 100 m. *See* Figure 2.31.

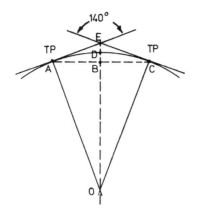

Figure 2.31

(*a*) Extend the line of each straight beyond the tangent points to meet at point E. The angle between the two intersecting lines may be measured and is, say, 140°.

(*b*) In the quadrilateral AECO

If AÊC = 140° and OÂE = OĈE = 90°
Then AÔC = 40°
(internal angles of a quadrilateral must total 360°).

(*c*) Consider the right-angled triangle OAB

AÔB = 20° (half of AÔC).
Use the ratio $\sin \text{A}\hat{\text{O}}\text{B} = \dfrac{\text{AB}}{\text{OA}}$
Then AB = OA sin AÔB
    = 100 sin 20°
    = 100 × 0.342.
Half chord = 34.2 m
so chord length AC = 68.4 m.

(*d*) Using Pythagoras in this triangle

OA² = AB² + OB²
100² = 32.2² + OB²
so OB² = 100² − 34.2²
    OB = √10 000 − 1 169.64
       = √8 830.36 = 93.069 m, say 93.970 m

Therefore the central off-set BD = 100 − 93.970 m (radius OD − OB) = 6.030 m.

(*e*) Now proceed as with Method I to set out the curve by halving and quartering.

Method IIIa

To establish a circular curve from one tangent point (working from either side of the curve): consider an example from a typical problem.
Using the formula,

$$\text{off-set} = \frac{\text{chord}^2}{2 \times \text{radius}}$$

describe how a horizontal curve of 50 m radius can be set out from one point of origin.

Answer (see Figure 2.32(a))
(*a*) Set out radius from the tangent point (TP), i.e. at right-angles to the kerb line.
(*b*) Calculate the off-sets '*y*' for various values of chord length '*c*': say, 4, 8, and 12 m.
(*c*) Using off-set $y = c^2/2r$, calculate the off-set '*y*' for each value of '*c*'.
(*d*) When

$$c = 4 \text{ m}, \quad y = \frac{4^2}{2 \times 50} = \frac{16}{100} = 0.160 \text{ m}$$

(*e*) Then measure 0.160 m from TP towards the centre of the circle and set up an off-set of 4 m at that point. This gives point A on the curve.

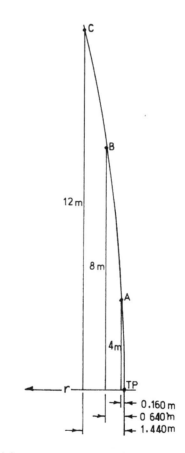

Figure 2.32(a)

(*f*) Repeat the procedure for increasing values of '*c*'.

$$\text{when } c = 8 \text{ m}, \quad y = \frac{8^2}{2 \times 50^2} = \frac{64}{100} = 0.640 \text{ m}$$

and when $c = 12$ m, $y = \dfrac{12^2}{2 \times 50} = \dfrac{144}{100} = 1.440$ m

(*g*) Note: This expression is an approximation and is only accurate up to angles of about 30° or where *c* is ¹/₄ (or less) of the value of the radius *r*. In this example *c* must be less than $\frac{50\,\text{m}}{4}$ or 12.5 m.

Method IIIb

To establish a circular curve from one tangent point (working from the outside of the curve on longer curves of larger radius): consider an example from a typical problem.

Using the formula,

$$\text{off-set} = \frac{\text{chord}^2}{2r}$$

describe how a horizontal curve of 120 m radius may be set out from one tangent point.

Answer (see *Figure 2.32(b)*)

(*a*) Choose a suitable chord length 'c' (but less than $\frac{r}{20}$): say 5 m, as $\frac{120}{20}$ m = 6 m.

(*b*) Extend straight past tangent point (TP) a distance of 'c'.

(*c*) Calculate off-set $\frac{5^2}{2 \times 120} = \frac{25}{240} = 0.104$ m.

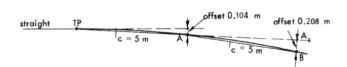

Figure 2.32(b)

(*d*) Intersection of 'c' and off-set 0.104 m gives point A.
(*e*) To establish point B and successive points, determine value of $\frac{c^2}{r}$ for further off-sets, i.e.

$$\frac{5^2}{120} = \frac{25}{120} = 0.208 \text{ m}$$

(*f*) Extend line from TP through point A for a further distance 'c' to point A_1 and fix point B by intersection of length 'c' (from point A) and off-set 0.208 m (from A_1).
(*g*) Continue as above to fix further points.
(*h*) Finally, as the closing chord is likely to be a fractional part of 5 m, say *p*, then a modification of the formula is used, i.e. off-set $= \frac{p(5 + p)}{2 \times 120}$.

2.3.3 Transition Curves

If a vehicle changes direction from a tangential line into a circular curve when the centrifugal force or side thrust

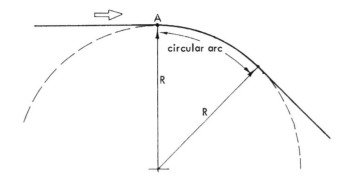

Figure 2.33

acting upon it will increase instantaneously from zero to its maximum value at point A (*see* Figure 2.33).

To prevent passengers in the vehicle experiencing a lateral shock as the tangent point is passed, a curve of variable radius is inserted between the tangential line and the circular curve. The effect of this is to allow the centrifugal force to be built up in a gradual and uniform manner and in this way the lateral shock is diminished. This variable radius curve is called a transition curve (*see* Figure 2.34).

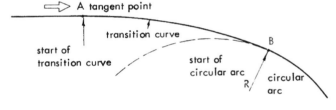

Figure 2.34

Figure 2.35(*a*) shows two transition curves joined together by a circular curve but in some cases the circular curve is replaced by two transition curves having one common tangent point (*see* Figure 2.35(*b*)). Setting out a transition curve is usually done by theodolite.

The setting out of transition curves has been greatly simplified by using computerized setting out data and a digital electronic distance measuring (EDM) theodolite

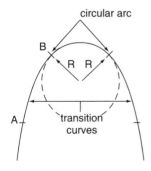

Figure 2.35(a) With circular arc

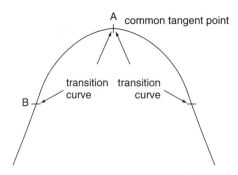

Figure 2.35(b) Without circular arc

via co-ordinates. However, this requires specialist surveying experience.

2.4 Surveying

The term 'surveying' describes the technique of taking measurements of the relative positions of man-made and natural features on the earth's surface and then drawing these measurements to some scale to make a map or plan. In highway surveying work the plan produced will be drawn from measurements made in two dimensions which form the horizontal plane, but surveying generally is divided into two branches; geodetic surveying and plane surveying.

In geodetic surveying the curvature of the earth's surface is taken into account and one such example is the Ordnance Survey of Great Britain. These maps are produced using the geometrical properties of a sphere. *See* Figure 2.36.

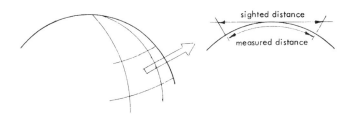

Figure 2.36 Geodetic survey principles

However, in plane surveying the highway site is taken to be a horizontal plane and the measurements plotted will represent the projection on the horizontal plane of the actual field measurements. If this technique is used to survey a large area, noticeable errors occur; however, surveying done in highway work involves too small an area for a discrepancy to be seen.

2.4.1 Plane Surveying

To survey a site it is necessary to choose two points and then measure the distance between them to form a base line AB, which can be drawn on the plan to scale. A point

C which may be on either side of this line can then be located relative to the line by taking two measurements, i.e. two distances, one distance and an angle, or two angles (*see* Figure 2.37). Which of these methods is used will depend on the equipment available.

The point C can then be drawn to a suitable scale on the paper and in this way a map is constructed.

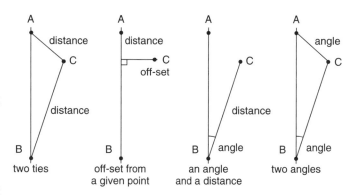

Figure 2.37

All surveys involve a series of measurements and so must have accumulated errors. The simpler and quicker methods in use are liable to have greater errors. Therefore, it is good practice to survey carefully a number of suitable points distributed throughout the area and then fill in the intervening detail by simpler and cheaper methods adequate for short distances.

Always:

(*a*) Work from the whole to the part. Minimize accumulated errors by making a rigid framework covering the whole area.

(*b*) Make sure the method used is appropriate as accuracy is very expensive in terms of time and equipment.

(*c*) Ensure that all important points are checked. Checking measurements should be taken, i.e. if three sides of a triangle are measured then the checking measurement should be from one corner to a point on the opposite side (an error in one of the triangle sides would still allow the triangle to be drawn, but this error would be shown by the checking measurement).

(*d*) Make a preliminary reconnaissance and decide what *you* are going to do before you start work.

2.4.2 Chain Surveying

Chain surveying uses simple apparatus and it is suitable for producing large-scale plans for a limited area. It is awkward in towns and hilly country and the accuracy is about 1 in 500. It is in fact almost always carried out using a measuring band rather than a chain.

Chain surveying is rarely carried out nowadays, but the principles, particularly of triangulation, remain valid.

(a) The Chain

This is made of wrought iron or steel wire links which are galvanized or black enamelled. Lengths are 20 m or 30 m.

The 20 m chain has one hundred 200 mm links with yellow tallies every metre and red numbered tallies every 5 m. Swivelling brass handles are fitted each end of the chain and the total length is measured over the handles. The chain is robust, easily read and easily repaired in the field if broken. It is liable to vary in length, however, due to wear on the metal-to-metal surfaces, bending of the links, mud between the bearing surfaces, etc. (see Figure 2.38(a)).

Prior to the metric chain there were other types of chain such as Gunter's chain (66 ft in length made up of 100 links of 7.92 inch) and the engineer's chain (100 ft in length and made up of 100 links). Gunter's chain was convenient for land measurement but the engineer's chain was used for construction work.

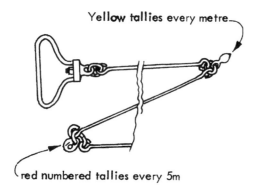

Figure 2.38(a)   Measuring chain

(b) The Measuring Band (Tape)

This is an alternative to the chain and may be made of steel strip or white steel band nylon-coated or even glass fibre. The length can be 20, 50 or 100 m and the width is usually 7 mm. One type with its mounting reel or open frame is shown in Figure 2.38(b).

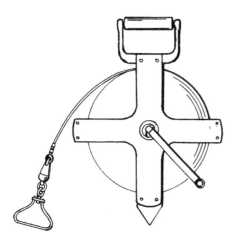

Figure 2.38(b)   Measuring band

The tape that has replaced the chain is more precise, less prone to incorrect reading, and lighter and easier to handle. Correct tension is obtained by using a spring balance (Figure 2.39).

Measuring tapes of Fibron, glass fibre and steel are not as strong or as true as bands and are consequently used for short measurements where a high degree of accuracy is not essential.

roller grip   spring balance   tension handle

Figure 2.39   Tensioning of chain/band

2.4.3   Chaining Procedure

This is basically measuring along a straight line with the chain, while fixing other points relative to the chain line by ties or offsets (see Figure 2.41). The following procedure should be undertaken when laying down the first chain line.

1 Reconnaissance: walk over the area to be surveyed note the general layout, shape and position of features.

2 Decide the location of each station.

  Drive in the station pegs.

  Check measurements should be taken to permanent objects to tie in the station peg (so that it can be replaced accurately if damaged or disturbed).

3 Sketch the layout of the survey on the last page of the field book (see Figure 2.40). Beware a sloping site (see Section 2.4.4).

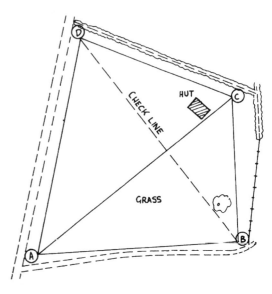

Figure 2.40   Sketch showing layout of chain lines (not to scale)

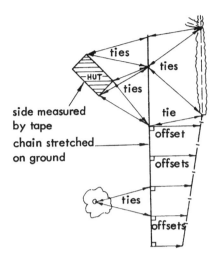

Figure 2.41 Method of measuring

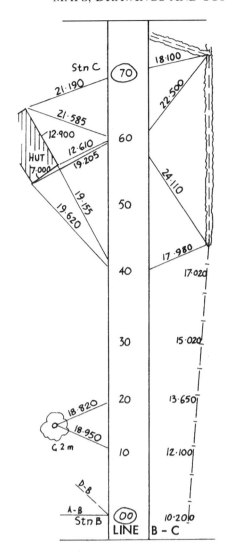

Figure 2.42 Page from a field book

4 Insert ranging rods at each station ensuring that they are on the line and upright.

5 Chainman then sets out from the first station towards the second station holding the leading end of the chain.

6 When the chain is fully extended, the surveyor lines in the chainman who places the first arrow.

7 The surveyor then walks along the chain measuring offsets or ties to each feature, using the tape, the end of which is held by the chainman; the chainage and offset are booked (*see* Figure 2.42).

8 This procedure is then repeated. As each chain length is completed, the surveyor collects the arrow from the 'rear' end.

9 At the completion of the chain line, the number of arrows held by the surveyor forms a cheek on the number of chains measured. N.B.: Offsets over 5 m in length are usually 'tied in' for greater accuracy.

2.4.4 Slopes and Obstructions

Chain lines sometimes have an obscured line of sight or perhaps a physical obstacle in their path. In the first case a line of ranging rods may be set up over a small hill by the technique of 'repeated alignment' already discussed earlier (*see* Figure 2.13). If chaining along a steep slope is to be undertaken, a method of carrying this out and continuing past an obstruction is as follows.

Chaining along a steep slope:
Length of chain line = sum of horizontal chain lengths.

The chain line is set out with vertical ranging rods at short intervals along the length. The horizontal distance is then checked by pulling the chain horizontally between adjacent rods as shown in Figure 2.43 (pull of about 4 kg max. which may be checked by pulling with a spring balance).

Add the series of horizontal distances so measured to determine the true horizontal length of the chain line.

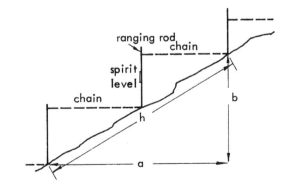

Figure 2.43 Chaining on a slope

Alternatively, if the slope is steep but even, the vertical distance between the stations can be found (*b* metres) and the slope chained (*h* metres).

Using Pythagoras' theorem $h^2 = a^2 + b^2$
so $a^2 = h^2 - b^2$

The true horizontal distance $a = \sqrt{h^2 - b^2}$

Chaining past an obstruction (*see* Figure 2.44):
Either (i) Set out AC to clear obstruction:
offset at D to original line (E)
offset from C to B.

By similar triangles $\dfrac{\text{side X}}{\text{side Y}} = \dfrac{\text{side } x}{\text{side } y}$

So $\dfrac{\text{AB}}{\text{AD}} = \dfrac{\text{AE}}{\text{AD}}$

Therefore $\text{AB} = \dfrac{\text{AE} \times \text{AC}}{\text{AD}}$

or (ii) Offset at P and measure distance PQ (using
equipment to measure off accurately at right-
angles (*see* Section 2.3.1).
Offset again at Q to R, measure QR.
Offset back on to original line at S.
Measure RS to equal PQ.
Then AB = AP + QR + SB.

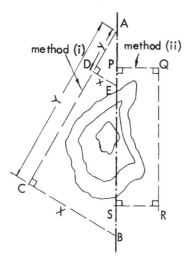

Figure 2.44 Chaining round obstacles

2.4.5 *Surveying by Theodolite*

This technique will only be used by a surveyor or
engineer, but it is of interest to the roadcraft
student. The theodolite is a precision instrument
which is capable of measuring horizontal and
vertical angles between objects

with great accuracy. Theodolites range from basic
models which measure angles to 20 seconds to high
accuracy versions which read direct to 0.5 and 0.1
second.

(a) *Traverse Survey*

A theodolite 'traverse survey' consists of measuring
the angles between successive traverse lines and then
finding the lengths of these lines. One method is to
set up the instrument and carry out a closed traverse
as in Figure 2.45(*a*). An alternative method is to
carry out an unclosed traverse shown in Figure
2.45(*b*), which although being more suitable where
the survey is comparatively long and narrow such as
pipeline or road construction, does not have the
advantage of being self-checking as with the closed
traverse. Yet another method is that of 'radiation'
where the instrument is placed at a fairly central
station and rays XA, XB, etc. are measured (*see*
Figure 2.45(*c*)). The traverse lines are then chained.

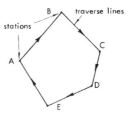

Figure 2.45(a) Closed traverse

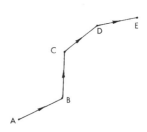

Figure 2.45(b) Open traverse

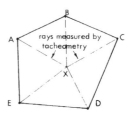

Figure 2.45(c) Radiation traverse

In the methods shown in Figures 2.45(*a*) and (*b*), the instrument progressively moves from station A to B, C, D, and E, the angle of the traverse lines being measured by the surveyor sighting on to the pole or staff held by his assistant at each station. The length of the traverse line may be measured by chain, band or the optical technique known as tacheometry. (*See* Section 2.4.6.)

(b) Modern Equipment

There are now two basic types of theodolite, the optical microptic and the electronic digital. Both of these are capable of the above tolerances.

Figure 2.46 shows a typical optical microptic theodolite from which all modern designs are developed.

The basic features are:

1 The trivet stage forming the base of the instrument and connecting it to the tripod head.
2 The tribrach which supports the rest of the instrument and, with reference to the plate bubble, can be levelled using the footscrews which act against the fixed trivet stage.
3 The lower plate carries the horizontal circle, made of glass, with graduations from 0° to 360° photographically etched around the perimeter. This process enables lines of only 0.004 mm thickness to be sharply defined on a 100 mm circle, thereby resulting in very compact instruments.
4 The upper plate carries the horizontal circle index and is fitted concentrically with the lower plate.
5 The plate bubble is attached to the upper plate and, when centred, using the foot screws, ensures that the instrument axis is vertical. Some digital and electronic theodolites are fitted with an electronic bubble instead of a spirit level.
6 The upper plate also carries the standards which support the telescope on its transit axis. The standards are tall enough to allow the telescope to be fully rotated about its transit axis.
7 The vertical circle is similar in construction to the horizontal one and is fixed to the telescope axis, rotating with the rotation of the telescope.
8 The vertical circle index, against which the vertical angles are measured, is set normal to gravity, either by (a) an altitude bubble attached to it, or (b) an automatic compensator, which is now standard for modern theodolites.
9 The lower plate clamp enables the horizontal circle to be fixed in position. The lower plate slow motion screw permits slow movement about the vertical axis when the lower plate is clamped. In most theodolites, the clamp and slow motion screw have been replaced by a horizontal circle-setting screw, which rotates the horizontal circle to any required reading.
10 Similarly, the upper plate clamp and slow motion screw allow rotation of the horizontal circle index.
11 The telescope clamp and slow motion screw allow fine movement of the telescope in the vertical plane.
12 The altitude bubble screw centres the altitude bubble, which, being attached to the vertical circle index, establishes it truly horizontal prior to reading the vertical circle.
13 The optical plummet, built into either the base of the instrument or into the tribrach, enables the instrument to be centred precisely over the survey point. The line of sight through the plummet coincides with the instrument's vertical axis.
14 The telescopes are similar to those of the optical level, but are usually shorter in length. They also possess rifle sights or collimators for initial pointing.

(c) Reading Systems

At the present time there are, basically, three different reading systems in use on theodolites. These are:

(1) Optical scale reading This system is generally used on theodolites which have a resolution of 20 seconds or less. Both horizontal and vertical scales are displayed simultaneously and are read directly with the aid of the auxiliary telescope (*see* Figure 2.46). The telescope used to give the direct reading may be a 'line' or 'scale' microscope. The former uses a fine line etched on to the graticule as an index against which to read the circle, whereas the latter has a scale on its image plane, the length of which corresponds to the line separation on the graduated circle (*see* Figure 2.47).

(2) Optical micrometer reading For this type of reading, an index must be set with the micrometer before taking or setting a reading. A line microscope, combined with an

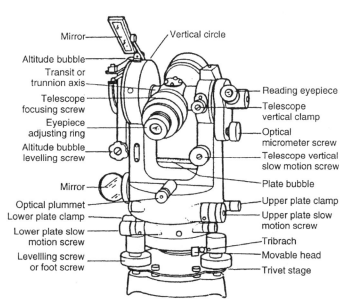

Mirror — Vertical circle
Altitude bubble
Transit or trunnion axis
Telescope focusing screw
Eyepiece adjusting ring
Altitude bubble levelling screw
Mirror
Optical plummet
Lower plate clamp
Lower plate slow motion screw
Levellling screw or foot screw

Reading eyepiece
Telescope vertical clamp
Optical micrometer screw
Telescope vertical slow motion screw
Plate bubble
Upper plate clamp
Upper plate slow motion screw
Tribrach
Movable head
Trivet stage

Figure 2.46 Optical microptic theodolite

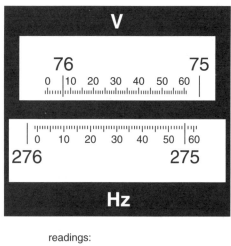

readings:

vertical	76° 06′ 30″
horizontal	275° 56′ 30″

Figure 2.47   Wild T16 optical scale theodolite

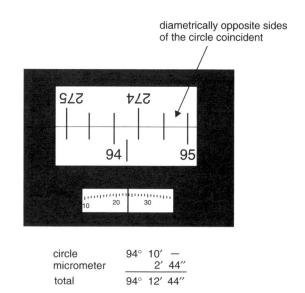

circle	94° 10′ —
micrometer	2′ 44″
total	94° 12′ 44″

Figure 2.49   Wild T2 theodolite (old pattern)

optical micrometer is used, on the same principle as the parallel plate micrometer on a precise level. The optical micrometer reads only one side of the horizontal circle, which is common practice in 20-second instruments (*see* Figure 2.48). On more precise models, reading to 1 second of arc and above, a coincidence microscope is used. This enables diametrically opposite sides of the circle to be combined and a single mean reading taken (*see* Figures 2.49 and 2.50).

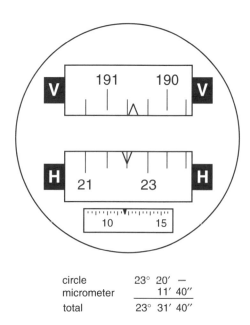

circle	23° 20′ —
micrometer	11′ 40″
total	23° 31′ 40″

Figure 2.48   Watts Microptic No. 1 theodolite showing the reading system

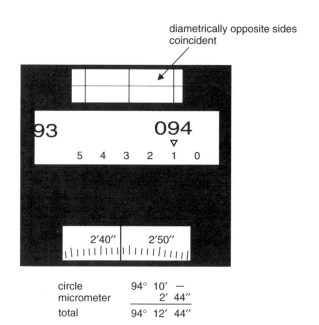

circle	94° 10′ —
micrometer	2′ 44″
total	94° 12′ 44″

Figure 2.50   Wild T2 theodolite (new pattern). N.B.: readings to nearest second

(3) Electronic digital display The new technology of electronic reading instruments, which are more easily read and more accurate are now competitive in price with the optical instruments and in common use. They are set up in much the same way as traditional theodolites, differing only in the external digital LED (light emitting diode) or LCD (liquid crystal display) displays derived from electronic scanning and counting of evenly spaced lines on the circles (protractors) and the ability to convert

received data into a form suited to the user at the touch of a button (*see* Figure 2.51).

Most electronic theodolites provide key setting of the circle to 0°, electronic hold of the circle reading, counting the angle either clockwise or anticlockwise and some provide key setting of any angular value. Compensation for maladjustment of the instrument by the application of corrections before displaying the reading may also be provided.

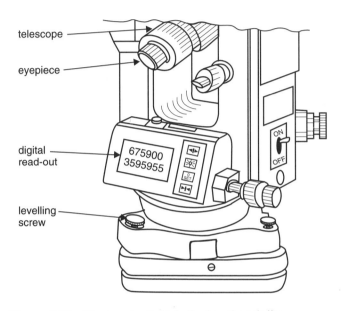

Figure 2.51 Electronic digital display theodolite

In the latest models, facilities can include automatic data recording and processing, storage on memory cards for downloading information into popular information technology systems and include computer aided design, mapping, geographical and land information systems (GIS/LIS) and global surveying technology applying global position fixing systems (GPS). Some are even linked to laptops or data tablets, to enable more effective co-ordinates and setting out.

2.4.6 Tacheometry

A tacheometer survey may be undertaken using one of several different techniques but the usual method is that described below. This requires a theodolite, having a measuring device inside it and hence sometimes referred to as a tacheometer, together with a normal graduated staff. By measuring distances in this manner, the chaining operation is eliminated and hence distances may be measured over broken ground or standing crops.

Stadia system tacheometry makes use of stadia line measurement together with a staff held in the vertical position. The theodolite has two horizontal stadia lines or hairs on the instrument's reticule, so placed that they define sight rays/lines which intersect as the centre of the instrument (the junction of the transit and vertical axes).

These rays/lines thus form the sides of an isosceles triangle the proportions of which are 1:100. Hence, when the theodolite (or level having stadia markings) is set horizontally the staff readings between the wires are 1/100th of the distance of the staff from the instrument (*see* Figure 2.52).

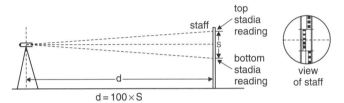

Figure 2.52 Tacheometry

This method remains a possible surveying technique, but it cannot compare with the degree of accuracy of modern data capture systems, being also extremely slow and requiring a great deal of processing. It should, therefore, only be used when there is no alternative.

2.4.7 Electromagnetic Distance Measuring (EDM)

EDM equipment has revolutionized surveying procedures. Distances can now be measured easily, quickly and with great accuracy, regardless of terrain conditions. For most engineering surveys, 'total stations' may be used. This requires a fully integrated instrument which captures all the spatial data necessary for a three dimensional position fix and displays it on digital readout systems, recorded at the press of a button. These 'total stations', combined with electronic data loggers are now virtually standard equipment on site.

Basic theodolites can be transformed into total stations by adding top mounted EDM modules. A standard measurement can now be taken in under 3 seconds.

EDMs all work on the same principle. They emit a carrier wave, usually infra red in the electromagnetic spectrum, this wave being reflected back to the instrument, usually from a corner cube prism. The distance which the beam has travelled is computed by changing the modulation frequencies of the carrier wave and comparing the resulting phase shift. The speed of the wave may be affected by meteorological conditions and corrections for this can be made either manually or automatically. Accuracy of readings to within 3 mm at distances of up to 2.5 km can be achieved for a single prism reflector or up to 3.5 km for a prism cluster.

There is a vast range of equipment now on the market and care must be taken to ensure that equipment is not out of date when purchased. The range includes purely slope indicators through to more complex ones which solve standard triangle equations and interface with input/output devices to receive, store or download data.

A recent development in EDM technology is the 'one man' system – called the Remote Positioning Unit (RPU) – in which the operator at the prism reflector end of the operation can control the functions of an unmanned servo-driven instrument. The 'magnetic memory card' which can store or provide data without the need for cables or other external devices is another development.

Setting Out by EDM

Due to the variety of EDMs available, only the general principles are dealt with here.

Lengths detailed on drawings are usually 'plan' or 'horizontal' and as EDM instruments measure slope lengths from instrument to prism, application of slope correction is necessary. In setting out, the prism reflector should be held as close as possible to the ground in order to avoid error due to the leaning of the single staff holding the prism. Total stations will automatically apply the vertical angle to display the slope corrected length, but 'semi-total' stations and top mounted EDMs require manual input of the vertical angle. This manual input can result in error and great care must be taken when making the corrections.

Most EDMs have a rapid measure facility, known as 'tracking' which measures to ± 10 mm, subject to a small error in the location of the peg, which may need adjustment. Some EDMs are capable of calculating this adjustment.

It is important that clear hand signals are given to the person holding the staff, if minor adjustments necessary for accurate readings are to be made.

The 'total' station allows input of co-ordinates of the markers over which the theodolite and base line target are set up, together with the co-ordinates of the point to be set out. The total station will then calculate the direction of pointing and length required. A servo-driven instrument can rotate to the required direction and some computer programmes display the forward/back and left/right adjustments required to the prism to achieve the stored co-ordinates of the point which is being set out.

2.4.8 Global Positioning System (GPS)

The GPS system is founded on space-based microwave technology and its primary use was for military purposes as a global navigation system. In recent years GPS surveying techniques have replaced conventional technologies, such as Doppler Satellite positioning and long range EDM for control surveys and is being used increasingly in cadestral, topographical and engineering surveying. The full constellation of satellites is now in place, and the GPS system will ultimately dominate the survey equipment field. The distances in GPS are called 'ranges' and are measured to satellites orbiting at nominally 20 183 km

above the Earth, instead of control points on the Earth. This system requires four distances, each from a different satellite, in order to achieve a positive fix and this is then complimented by a pseudo-random noise code (PRN). PRN enables point accuracy to be achieved.

A disadvantage of GPS is the need for a direct line of sight. A good example of this would be when working in a large city having high rise buildings, which could obscure a satellite.

2.5 Transferring a Level

Having set out the work, it is necessary to fix levels.

Perhaps one of the oldest methods of transferring a level is by using a flexible tube filled with water. When in this state, i.e. completely full, both ends are at the same level.

Several other methods, making use of both simple and sophisticated equipment, are outlined in this section and typical uses are given in Chapter 3, Earthworks.

2.5.1 Straight Edge and Spirit Level

A straight edge is usually made of hardwood with planed parallel edges. Used with a spirit level it is possible to transfer a level within the length of the straight edge. Figure 2.53 shows a level transferred from a peg to a pin.

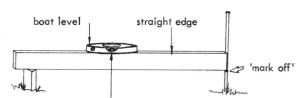

Figure 2.53 Transferring a level

2.5.2 Boning Rods

These are portable 'T' shaped rods (in sets of three) having the same dimensions and used to furnish a line of sight, whereby from two given points, other points at the same level or on the same gradient, may be established (*see* Figure 2.54).

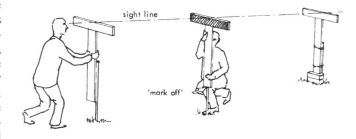

Figure 2.54 Fixing intermediate levels

2.5.3 Sight Rails and Travellers

These are similar to boning rods and are used in a fixed position for construction work. The sight rails are erected as horizontal rails, fixed to a post and the traveller is cut to a length which will give the required level below that of the sight line between the sight rails (*see* Figures 3.1 and 4.4).

2.6 Levelling

The term 'levelling' is used to describe the operation of finding the difference in height between points on the earth's surface. These heights are given relative to a plane or datum which is known as the Ordnance Datum (OD). This is the mean level of the sea and is measured at Newlyn, Cornwall. Levelling is only accurate over relatively short distances as the line of sight is a tangent to the earth's surface and longer sights would need to be corrected to take account of the earth's curvature as in Figure 2.55.

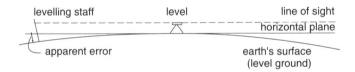

Figure 2.55

Site surveying is normally so limited in distance that no correction for curvature need be made. Corrections for curvature are beyond the scope of this section.

2.6.1 Uses of Levelling

In highway work, levelling is used to:

(*a*) find the difference in level between two points (often known as 'check levelling');

(*b*) take the longitudinal section and the cross-sections of the ground along a given line; and

(*c*) set out levels to control work in progress (often known as 'construction levelling').

2.6.2 Equipment

Although advanced levelling equipment such as the electronic levelling head is now in use by specialist surveyors, it is thought that most highway personnel employed in setting out will use simple laser levelling equipment (*see* Section 2.6.8) or the older and more basic equipment, which is:

(*a*) a device which gives a truly horizontal line, such as a Dumpy level, Tilting level or a Precise level; and

(*b*) suitably graduated staff for reading vertical heights. This may be designed as a telescopic wooden or aluminium staff and can be 3 m to 5 m in length. The scale markings on the face are in metres, 100 mm and 10 mm units or in metres, 100 mm and 5 mm units.

2.6.3 The Dumpy Level

This is a common form of level and will be considered in detail.

(a) Description of Dumpy Level

The levelling head consists essentially of two plates, the telescope being mounted on the upper plate while the lower plate screws directly on to a tripod. A small circular bubble is mounted on the upper plate and this enables a preliminary levelling up of the instrument to be made.

The two plates are held apart by three levelling screws and adjustments to these bring the main bubble in the bubble tube to the central position, which accurately levels the instrument (*see* Figure 2.58).

The internal focusing telescope has a rack and pinion focusing device together with an adjustable eyepiece (*see* Figure 2.56).

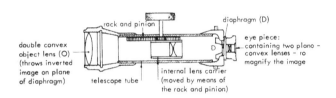

Figure 2.56 Section through Dumpy level

The distance between the object lens (O) and diaphragm (D) is fixed. The diaphragm has either fine wires or an etched glass plate. These wires are one horizontal and two vertical as shown in Figure 2.57. The image (inverted) of the staff is viewed between the two

Figure 2.57 Dumpy level cross wires

vertical lines. The reading is taken at the horizontal line.

(b) Setting Up the Dumpy Level

To set up the level a procedure is followed: First, place the tripod legs firmly in the ground and screw on the instrument. The circular bubble should be roughly central and the plumb-bob over the station mark. Secondly, the

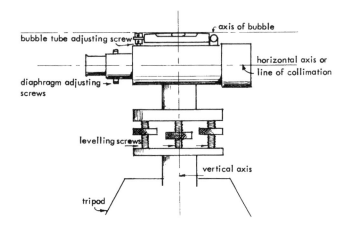

Figure 2.58 Dumpy level

telescope is turned so that it is parallel to two of the levelling or foot screws and by adjusting these the main bubble is brought to the centre of the tube. The telescope is then turned through 90° and the bubble once more brought to the centre of the tube by adjusting the third screw. It is necessary to repeat the operations once more to ensure that the bubble will remain central wherever the telescope is pointed. Levelling really means setting the vertical axis vertical. Finally, point the telescope at a piece of blank paper and focus the eyepiece so that the cross hairs are clearly seen. (This adjustment depends only on your eyesight and will differ between individuals.)

Then, sight on to the levelling staff and focus its image with the focusing screw so that when the eye is moved slightly from side to side, there is no apparent relative movement between the image and the cross hairs, i.e. no parallax.

N.B.: If the main bubble cannot be maintained in a central position wherever the telescope is pointed, the instrument may need adjustment.

(c) Levelling

The staff is seen upside down in the telescope; do not correct this by holding the staff upside-down (*see* Figure 2.59). Before starting work, study the scale carefully, both with the naked eye and through the telescope.

The staff-man stands facing the instrument and holds the staff in front of him with one hand on each side so as not to hide the scale. It is very difficult to hold the staff vertical. The observer can see if it is leaning sideways, and signal to the staff-man accordingly, but he cannot see if it is leaning forwards or backwards. The staff-man should therefore swing the top of the staff slowly towards and away from the instrument, passing through the vertical position. The observer records the LOWEST reading, since this corresponds with the vertical position of the staff. At change points rest the foot of the staff on something firm; if necessary drive a peg.

To find the difference in level between two points A and B, the observer sets up and levels the instrument at a third point I_1. An assistant holds the staff upright with its foot resting on A (*see* Figure 2.60).

The observer traverses the telescope until the staff appears in the centre of the field of view and then reads the scale of the staff against the horizontal cross hair. Call this reading '*a*'. The staff is then moved to B and the observer again directs the telescope on to it, obtaining a reading '*b*', the difference in level between A and B is then $(a - b)$ since, if the instrument is correctly adjusted, the 'line of collimation' (sighting line) is horizontal. The height of the instrument at I_1 does not affect the calculation.

If the height of a third point C, beyond B, is required, the instrument is moved to I_2 between B and C. The difference in level between B and C is then found in the same way. By repeating this process the difference in level between points at any distance apart can be found.

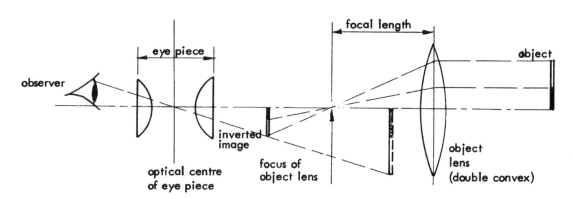

Figure 2.59 Optical system of dumpy level

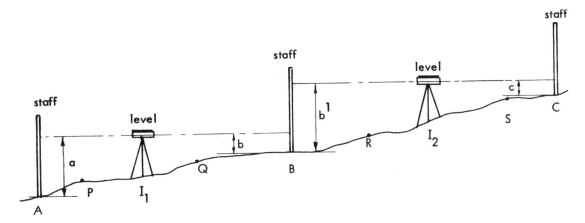

Figure 2.60

If the heights of intermediate points are required, the staff is held say, at P and Q and at R and S and readings are taken from I_1 and I_2 respectively. If the horizontal distances between the staff stations are chained, a vertical section of the ground (a 'level section') can be plotted. The instrument positions I_1, I_2, etc. may be offset a little to one side of the line being levelled.

In levelling from A the observations I_1A, I_2B, etc. are called 'backsights' and the observations to P, Q, R, S, etc. are called 'intermediate sights'. I_1B and I_2C are called 'foresights'. B, C, etc. are called 'change points'. Notice that when the staff is moved the instrument remains stationary and vice versa.

The carrying forward of the height depends on the backsights and foresights. Each position of the instrument should therefore be nearly equidistant from the corresponding change-points of the staff, e.g. I_1A should equal I_1B. Then if the line of collimation is not perpendicular to the vertical axis, and therefore always points up (or down) slightly when the instrument is levelled, the resulting error will be the same on backsight and foresight, and will therefore cancel out. For clear reading, with ordinary instruments, no observation should be taken over a greater range than 100 m. Always check work of any importance by levelling back from finish to start, unless both ends of the line can be checked against 'Bench Marks'.

In levelling, an accuracy of 20 mm vertically per kilometre horizontally should be obtainable.

(d) Booking

Either of two common methods of booking may be used. Table 2.1 shows examples. In both methods successive lines across the field book page refer to successive STAFF stations. In the Rise and Fall method the rises and falls are the differences between successive staff readings and successive 'reduced levels' (heights above datum) are calculated from the rise or fall. In the Line of Collimation method, at each instrument station the height of the line of collimation is found from the backsight and hence the reduced levels of all other staff stations by subtracting the intermediate sights or the foresight. The Rise and Fall method takes rather longer but contains a check on all arithmetic. By the Line of Collimation method only the calculation of the last reduced level is checked. If the reduced level of first staff station is not known, a value is assumed such that all reduced levels will be positive.

The line of collimation method is the technique more commonly used by Highway Surveyors.

2.6.4 Worked Example on Booking

The answer to a typical problem will further illustrate a method of booking (*see* Table 2.2).

The following staff readings, in metres, were obtained at ground level along the line of a proposed new sewer:

1.500 1.700 2.300 3.520 1.240 1.855 3.750 2.805

The first reading is on the cover of a manhole which is to be the outfall of the sewer and is 30.000 m above OD. The third reading is a change point. All other readings are taken at 25.000 m intervals.

Example

1 Make out a typical level book and reduce the levels.
2 If the depth to invert at the outfall manhole is 3.000 m and the overall gradient 1 in 200, what is the invert level of the sewer at the last reading?

Rise and fall method							
Back-sight	Inter-mediate	Fore-sight	Rise	Fall	Reduced level	Distance (m)	Remarks
0.185					20.560		B.M.
	0.745			0.560	20.000	0	Manhole cover at A
	0.640		0.105		20.105	15	
	0.590		0.050		20.155	30	
	0.540		0.050		20.205	42	Centre of road
0.320		0.530	0.010		20.215	50	
	0.285		0.035		20.250	65	
		0.420		0.135	20.115	72	Front of wall
0.505		0.950	0.250	0.695	20.560		
		0.505		0.250	20.115		
		0.445		0.445	0.445		

Line of collimation method						
Back-sight	Inter-mediate	Fore-sight	Line of collimation	Reduced level	Distance (m)	Remarks
0.185			20.745	20.560		B.M.
	0.745			20.000	0	Manhole cover at A
	0.640			20.105	15	
	0.590			20.155	30	
	0.540			20.205	42	Centre of road
0.320		0.530	20.535	20.215	50	
	0.285			20.250	65	
		0.420		20.115	72	Front of wall
0.505		0.950		20.560		
		0.505		20.115		
		0.445		0.445		

Table 2.1

Back-sight	Inter-mediate	Fore-sight	Height of collimation	Reduced level	Chainage (m)	Remarks
1.500			31.500	30.000	0	Manhole cover (outfall)
	1.700			29.800	25	
3.520		2.300	32.720	29.200	50	C.P.
	1.240			31.480	75	
	1.855			30.865	100	
	3.750			28.970	125	
		2.805		29.915	150	
5.020		5.105		0.085		
		5.020				
		0.085				

Table 2.2

Level at invert of outfall manhole = 30.000 − 3.000
= 27.000 m
Length of sewer = 150 m
Gradient = 1 in 200
Therefore rise of sewer over this length

$$= 150 \times \frac{1}{200} = 0.750 \text{ m}$$

Therefore invert level at upper end of sewer
= 27.000 + 0.750
= **27.750 m**

2.6.5 Tilting Level

This type of level has become increasingly popular as a replacement for the Dumpy level. It is a smaller, more compact instrument and the telescope is not rigidly fixed to the vertical spindle as with the Dumpy level, but is capable of a slight tilt in the vertical plane about the axis placed immediately below the telescope. This movement is made by a fine setting screw at the eyepiece end, which brings the bubble to the centre of the bubble tube, for each reading of the level. So the line of collimation need not be perpendicular to the vertical axis.

The tilting principle is used in most modern levels, giving great accuracy as in an Engineer's Microptic level used for precise levelling or in a Quickset level which may be used for ordinary work.

The Quickset Level (see Figure 2.61)

The main feature of this instrument is the rapidity with which it can be attached to the tripod and levelled. The domed head of the tripod carries three bearing pads which support the concave base of the level, an arrangement which enables the instrument to be easily set to an approximately level position and clamped (using a small circular bubble). The final precise levelling is quickly done with the aid of a slow motion tilting screw.

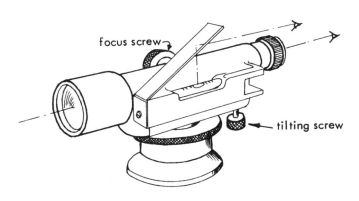

Figure 2.61 Quickset level

Setting up a Quickset level

1 Unstrap tripod legs, extend them and lock in position.
2 Erect the tripod approximately level and press each foot firmly into the ground.
3 Remove the cover from the mounting head on the tripod.
4 Remove instrument from its carrying case and place on tripod, partially tightening the securing bolt. Adjust the position of the level on the domed head to centre the circular bubble and then clamp firmly with the securing screw.
5 Focus the eyepiece; in order to do this, point the telescope at a light background and turn the eyepiece fully anti-clockwise, then rotate it clockwise until the graticule lines are in sharp focus.
6 Undo the azimuth clamp, point the telescope at the staff and use the focusing knob to bring it into sharp focus. Lightly tighten the azimuth clamp and then use the slow-motion screw for the fine directional adjustment.
7 Use the tilt screw to level the telescope, bringing the main bubble to the centre of the vial.
8 Check that no parallax exists by moving the eye from side to side (this should not cause apparent relative movement between the graticule and the staff image). If parallax exists, re-adjust focus.
9 Read the staff.
10 Re-level the telescope by means of the tilt screw every time the telescope is pointed in a different direction.

2.6.6 Automatic and Semi-automatic Levels

These are replacing levels which have to be adjusted before readings are taken. Once the level is set up, no further adjustment is necessary, making them more simple to use and giving greater accuracy than the older non-automatic levels, such as the Quickset type.

2.6.7 Laser Levels

This type of level may also be referred to as a 'diode laser', 'electronic level' or reference plane laser.

The principle of operation is simple. A beam, usually infra-red, is produced by the laser and is rotated at great speed by means of a spinning penta prism, thereby simulating a constant plane. All the rotating beam laser levels incorporate automatic self-levelling systems and can generate horizontal planes. Many types can also produce vertical and tilting planes when required.

The range of use of a laser level is much greater than that of the optical type instruments. Some are capable of giving precise readings up to 500 m from the instrument.

The main advantage of a reference plane laser over optical levels is that it can be used by a single person. The operator can set up the instrument in the middle of a large site and then take levels at various points around the site by using a detector. Several operators, each using a detector, can use the same laser at the same time. See Figure 2.62.

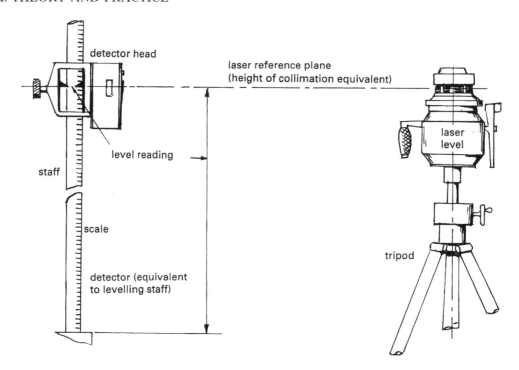

Figure 2.62   Laser level and detector

Figure 2.63   1 Rotating laser 2 Laying sub-grade to laser control 3 Checking formation level 4 Fixing wall levels
5 Staff with photoelectric detector fixing foundation levels 6 Taking ground levels 7 Laser plane of reference

This type of equipment has a very high initial cost, but it must be borne in mind that labour resources are immediately reduced, particularly when a number of operatives are 'levelling' simultaneously from the one laser source (*see* Figure 2.63).

2.6.8   Worked Example on Fixing Sight Rails

Sight rails are to be fixed for a new sewer construction. The length of the sewer is 100 m and the gradient is to be 1 in 125. Sight rails are to be fixed at each end and the centre of the proposed sewer line. The staff reading on the lower rail is 2.630 m and optical instruments are to be used.

1   The required readings at each end and the centre of the length must be calculated (i.e. at 50 m intervals).
2   Drive in stakes to support the sight rails at each of these three points (*see* Figure 2.64). These must be offset from the line of trench sufficiently to avoid interfering with the work of the machine excavating the trench.
3   Set up the level at a point reasonably central to this length of trench, but offset to the side to give approximately equal lengths of sight to each rail. (Equal sight distances will eliminate any error in collimation of the instrument.)
4   Sight on to the lower rail (at the reading on the staff of 2.630 m).

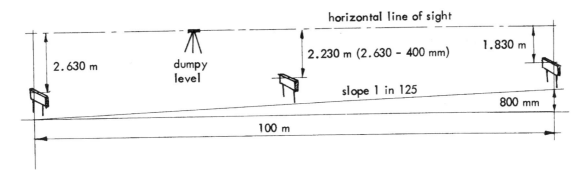

Figure 2.64

5 Rotate the telescope of the instrument to sight on to the centre rail position.
6 Adjust the staff, held at the sight rail upright, until the reading is 2.230 m. Mark this position (the bottom of the staff) on the stake.
7 Fix the cross piece on the sight rail with the top level with the mark.
8 Repeat this procedure for the top sight rail, but with a staff reading of 1.830 m.

The calculations for these figures are shown in Figure 2.64.

Revision Questions

1 Several years ago most engineering drawings were hand drawn. What methods do we use today.
2 How would you set out a right-angle using pins and a tape only?
3 What surveying instrument would you use to undertake a closed traverse?
4 What problems might you encounter when undertaking surveying in a dense urban area using a GPS system?

Note: See page 305 for answers to these revision questions

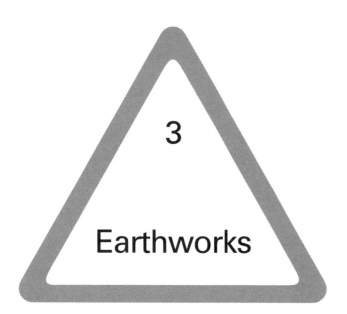

3
Earthworks

3.1 Setting Out Prior to Starting Earthworks

Before the earthworks necessary to a road improvement or new construction can be started, the site for the intended work must be surveyed and set out. This survey would not normally be part of the work of the roadwork craftsman.

3.1.1 Procedure (see Figure 3.1)

Prior to the start of construction work, pegs will be driven into the ground by surveyors at key points and a schedule giving the levels of these pegs will be handed over to the site supervisor. The level pegs will have been placed at strategic and significant points and would in most cases include a set of pegs down the centre line of a new carriageway. Further setting out pegs would be placed at manholes and other drainage points. Different practices

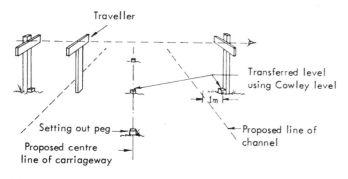

Figure 3.1 Use of simple sight rails and traveller

apply in different areas, as a line of pegs is sometimes inserted 1 m back from the proposed channel edge or, alternatively, the top kerb face. Improvements based on an existing road will often have steel pins driven into the surface 1 m inwards from the proposed kerb line.

The site supervisor would normally be expected to erect site rails from the setting out pegs, in order that travellers or travelling rods may be used to check the depths of excavation as the work proceeds.

In setting out the crown and channel levels, either two different travellers or one traveller and double sight rails may be used (see Figures 3.2 and 3.3(a)).

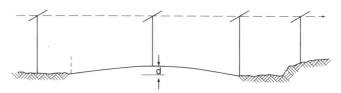

Figure 3.2 Two traveller method

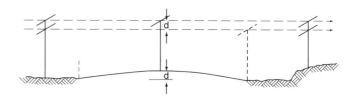

Figure 3.3(a) Single traveller/two sight rail method

N.B.: Care must be taken to avoid errors through sighting on to the wrong rail. Complementary rails should be painted in the same colour and each pair of rails should be of different colours in order to avoid confusion. The camber of the carriageway must be checked by means of a specially shaped template.

An alternative method is shown in Figure 3.3(b). This method can be used for a carriageway with crossfalls (i.e. a hanging section) or a super-elevation. In this, by boning between the lower rail (A2) on the left-hand side to the higher rail (B1) on the right, the crossfall can be set out by sighting in the traveller at points T1, T2, and T3 in turn. Then reverse the sighting, using rails B2 and A1 to obtain the crossfall on the other side of the carriageway.

On small works, such as a road realignment to an existing carriageway, it may be possible to check the excavation

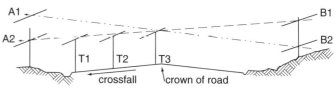

Figure 3.3(b) Sighting for crossfall setting out

depths by using a set of boning rods and working off the existing carriageway (*see* Figure 3.4). This at best, is a very approximate method and the establishment of level pegs and sight rails is a safer and much more accurate method of checking the earthworks as they proceed.

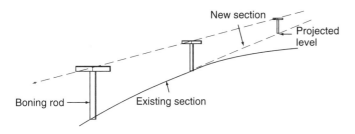

Figure 3.4   Use of boning rods

3.2   Earthmoving or Muck Shifting

3.2.1   Site Clearance and Preparation

At the start of the work in the construction of a new road or the improvement of an existing road there is usually a problem of site clearance and preparation. This is basically the preparation of the entire site to allow subsequent work to 'flow' without annoying stoppages caused by known obstacles. The problems are quite different in urban and rural areas.

In rural areas, trees may have to be felled, and removed from the site. Hedges may need to be grubbed up and burnt on site, but a tangle of roots and topsoil must not be allowed to accumulate as it is very difficult to dispose of at a later stage in the work. Temporary fences may need to be erected and access points provided to re-route the public past the construction site.

Temporary accommodation and storage huts will need to be carefully sited and compounds erected to take materials and equipment. In urban areas it may be very difficult to avoid obstructing public access. The problems of demolition and of locating plant belonging to statutory undertakers, together with the signing and protection of roadworks, are dealt with elsewhere in this book. (*See* Chapter 15: Section 15.1).

Much thought needs to be given in the planning stages of scheduled work for arranging that items such as telegraph poles, street lighting lamp standards and services are repositioned prior to work commencing on site. This is not always possible, even with careful planning, in cases where the resited positions require appreciable changes in level, etc., which requires the substantial completion of the earthworks prior to re-erection of the telegraph poles and lighting standards.

3.2.2   Soil Stripping

In almost all cases the topsoil is first removed, even in cases where the ground is to be built-up.

For a high embankment, the topsoil is not usually removed prior to the build-up of the ground. A high embankment may be defined as one which is more than 2 m high.

If the topsoil is left under a fill of less than this amount, there will always be a risk of vegetation growing through it. There is also a risk of this sub-grade being 'spongy' due to the humus content of the ground.

When the topsoil is stripped, it should be stored in soil heaps for re-use after the completion of the road construction on verges and embankments, etc. These soil heaps should be carefully shaped to shed water and care must be taken that the soil from them does not obstruct land drainage. When the topsoil is replaced, it should be laid only to a thickness of about 100 mm.

On jobs where there is to be an entirely new carriage-way or dual carriageways and particularly if the site had been under the plough, the volume of topsoil stripped may be considerably in excess of that required for re-soiling later and the estimated surplus should there-fore be taken off site immediately to prevent double handling.

Crawler tractors and/or scrapers are normally used for topsoil removal. For small-scale jobs, a tracked or wheeled excavator/loader is suitable.

3.2.3   Excavation

This should preferably be carried out immediately after topsoil removal. If possible, the optimum moisture content of the soil should be retained. The optimum content is usually the water content of the soil at a depth of about 1 m below the surface (within ±2%) during periods of normal summer weather. In some cases, the soil will tend to have a moisture content higher than the optimum and may even be saturated.

When the excavated material is to be used for forming embankments elsewhere, the following comments about moisture content are most important.

(*a*) If the ground is too wet (i.e. if it contains more water than when at the optimum moisture content) it becomes plastic and even semi-fluid.

(*b*) If the ground is too dry, it may crack and crumble.

(*c*) A soil sample, when gripped in the hand, should just stick together in a single cohesive lump. If the sample falls apart when the hand is opened, if a cohesive soil, it is too dry. Non-cohesive soils will always fall apart.

Every effort must be made to keep the excavated roadworks dry (i.e. free of surface water) whatever the weather conditions and for this reason it may be necessary first to excavate and lay a drainage system prior to the general site excavation commencing.

3.2.4 Formation

If new roadworks have been excavated to formation level, this surface should be protected. The formation may best be protected by immediately laying and consolidating the sub-base material. If this is not possible, the excavation should be left at about 150 mm above the formation level, so that the natural ground is used to provide the desired protection. This latter procedure is to be preferred. Loss or contamination of the sub-base material can and does occur during use by construction traffic.

In some cases, where the formation is not to be used by construction traffic, it can be protected from the weather either by surface dressing or by sheets of polythene/waterproof paper.

3.2.5 Excavation in Cutting

Initial Excavation

In general, for all types of earthworks, the rule is to work uphill, thus allowing drainage from the working face, but in certain cases such as well-drained sites or in very firm ground, the scrapers should work 'downhill'. The reason for this is to avoid any excessive pull on the scraper (particularly when full) which could easily result in 'snatching' or 'digging-in' of the scraper (*see* Figure 3.5). When running downhill, the tractor pull is assisted by gravity thus reducing the draw-bar pull and avoiding 'digging-in'.

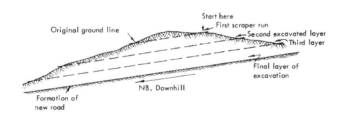

Figure 3.5 Excavation by scraper in cut

Cuttings are normally left with the final 300 mm of excavation above the formation level *in situ* until the last possible moment, as earth is its own best protector.

Drainage

In cut, the excavation to formation level may well be lower than the natural water table at that point. In this case the road side drains should be laid before excavation is started. The provision of this drainage can be most useful in lowering, or depressing, the water table so that the ground water does not affect the road construction (*see* Figure 3.6).

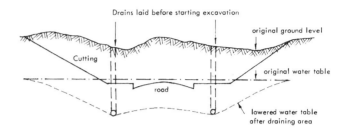

Figure 3.6 Excavation lower than water table

Springs

If springs or other underground waters are encountered at or about formation level, a system of land drains should be laid herringbone fashion and connected to the main drains (*see* Figure 3.7). A land drain is a plain porous earthenware pipe, usually of 100 or 150 mm diameter or alternatively of 100 mm flexible plastic pipe with multiple holes.

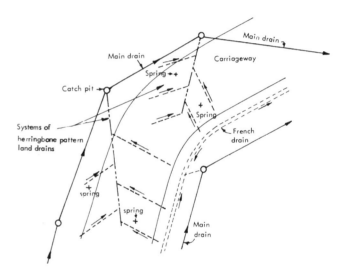

Figure 3.7 Method of draining ground water from below formation

Where springs or land drains are encountered part way down the cutting slope these should be connected to the side drains by counterfort or block stone drainage channels cut into the sloping face. In the case of a deep cutting (5 m or more) these channels may be arranged in a herringbone fashion on the sides of the cutting.

The cutting slopes should be left with a rough face to form a key for the subsequent topsoiling. The practice of neatly planing the face with a grader will lead to topsoil slips later (*see* Figure 3.10).

3.2.6 Earthwork in Widening an Existing Carriageway

In this case, a temporary bituminous bound material dam is constructed along the existing channel to prevent water from draining off the existing road into the excavation for the widened road (*see* Figure 3.8). However, care must be taken to ensure that flooding of the existing carriageway does not become a hazard to road users.

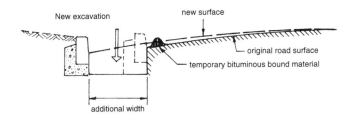

Figure 3.8 Widening a carriageway

3.2.7 Embankments

When the earthwork is to be left overnight, the work should be left with an exaggerated cross-fall to prevent water from standing on the newly exposed ground.

Embankments should be built higher, by about 300 mm, than the final formation level. If an embankment is to be left for a considerable period of time, say several months, before the roadworks are done, care must be taken to preserve the top surface.

This is particularly important in clay ground. In this type of ground, if the surface is left exposed, it may dry out and crack. It should be well compacted and covered with an insulating material, such as earth. If allowed to dry out, deep cracks occur, which in wet weather allow water to penetrate and form soft areas at the bottom of the cracks. This causes instability and plasticity of the embankment (*see* Figure 3.9).

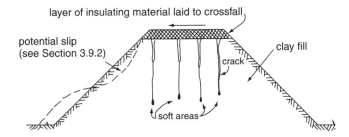

Figure 3.9 Protection of clay fill embankment

Embankments are often constructed with terraced slopes. This helps both the stability of the embankment and the retention of the topsoil on the slope (Figure 3.10). N.B.: To clarify Figures 3.9 and 3.10, the vertical scale is greater than the horizontal.

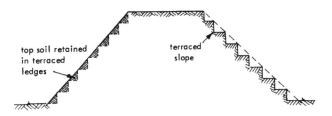

Figure 3.10 Terracing embankment to retain topsoil

3.3 Compaction

It is essential that fill material used to form embankments is compacted to the maximum dry density and to an accurate surface profile. The factors that determine this are:

(*a*) material characteristics of the fill
(*b*) moisture content of the material
(*c*) type of compaction equipment being used
(*d*) mass (weight) of the equipment in relation to its width of roller or base plate
(*e*) thickness of the layer being compacted
(*f*) number of passes required.

The DfT Specification, Clause 612, provides detailed guidance to the use of a range of materials. Compaction

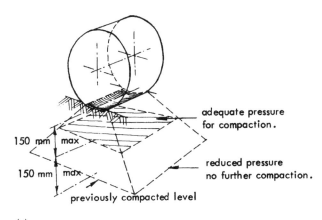

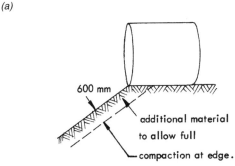

Figure 3.11 (a) and (b) Compaction of top edge of embankment

plant (*see* Section 11.14), which should be selected carefully to give the best results on the material used, consists of:

(*a*) rollers of various types – smooth wheel with or without vibration, grid, tamping, sheepsfoot, pneumatic

(*b*) vibrating plates – vibro tampers, power rammers, dropping weight compactors.

The final compaction is specified either by method (type of fill, thickness of layer, type of equipment, number of passes) or by performance requirement – a stated dry density which can be achieved by any method and is determined by site tests.

Specification by performance is usually applied to sites large enough to justify the cost of special equipment and laboratory analysis of the fill. On smaller scale works, it is prudent to fill in layers of not more than 150 mm compacted thickness and to roll with a minimum of four passes or until compaction marks disappear. Figure 3.11(a) shows compaction of 150 mm layers and Figure 3.11(b) the precautions taken at the edge of an embankment.

Shaping of the fill should be checked continuously so that, in the course of filling, the surface develops an accurate profile.

3.4   Replacement of Topsoil and Grassing

This is the final operation in the earthworks.

On an embankment, the less the quantity of topsoil that is to be replaced, the less chance there is of slips occurring. The standard thickness is 100 mm. The growth of grass roots helps to consolidate and bind the topsoil. Turfing, seeding, glass fibre strands, bitumen spray, chopped straw, etc. are all useful ways of binding together the sloping surface of the side of an embankment. For seeding *see* Chapter 8.

In recent years many other techniques have evolved, which include laying geotextiles or geosynthetic materials, some being in mats or grids and other types in pre-formed shapes. These methods address the problems in providing reinforcement and protection from erosion with embankment slopes. *See* Sections 3.9.2 and 3.9.3 for some details of methods.

3.5   Cut and Fill

These terms are used in earthworks where, for economic reasons, the excavated earth in cuttings (cut) should, if possible, balance the amount needed to form embankments (fill). Disposing of surplus cut or importing fill usually involves extra work and expense.

In new highway construction, 'cut and fill' refers to excavation in cuttings down to formation level for the new highway, this excavated material then being carted to adjacent areas where the formation is to be above the natural ground level and where 'fill' is required to form an embankment up to the necessary formation level (*see* Figure 3.12).

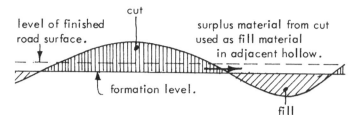

Figure 3.12   Longitudinal section of road to show areas of cut and fill

Ideal design uses all the excavated material from the cuttings in exactly forming the embankments. Care must be taken to ensure that the excavated material is suitable for use in the fill areas. Topsoil should be stored on one side until the fill is virtually completed and can then be used on the sides of both cuttings and embankments.

Suitable machines include: scraper; bulldozer, grader, dragline excavator, dump trucks/lorries and tractor shovels.

3.6   Capping Layer

In situations where the sub-grade soil has a CBR (California Bearing Ratio) of less than 5% if the road is to be of flexible construction, or less than 3% for a rigid pavement construction, a capping layer should be included. (The CBR test is described in Section 10.35.) This capping layer should consist of material of CBR value at least 5% above the normal sub-grade value.

One of the main functions of the capping layer is to provide a firm foundation on which to lay and compact the sub-base and subsequent pavement layers. By this method, a cost saving in the more expensive sub-base materials should be possible.

3.7   Methods of Drying Out Waterlogged Ground

The normal way of dealing with wet ground is to improve the drainage, usually by laying a system of shallow drains of agricultural pipes or porous concrete pipes, to carry the water away from the area where the earthworks are to be carried out.

An alternative is to fill the drainage trenches with broken stone or gravel to form filter drains in order to lower the water table.

This may also be achieved by laying fin drains. These consist of pervious fabric at each side, separated by spacers to form a type of honeycomb structure. This is laid vertically at one side of a very narrow trench laid to falls and held in position by a backfill of excavated soil. Fin drains may be used in situations where French drains were formerly laid and are useful for draining soft ground (*see* Figure 3.13).

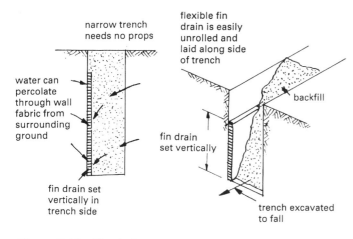

Figure 3.13   Fin drain

Further information about fin drains is given in Chapter 4, Section 4.19.3.

In clay soils one suitable treatment is to use quicklime (CaO: calcium oxide). This immediately changes the consistency of the clay with the clay particles tending to flocculate into silt-sized aggregations, the soil becoming more friable, apparently drier and more workable. The clay becomes less sticky and its shearing resistance increases.

Quicklime is better than hydrated lime as it takes up a third of its own weight in hydrating (to CaOH). In addition, heat is generated in this chemical reaction and this can be sufficient to drive off water from wet soil (as steam and water vapour). The addition of 2–5% of quicklime (this percentage being by weight of wet soil) to a very wet, soft, clayey soil will stiffen it sufficiently for site plant to be able to travel across it and for the soil to be handled (e.g. loaded and carted away or bulldozed aside, etc.).

3.8   Fascine Construction

This is used where the bearing capacity of the natural ground is inadequate to support the road structure, i.e. to improve the bearing capacity of that ground, but is suitable only for lighter traffic and may not be able to stand up to heavy main road traffic.

This type of construction is used across marshy ground, swamps, peat bogs, etc. The alternative is to blast the peat bog along the line of the road and replace the peat by an additional depth of sub-base stone.

Fascines consisting of bundles of brushwood, cut to approximately the same length and in equal-sized bundles. These are laid, close packed, on the ground to be crossed and at an angle to the line of the intended road, to form a mat. Additional layers of fascines are added, each at a different angle, and intertwined, until the ground is firm. The normal type of road structure can then be laid on this mat of brushwood.

Care must be taken to ensure that the fascines are laid wider than the intended road, in order to provide adequate support for the road edges or hard shoulder and to provide a base for a verge.

3.9   Stabilizing Ground

Geotextile synthetic materials, often referred to as 'plastic mesh', have many advantages over earlier fabrics. They are usually made of high-strength filaments encased in a polyolefin sheath in flat form, and are non-corrodible, lightweight and flexible.

Plastic mesh structures can be used to stabilize and improve the load-bearing capacity of soft ground. These net structures allow free passage of water and do not clog. The mesh also acts as a load-distributing layer, thus reducing the stress in underlying weak soils.

Plastic mesh sheet has a basic advantage over solid materials when used as a horizontal sheet to improve the load-bearing capacity of natural soils. The solid materials, such as fibrous felt sheeting, caused problems due to the clogging of the sheet. This resulted in a build-up of pore water in the sheeting which in turn led to separation of the sheeting from the soil and the development of slip surfaces, which weakened the ground structure. Polythene sheeting also causes water retention with resulting weakness due to the excess of water in the ground above the sheeting. However, with the plastic mesh, or net, the water can drain through the mesh and so avoid the clogging and weakening effect.

The mesh can be used to reduce the thickness of road construction required and allow traffic to use ground of very low CBR value with a minimum of surface material. Care must, however, be taken to leave sufficient thickness of material for frost protection, many low CBR sub-grades being frost susceptible.

The mesh helps to produce a high-density, shear-resistant layer of soil which distributes vertical loads (*see* Figure 3.14) and restrains soft ground in the lower layers of the sub-grade – it is reputed to make the soil more cohesive.

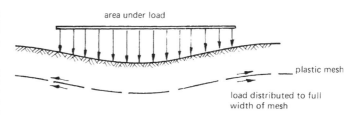

Figure 3.14   Plastic mesh used as shear resistant layer in soft ground

3.9.1 Geotextile Separation

On certain types of pervious soil, geotextile fabric may be used:

(a) to reduce the possibility of loss of granular material from the sub-base as a result of water seeping through the pavement and carrying it into the sub-grade;

(b) to prevent the intermixing of the sub-grade and sub-base (capping) during construction and in use;

(c) to allow work to proceed during adverse weather conditions;

(d) to allow construction to be carried out over ground conditions where traditional methods would not be economically viable.

There is a variety of fabrics available to meet specific requirements. Figure 3.15 shows the use of Terram as a separation membrane.

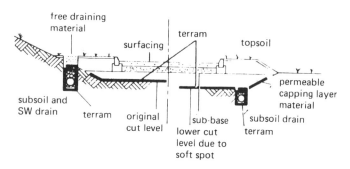

Figure 3.15 Typical surfaced road construction

3.9.2 Geotextile Reinforcement

Soils can only withstand limited shear stress and cannot generally be used for steeply angled profiles without external support or internal tensioning. Soil structures have been designed so that they are either in compression due to their geometrical shape, or have external support, such as retaining walls. Now, with the use of plastic mesh, slope point reinforcement and anchorages can provide a cheaper solution. Figure 3.16(a) illustrates this. Figure 3.15 shows

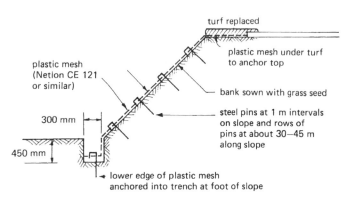

Figure 3.16(a) Plastic mesh as embankment slope reinforcement

how grids of plastic mesh can be used in a different, in this case horizontal, application.

Where embankments are constructed of piled and soft soil, soil reinforcement may be used. Sheets of geosynthetic material are put down in layers (see Figure 3.16(b)). This can significantly improve the strength of the soil embankment.

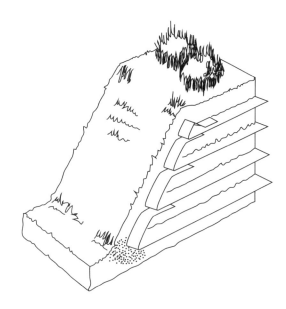

Figure 3.16(b) Geosynthetic sheet reinforcement on steep bank

3.9.3 Geotextile Erosion Control

To prevent erosion, banks can be covered with a geomat or grid consisting of two layers of extruded and bi-orientated material. This is designed to protect and help establish rapid grass growth on a slope likely to be subjected to surface erosion. The grid will protect the top soil and anchor the roots of the grass to the soil (see Figure 3.17(a)). Where the ground is dry and stony due to lack of organic material, then a grid of honeycombed geocells may be laid in mat

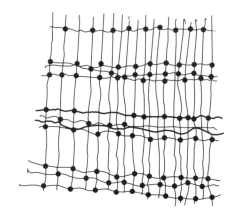

Figure 3.17(a) Polypropylene geomat

form (*see* Figure 3.17(*b*)). Once laid it is filled with lightly compacted soil and the area seeded.

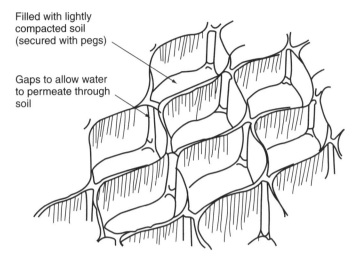

Figure 3.17(b) Honeycombed geocells (height 25–100 mm)

Coast, river and canal banks are subject to erosion. The defence structures designed to combat this erosion can take various forms, but usually require a filter to prevent soil outwash from a sloping soil face. Traditional filters consist of graded aggregate, whereas geotextiles can easily be placed under large rocks to give the same filtering effect (*see* Figure 3.17(*c*)).

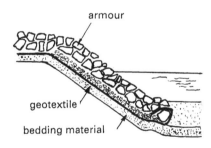

Figure 3.17(c) Geotextile erosion control

3.9.4 Gabion Walls

Gabion walls are built from layers of stone-filled 'boxes'. They have numerous uses as retaining walls, support for river banks, etc. The wall is formed by placing galvanized mesh rectangular boxes in position and filling them with hard, durable stone, not larger than 250 mm and not smaller than the size of the mesh. The best range is between 125 mm and 250 mm.

To be effective, the mesh boxes must be carefully formed from the flat packs or bundles in which they are delivered, using binding wire laced through adjacent panels. The gabions are wired together in small groups and carried to the levelled ground where they are joined to other completed sections (*see* Figure 3.18). The end gabions are anchored with rods driven through the end corners and the empty gabions are stretched until filled with stone. Horizontal bracing wires are fitted as the gabions are filled and finally the 'lids' are secured.

When placed, the gabions are wired down to the course below, thus forming a solid structure.

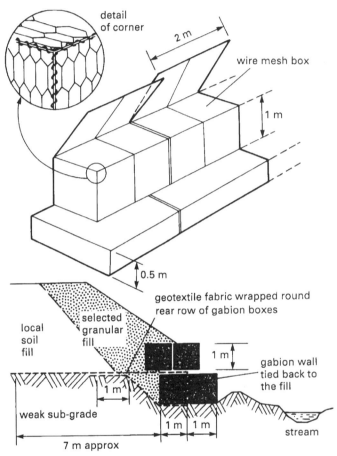

Figure 3.18 Gabion wall and geotextile application

3.9.5 Cement or Lime Stabilization

Given suitable ground conditions, the addition of cement or lime can give increased strength. A content of more than 2% cement or $2\frac{1}{2}$% lime is recommended, enough water being added to hydrate the cement or slake the lime. The exact content required is determined by laboratory test.

Mixing is carried out *in situ*, using a suitable stabilizing machine, in layers not less than 130 mm or more than 250 mm thick. Compaction must take place as soon as possible. The DfT Specifications for Highways Works Clauses 614 and 615 provide detailed guidance.

3.10 Load-bearing Fill

Many materials are suitable for use as load-bearing fill in roadworks.

These materials include gravels, crushed rock, colliery shale, and hardcore (such as broken brick, etc.). Pulverized fuel ash (PFA) is useful as it possesses excellent properties as load-bearing fill material.

3.11 Soils

Although soil mechanics is a complex subject, it is considered that a very elementary and, therefore, rather over-simplified introduction to the subject will help the road worker better appreciate the difficulties experienced in dealing with various types of ground and understand some of the technical terms used.

Soils may be divided into three main classes:

coarse-grained soils, also known as non-cohesive;
fine-grained soils, also known as cohesive;
organic soils.

Coarse-grained soils vary in size from boulders down to gravel and fine sand. Generally these materials do not shrink or swell and since they are stable under loads they form an excellent foundation.

Fine-grained soils include clays, marls, and silts. These materials show varying degrees of plasticity and are liable to considerable changes in volume following frost action and changes in moisture content.

Organic soils consist of topsoil, peat and other soils with a high content of decomposed vegetable matter. Organic soils are highly compressible.

A typical soil strata is shown in Figure 3.19.

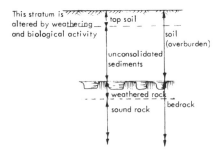

Figure 3.19 Typical soil strata

3.12 Soil Phases

Soil can be considered as an assemblage of solid particles interspersed with void or pore spaces which may or may not contain water.

In highway engineering, the aim is to keep void spaces to a minimum and make the volume occupied by mineral particles as great as possible.

A soil consists of three main items or phases: air, water, and solids. A soil in its most compact state will have no air voids, while one in its loosest state will have a high degree of air voids. In practice 3–5% of air voids will always be left in the soil, even at maximum compaction.

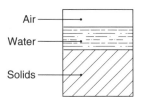

Figure 3.20 Soil phases

3.13 Compaction Testing

The maximum dry unit weight obtained in this test is regarded as standard for comparison purposes and is known as 100% compaction.

Actual field measurements of compacted soil are compared with this standard. It should be noted that soils which have been very well rolled may have a compaction figure in excess of 100%. Compaction may continue until virtually all the air voids have been removed, at which stage the soil is said to be 'saturated'.

As the moisture content of a soil increases, the unit weight increases, up to a maximum value, after which the unit weight decreases. This maximum value is known as the Optimum Moisture Content.

3.13.1 Optimum Moisture Content

The density of soil which is achieved by compaction is usually expressed as 'dry density' generally in kg/m³.

The optimum moisture content of a soil, which is the moisture content at which maximum dry density is obtained for a given amount of compaction, is normally arrived at by means of the BS Standard Compaction Test (BS 1377), which is described in Chapter 10, Section 10.36. It can vary from less than 10% to about 30%.

In general, the greater the density of a soil the lower is its optimum moisture content. This is illustrated by the examples given in Table 3.1.

Type of soil	Maximum dry density (kg/m³)	Optimum moisture content (%)
Gravel/sand/clay (mix)	2060	9
Sand	1930	11
Sandy clay	1840	14
Silty clay	1660	21
Heavy clay	1550	28
PFA*	1280	25

*Pulverized Fuel Ash (PFA), obtained from power stations, is the exception in the list of 'soils'

Table 3.1 Soil densities and optimum moisture content

3.14 Soil Water

The properties of a soil are greatly affected by its moisture content.

Soil moisture may be considered as follows:

1 *Gravitational water*. This is not held by the soil, but drains from it under the influence of gravity.

2 *Capillary water*. This is held by surface tension forces as a continuous film about the soil particles and in the capillary pores.

3 *Hygroscopic water*. This is absorbed (or, more correctly, 'adsorbed') from water vapour in the atmosphere as a result of attractive forces within the surface of the particles.

3.14.1 Gravitational Water

When moisture enters a soil as a result of rain, the tendency is for it to drain downwards through the soil due to the pull of gravity. The rate at which drainage occurs is mainly dependent upon the texture of the soil and its structure. Fine-grained soils, although having considerable porosity, are noted for the long period of retention of gravitational water (due to longer perco-lating passages and higher side wall frictions than most other soils).

The downward percolation of gravitational water continues until a depth is reached below which the soil pores are saturated with water. The water in this zone is known as groundwater and the upper surface of this is called the water table. Groundwater is not at rest in a soil, but tends to flow laterally and it should be noted that the water table is not horizontal, but tends to be a subsurface replica of the ground surface.

3.14.2 Capillary Water

If soil contained only gravitational water, then the pores in the material above the water table would be dry at all times other than when gravitational water was flowing through. However, if a cross-section is taken through the ground, it is found that a zone exists, above the water table, within which the pores are wholly or partially filled with moisture. This zone is often termed the 'Capillary fringe' and its thickness in various materials is as shown in Table 3.2.

Material	Capillary fringe thickness
Coarse sands	20–50 mm
Medium sands	120–350 mm
Fine sands	350–700 mm
Silts	700–1500 mm
Clays	2.000–4.000 m

Table 3.2

The pores in a soil above the water table can be considered to form an irregular mass of inter-related capillary tubes; the smaller the size of the pores, the greater is the height through which the water will rise due to the capillary action.

Capillary flow of moisture can occur in any direction, not just vertically.

The stability of fine grained soils, particularly clays, is very heavily dependent upon the moisture content. For a soil above the water table and within the capillary fringe, the cohesion of the material is mainly dependent upon the capillary water present, this moisture being held between adjacent soil particles.

When a soil is compacted, the soil particles are pushed more closely together, the water tending to be pushed out from between them; if sufficient compaction is applied, it is possible to reach the condition where the soil pores are full of water, all the air having been squeezed out, and the soil cohesion between particles is lost.

3.14.3 Hygroscopic Water

When a soil is dried in air, say in a laboratory, moisture is removed by evaporation until the moisture remaining in the soil is in equilibrium with the moisture vapour in the air.

If a sample of the soil is then dried at 105–110°C and the moisture content driven off is determined, the value obtained is called the hygroscopic moisture content. Knowledge of this hygroscopic moisture content is required for laboratory tests involving the addition of controlled amounts of water to the soil.

On site, it must be remembered that the hygroscopic content can be a variable quantity for a given soil, as its value depends on the temperature and humidity of the air above the soil at the time the test is carried out.

3.15 Consistency of a Soil

The consistency of a soil is a measure of its resistance to flow and is affected considerably by the moisture content of the soil.

The consistency is specified according to one of six stages of cohesiveness, based on moisture content limits, as follows.

3.15.1 Upper Limit

The upper limit of viscous flow is the point above which the mixture of soil and water flows like a liquid.

3.15.2 Liquid Limit

This is also called the lower limit of viscous flow, above which the soil and water flows as a viscous liquid and below which the mixture is plastic.

3.15.3 Sticky Limit

Above this the mixture of soil and water will adhere, or stick, to a steel spatula or other object that is wetted by water.

3.15.4 Cohesion Limit

At this limit crumbs of the soil cease to adhere when placed in contact with each other.

3.15.5 Plastic Limit

This is the lower limit of the plastic range, at which the soil starts to crumble when rolled into a thread under the palm of the hand.

3.15.6 Shrinkage Limit

This can also be called the lower limit of volume change, at which there is no further decrease in volume as water is evaporated from the soil.

3.16 Liquid Limit and Plastic Limit

Of these limits, two are of importance to highway engineers. These are the liquid and plastic limits.

3.16.1 Liquid Limit

The liquid limit of a soil is the moisture content, expressed as a percentage by weight of the oven dry soil, at the limit, or boundary, between the soil being plastic or liquid. There is a special apparatus for determining the liquid limit, but the concern, in these notes, is only for the actual results, some of which are as follows:

(a) Clays   40–60%
(b) Silty soil   25–50%
(c) Sands   Sandy soils do not have liquid limits and are considered to be non-plastic.

3.16.2 Plastic Limit

This is defined as the moisture content of the soil at the boundary between plastic and semi-solid states, i.e. it is the moisture content at which a sample of soil begins to crumble when rolled into a thread under the palm of the hand.

Pure sands cannot be rolled into threads – hence their non-plastic definition.

Silty soils have a plastic limit of 5–15%
Clay soils have a plastic limit of 10–30%

Plasticity index = Liquid limit – plastic limit.

Characteristic of the soil	Soils of equal liquid limit with plasticity index increasing	Soils of equal plasticity index with liquid limit increasing
Compressibility	about the same	increases
Permeability	decreases	increases
Rate of volume change	increases	—
Dry strength	increases	decreases

Table 3.3 Comparison of soil characteristics

3.17 Backfill

In the New Roads and Streetworks Act, the Specification for the Reinstatement of Openings in Highways contains requirements for material used in backfill, some of which can be classed as soil.

Information contained in this chapter is intended to provide a general understanding of activities connected with earthworks. The specific requirements of the New Roads and Streetworks Act are dealt with more fully in Chapter 15.

Revision Questions

1 What pattern of land drains are laid when springs or other underground water features are encountered during excavation?
2 For a rigid pavement construction, under what circumstances would a 'capping layer' be included?
3 What are the 3 main classes into which soil can be divided?
4 What is meant by the 'LIQUID LIMIT' of a soil?

Note: See page 305 for answers to these revision questions

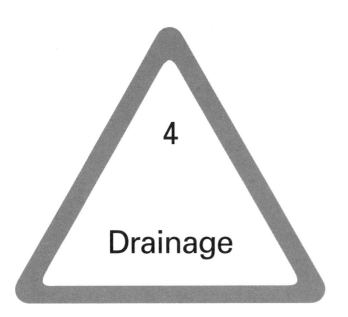

4

Drainage

4.1 General

The main drainage system consists of:

(a) Foul water drainage to carry away safely, and without causing unacceptable pollution, domestic sewage, and factory liquid wastes.

(b) Storm water drainage to limit or prevent flooding caused by rainfall.

In a drainage system, for reasons of economy, as much use as possible must be made of gravity. (The alternative is continuous pumping of the sewage, which is costly.) This requires, as the first step in the design of a drainage system, a close study of the contours of the area to be drained so that the system flows, to the greatest possible extent, downhill to the final point of disposal.

In the case of foul water, the drains will lead to a sewage treatment plant and, for storm water, usually directly to a ditch or stream.

4.2 Foul Water Drainage

Three systems are in existence as follows:

4.2.1 Totally Separate

Two sewer networks are required, one carrying foul sewage and the other storm water. This is the standard system for all new development, though it is also the most expensive as it requires two separate drainage systems

throughout, including around houses which the system serves. It has the great value, however, of reducing pollution of watercourses to a minimum.

4.2.2 Partially Separate

This is a compromise system in which two separate networks are provided, one carrying storm water and the other foul water plus the storm water off house roofs and backyards. This has the merit of providing for a flushing of the foul sewer by the roof water, etc. each time it rains. To allow for very exceptional periods of rainfall, the foul sewer must be provided with storm overflows, but they very seldom come into use. The possibility of pollution of watercourses remains, and the system is not favoured when compared with the totally separate one.

4.2.3 Combined System

Only one sewer network is provided, to carry both foul and storm water. It is to be found in many old built-up areas, as it was the cheapest and quickest way of providing the drainage required under the Public Health Acts.

The storm water flow is many times greater than the quantity of foul sewage carried and, in fact the sewer is designed as a storm water drain. The size of pipe required tends to be very large in serving a district and to reduce this 'storm overflows' are provided. When the storm flow exceeds a predetermined limit, these overflows discharge the excess flow to a convenient watercourse, and permit the use of a smaller pipe size. The overflows are designed to operate when the flow in the sewer exceeds six times the dry weather flow. However, the discharge of the mixture of foul and storm water causes pollution of the watercourses and is a serious, and nowadays unacceptable, fault in the system.

Where the combined system is encountered in redeveloping old areas, it is necessary to convert it to a separate or partially separate system.

4.3 Foul Drains

The emphasis in roadwork is, obviously, on storm drainage, but some basic knowledge of foul drainage is needed as these systems will be encountered from time to time in carrying out road building and improvement schemes.

A foul sewer is normally designed to carry the foul sewage of the area which it serves; flow is based on the number of inhabitants and the normal amount of water which they use each day, with an allowance for future development of the area. Also it should be noted that, even with a totally separate system there will be some storm water getting into the foul drain, because yard gullies in factory and garage areas, etc. should be connected to it, not to the storm sewer. The quantity of liquid wastes from industry must also be allowed for in designing the foul sewer; these are determined by the actual amounts of liquid disposed of during the working hours of each industrial process.

4.4 Storm Water Drains

The estimation of the quantity of storm water to be carried in a sewer system is a difficult matter to determine as it is affected by:

(*a*) The unpredictable variation of rate of intensity of rainfall during a storm.

(*b*) The unpredictable direction of movement of the storm. (For instance a storm moving along a valley will produce more water than one moving across it.)

(*c*) The degree of impermeability of the area on which the rain falls (impermeable ground will not soak up any of the rain, which will, therefore, run off into the drains via the collecting gullies).

(*d*) The storage capacity of the sewer system itself.

(*e*) The time taken for rain water to get into the sewer after falling on the ground.

Although sewer design does not fall within the scope of this book, these basic factors affecting size of pipe used, etc. are included to provide a greater understanding of the function of the storm drain and foul sewer.

4.5 Road Drainage Systems

In roadwork, the main drainage requirements fall into two categories.

4.5.1 Sub-soil Drainage

There must be a provision of sub-soil drainage to cope with water in the ground and to ensure that the water table can be kept low enough not to allow saturation of the road structure and its surrounding soil. Excess water of this nature causes the structure and ground to become plastic and incapable of withstanding traffic loads. Subsoil drainage also helps to prevent frost damage to the road structure by keeping it drained.

4.5.2 Drainage of the Carriageway

The surface of the road must be kept clear of standing water in order not to increase the dangers to road users. To accomplish this, roads should be cambered when straight, laid to crossfalls on bends, and adequately provided with gullies or grips to dispose of rain water which falls on the carriageway.

In many cases, the drainage of this surface water and the sub-soil water is connected to the same systems of piped drains or open ditches through which the water is removed to a safe distance from the carriageway itself.

4.6 Sub-soil Drainage

This is usually associated with the construction of new roads, but it may also become necessary to lay a new drainage system for an existing road system, particularly where there have been other works, such as large-scale building projects, adjacent to existing roads.

In new works, great care must be taken to keep the water table below the road formation level. This, in low-lying wet areas, may require the provision of a system of land drains, collector drains and a long outfall pipe or open ditch before the ground can be drained sufficiently for the structural work of the road building to be started.

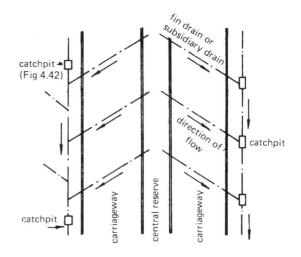

Figure 4.1 Sub-soil drainage

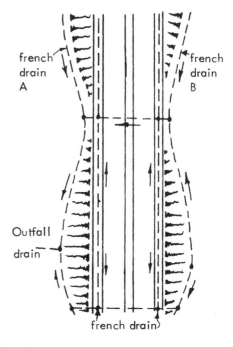

Figure 4.2 Cutting and embankment design

Springs may be encountered in a position close to or under the proposed road structure. These must be piped clear of the roadworks.

With most new works in rural areas, the construction alternates between cut and fill, with the design, where possible, using the surplus from the cutting to form the embankments.

Before the work of cutting and filling begins, it is good practice to lay French drains (*see* Figures 4.56*a* and *b*) along the limits of the roadworks (drains A and B in Figure 4.2). On completion of earthworks, similar drains should immediately be laid in the verges and central reserve, in order to keep the formation of the new road clear of surface water.

All these drains are laid to falls and interconnected and linked to a suitable natural watercourse by an outfall drain.

4.7 Road Drainage

For new roadworks in urban areas, which tend to be of the estate road development type, there is generally an existing sewerage system associated with the existing storm sewers to which the drainage from the new roads must be connected. If, however, the ground surface levels do not permit this to be done, a new outfall storm sewer must be provided to discharge either to a natural watercourse or into the same or another storm sewerage system at a point away from the site.

In this type of construction, an adequate system of road gullies will be needed with new storm sewers, usually laid in the road or alongside it, into which each gully can be connected.

4.8 The Theory of Drainage

4.8.1 Aim of Drainage

The objective in designing a road drainage system is to provide drains/sewers with sufficient capacity to deal with the most severe storm conditions and laid to falls so that the water entering the sewers will be conducted away from the road by gravity, eventually discharging to a natural watercourse.

Care must be taken that this outfall to the watercourse is high enough to be clear of the highest normal water level of that river or stream, to avoid any risk of excess water from the river backing up into the storm sewer. This is the work of the engineer, the roadman taking over when the actual work of drain laying is to start.

4.8.2 Self-cleansing Velocity

The pipes are laid uphill from the outfall point with the gradient of the pipeline being sufficient for any water flowing through it to move with at least 'self-cleansing velocity'. This is the minimum velocity of flow at which solid gritty particles will remain in suspension in the water and not settle out on the invert of the pipe. Any settlement of this suspended matter will tend gradually to build up and block the pipe.

4.8.3 Gradient

The gradients at which the various sizes of pipes are self-cleansing are approximately as follows:

Pipe size diameter (mm)	Self-cleansing minimum gradient
100	1 in 50
150	1 in 75
225	1 in 112
300	1 in 150

All these gradients are safely in excess of the absolute limit and can be determined simply by dividing the pipe diameter (measured in millimetres) by 2.

Hence 150 mm diameter pipe has a self-cleansing gradient of 1 in 150/2 or 1 in 75 (as above). The real limiting gradient for this size of pipe is 1 in 120, but the pipe must be running at least half full before the minimum gradient conditions apply. However, for this to be true, the pipe must be truly straight, all joints must be accurately aligned at the invert and the whole pipeline must be accurately straight. This is impossible to achieve under real working conditions, so the steeper gradient of 1 in 75 allows a margin of safety and ensures that the pipeline, as laid, is self-cleansing. This still means that the greatest care needs to be taken to lay the pipeline as accurately as possible to line and level.

Where there is little natural fall available on a site, the pipe size can be increased above that required to deal with the actual flow, so that a 'slacker' gradient may be used.

4.8.4 Manholes

All sewer pipelines must be laid in straight lengths to a constant gradient over each length.

At every change of:

(a) direction;
(b) gradient;
(c) pipe size;
(d) pipe type;

a manhole is required. Where several lengths of sewer join together, a manhole or inspection chamber should be built. For long straight lengths of sewer, manhole or inspection chambers are often provided at not more than 100 m intervals.

Manholes are provided to enable the sewer to be rodded in the event of it becoming blocked.

4.8.5 Gullies

Gully connections to storm sewers may be made by laying junction pipes at suitable points along the sewer pipeline. The branch pipe for the gully is not less than 100 mm diameter.

If it is not possible to predict the positions of the gullies and thus include the junction fittings at the time of laying the sewer, connection to the sewer can be made later by means of a saddle placed over a hole carefully cut on the existing sewer crown (*see* Figure 4.52*a*).

4.9 Drainage Work

Some procedures for setting out, excavating trenches and laying pipe lines are given under the following headings.

1 Reference pegs for alignment and invert levels of pipe lines.
2 Controlling the alignment of the excavation.
3 Controlling the depth of excavation.
4 Setting out for pipelaying.
5 Pipe laying.

4.9.1 Reference Pegs

Primary reference points should be provided by a surveyor in the form of pegs (*see* Section 3.1.1) which give the alignment of the pipe line (or distance offset to the centre of the pipe line) and the invert level of the pipe.

Pegs (or studs when on an existing road surface) should be driven in at

(*a*) each end of the pipe line;
(*b*) the centre of each intermediate manhole, inspection chamber, etc.;
(*c*) each change of direction or gradient not having a manhole, etc.; and
(*d*) additionally, where the distance between two pegs exceeds 50 m.

The drawing/schedule left with the site supervisor should indicate the

(*a*) location of offset pegs (numbered);
(*b*) offset dimensions;
(*c*) distance between pegs;
(*d*) depth below peg to pipe invert level;
(*e*) location of junctions, manholes, etc.

together with general information as to the type and size of pipe, bedding materials and any haunching requirements, etc.

4.9.2 Controlling the Alignment of the Excavation

Two common methods are in use when machine excavating a trench. N.B.: The excavator must be horizontal to ensure vertical trench sides are cut (*see* Figure 4.3*a*).

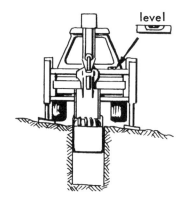

Figure 4.3(a)   Trench excavation across a slope

1   Pin and Line Method

(*a*) The centre line of the trench is first determined from the peg offset distances.
(*b*) The distance from the machine kingpost to the outside of the stabilizer is measured (*see* dimension 'w' on Figure 4.3*b*).
(*c*) A nylon line on pins is set out at distance 'w' from the centre line of the trench.
(*d*) The excavator is positioned so that the stabilizer is just clear of the nylon line.

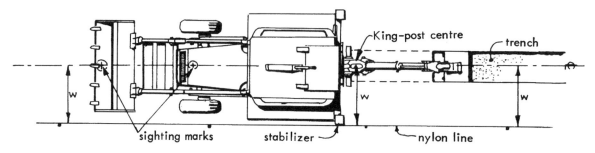

Figure 4.3(b)   Trench excavation: use of pin and line

2 Chalk Line (or Sand, etc.) Method

(a) As above in 1(a) the centre line is determined and marked with a contrasting material to that of the ground.

(b) The excavator is positioned so that the kingpost is directly over the marked centre line.

(c) Sighting marks are then made with chalk on the machine (see Figure 4.3b).

(d) The sighting marks must be lined up with the centre line every time the excavator is moved.

4.9.3 Controlling the Depth of Excavation

Care in this operation can save labour in 'bottoming out' the trench by hand or using extra bedding materials prior to laying the pipes. One method is to erect sight rails and use a traveller to determine the level of the trench bottom and thus control the depth of excavation.

1 Sight Rails and Traveller Method (see Figure 4.4a)

(a) Sight rails are erected at each manhole or inspection chamber and if the length is over 50 m, an inter-mediate rail is set up at the mid-point of this length. The intermediate rail helps in sighting over the length and also provides a check in case of damage to one of the rails during construction.

(b) The rail may straddle the centre line of the trench or it may be offset, but the vertical posts should not be placed in the path of the excavator.

(c) The height of the sight rail is found by transferring the level on the offset reference peg to the vertical post and measuring up to give a convenient height 'd' above trench bottom level. (The trench bottom level will be slightly deeper than the pipe invert level to allow for bedding of pipe.) A traveller is then constructed for use with the sight rails and a banksman can control the depth of excavation by holding the traveller vertically, whilst it is 'sighted in'.

(d) The trench is excavated to the correct level, the topsoil usually being piled to one side and the spoil to the other side. If the trench has a vertical face then timbering or other forms of trench support must be provided as the excavation proceeds (see Section 4.10).

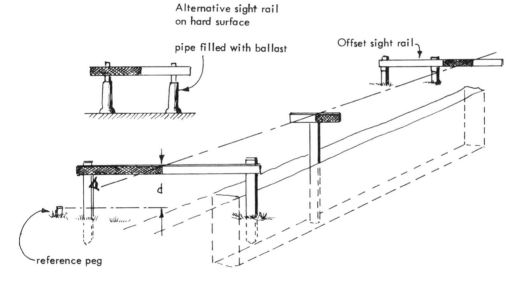

Figure 4.4(a)   Use of sight rails and traveller

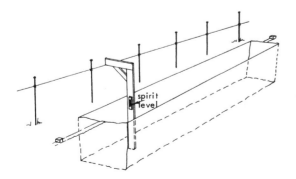

Figure 4.4(b)   Use of pins and line with special traveller

2 Pin and Line Method

Where the excavation is in open ground, the nylon line erected on pins to control the alignment of the excavation may also be used to control the depth.

(a) The nylon line must be pulled taut and tied off at each pin, then from the levels marked on the end pins, the gradient is boned in on to the intermediate pins.

(b) A traveller is constructed as in the previous method, but is designed to touch the line and hence control the depth and check the vertical cut of the trench (see Figure 4.4b).

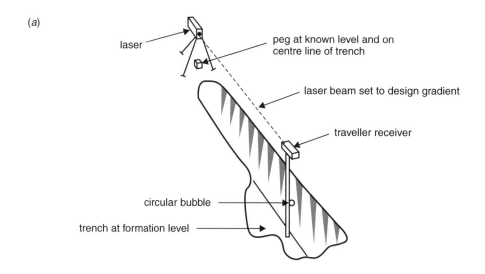

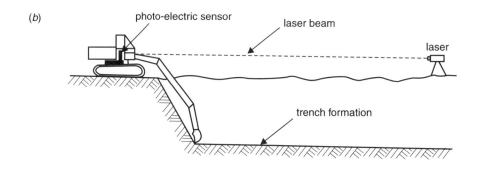

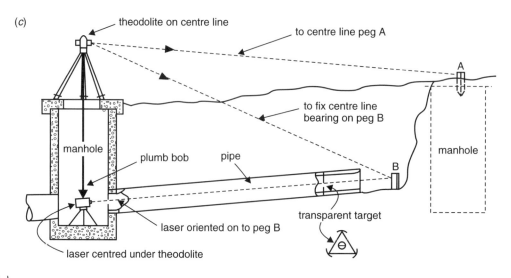

Figure 4.5(a)–(c)

3 Laser Method

A laser is an instrument which produces a powerful beam of light which can be concentrated into a narrow beam which does not scatter or become diffused like ordinary light. A laser is multi-functional, being capable of controlling trench excavation as well as the actual process of laying the pipes. In trench excavation, the laser replaces the sight rails and traveller method (Figure 4.4a) and the pin and line method (Figure 4.4b) by being set up on the centre line of the trench, over a level peg. The height above the peg is measured to obtain the reduced level of the beam. (*See* Chapter 2, Section 2.6.4.)

The laser is set to the required grade and is used in conjunction with an extendable traveller, set to the same height as that of the laser above the formation level. When the trench excavation is at the correct level, the laser spot will shine on the centre line of the traveller receiver, or alternatively, on to a levelling staff (*see* Figure 4.5a). With machine excavation, the laser may be used to control the depth of excavation by fitting a photo-electric cell on the machine at the appropriate height. The information can be relayed to the operator by way of a console in order to confirm that the trench has been excavated to the correct level and grade. *See* Figure 4.5b.

4.9.4 Setting Out for Pipe Laying

There are several techniques used for ensuring correct line and level when laying pipes. Four ways are described below.

1 Laser Use in Pipelaying

In this method, a target is fixed in the pipe and the laser installed on the centre line of the trench. The laser is then orientated in the correct direction and angled to the required gradient of the pipe. By centring on a target up to 150 m away, line and level can be maintained.

Where a manhole has already been constructed and invert levels are already determined (*see* Figure 4.5c), the centre line can be transferred from peg A to peg B in the trench, to orientate the direction of the laser beam.

2 Centre Line

In this method, pipes are laid below a nylon line on pins placed centrally along the trench bottom.

(a) From primary reference points, steel pins are positioned in the bottom of the trench to mark the centre line of pipelines at centres of manholes, changes of gradient, etc.
(b) The pins are marked at invert level from primary reference points or sight rails.
(c) The height of the line is determined by measuring up from the invert level the distance 'd' (*see* Figure 4.6), which includes 3 mm tolerance.
(d) The intermediate pins are placed at approximately 10 m and the gradient boned in on to them. The line is then 'eyed in'.

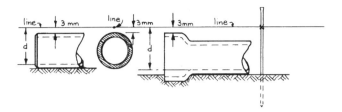

Figure 4.6 Pipe alignment: centre line method

3 Offset Line

This is an alternative to the centre line method.

(a) The nylon line and pins can be erected as previously described but offset at a suitable distance from the pipe line. The distance 'd' is from the invert to the top of the pipe.
(b) A simple gauge is then used to lay the pipes to line and level (*see* Figure 4.7). This allows the line to remain undisturbed as pins do not have to be removed as laying proceeds.

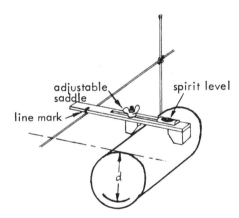

Figure 4.7 Pipe alignment: offset line method

4 Traveller

When laying pipes the line and pins can be an obstruction, hence pipes can be checked individually for line and level using a traveller in conjunction with sight rails. The traveller used in excavating the trench may be adapted by fitting a 'shoe' to sit in the invert of the pipe thus giving line and level (*see* Figure 4.8). The length of the traveller is easily determined. For instance, if the level peg at the one end of the sewer is 2.750 m above invert level, the rail may be set up at 1.250 m above the level peg and a 4.000 m traveller constructed, as shown in Figure 4.8.

If the level peg at the other end of the length of sewer is 3.100 m above the invert level, then the rail at that end should be erected 0.900 m above the peg, for the 4.000 m traveller to be used.

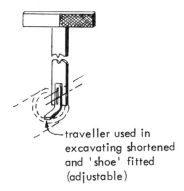

traveller used in
excavating shortened
and 'shoe' fitted
(adjustable)

Figure 4.8 Pipe traveller

4.9.5 *Laying the Pipes*

1 Placement of Bedding Material

(*a*) Check from established levels that the correct thickness of bedding can be laid. If not, the trench bottom must be trimmed.

(*b*) Place bedding material in trench and spread to correct level.

2 Pipe Selection and Handling

(*a*) Check that the correct size and type of pipe is available and inspect for defects, i.e. chips, cracks, damage to joints, etc. Reject where necessary.

(*b*) Place pipes ready for laying alongside the trench at 45° to the edge.

(*c*) Lower larger pipes onto bedding material using a sling or pipe hook.

3 Laying

(*a*) Clean pipe ends.

(*b*) Lay the pipe to line and level working uphill. If spigot and socket pipes are being used, they should be laid with the socket uphill. When laying pipes, holes to accommodate the socket or joint should be cut into the trench bottom or bedding, so that the pipes will lie with the length of the pipe barrel supported by the ground.

(*c*) Check that the highest point of the pipe is 1–3 mm below the line for the centre line method, or that the laser beam is correct for line and level.

(*d*) Lay subsequent pipes and form joints.

(*e*) Check the complete pipe line for line and level and, if used, remove line and pins.

4 Backfilling

(*a*) Supervisor to give permission prior to backfilling.

(*b*) This backfill should be carefully compacted, to avoid disturbance of the pipes, and to a depth of 300 mm above the barrel of the pipes.

(*c*) Thereafter, backfill should continue, layers of about 150 mm being compacted successively until the backfill is completed.

(*d*) If turf was removed prior to excavating then relay on a bed of about 100 mm of topsoil which is only lightly compacted. (The construction of manholes has been left out at this stage, but would normally proceed at the same time as the drain laying.) All surplus materials and equipment are then removed and the site left clean and tidy.

4.10 Notes on Supporting Excavations

Two methods of excavating trenches were previously outlined. The techniques by which excavations can then be made safe for operatives to work in should now be carefully considered.

If an excavation has its sides sloped back to a safe angle (which should be less than the angle of repose for the material), then there is no danger of collapse of earth on to operatives working in the trench. However, few excavations can be treated in this manner as there is often insufficient site space available for cutting back the sides. Consequently trenches are usually cut having vertical sides or faces.

A vertical face can under certain conditions, collapse, whether it be earth. rock or any other natural material. Hence, support is necessary and this may be in the form of timbering or steel sheeting.

4.10.1 *Reasons for Collapse of Face*

There are numerous reasons for collapse and it is often a combination of two or more factors which bring about this dangerous situation. The usual reasons are:

1 Failure of the soil to support its own weight.

2 Steeply angled bedding planes which encourage slips (Figure 4.9*a*).

3 Water sheds or seepages (Figure 4.9*b*).

4 Breakdown of the cohesion of the soil by frost or heavy rain.

5 Changes in the type of soil such as weak material underlying sound rock or layers of sand and clay (Figure 4.9*c*).

6 Failure due to trenching on or near the position of an earlier excavation.

7 Vibration due to the close passage of vehicles and plant.

8 Failure due to loads placed too near the edge of the face.

9 Impact of heavy loads such as pipes, striking the sides of a trench when being lowered.

10 Inadequate timbering in supporting a face.

11 Lack of cohesion (or shearing) of soil (Figure 4.9*d*).

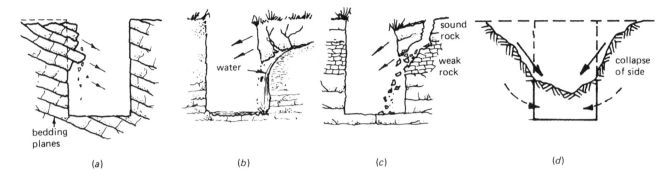

Figure 4.9(a)–(d) Collapse of trench face

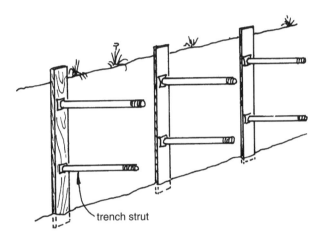

Figure 4.10 An example of a designed trench support system for shallow trenches

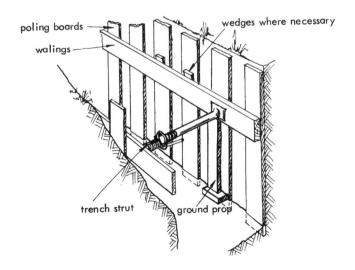

Figure 4.11 An example of a designed trench support system where greater support is required

4.10.2 Supporting a Face

The method used and the amount of support necessary must depend on the nature of the ground being excavated, the depth of working, and the presence of other factors such as water, or vibration due to heavy traffic flow in the vicinity.

Side support systems should be properly designed to withstand the loading forces likely to be encountered and should be installed, inspected, altered and removed by a trained and competent team. All excavations in unstable ground should be provided with a side support system, irrespective of excavation depth.

The type of support used can vary widely as different techniques exist making use of various items of equipment. Some of the more common methods are illustrated.

Close Boarding (Figure 4.12)

This was used in deeper trenches with variable ground conditions, but this method has been largely replaced by steel sheeting.

Steel Sheeting (Figure 4.13)

This technique is used both for deep excavations and also where poor ground is likely to be found, such as running sand. The sheets are sometimes driven before excavation commences.

Various section profiles are available in trench sheeting. Typical sections are shown in Figures 4.14(a) and (b).

The lap-jointed section can be supplied in standard mild steel (3.57 mm thickness) or high tensile steel (2.64 mm thickness) and lengths are 2–6 m in 0.5 m increments.

Steel trench sheeting can be driven and withdrawn much more easily than timber sheeting and it also provides reasonable water-tightness when under pressure. A special driving cap may be used to protect the top of the sheeting when driving by hand or by an automatic hammer. When withdrawing the sheeting, a pin and shackle can be fitted in

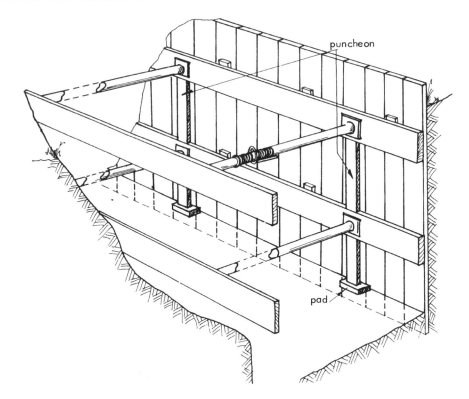

Figure 4.12   An example of a close boarded trench support system

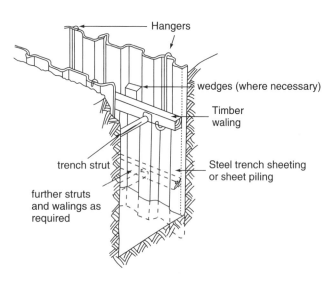

Figure 4.13   Steel sheet trench support

Figure 4.14(a)   Steel sheet piles: interlocking type

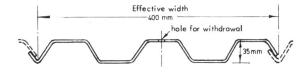

Figure 4.14(b)   Steel sheet piles: lap-jointed type

the hole provided, for hoisting by crane or winch. A profile such as this one will resist distortion that is liable to occur with much use.

Drag-Box (Figure 4.16(a))

This is a robust trench support unit having high strength and low weight (700 kg) which can be dragged forward in a newly excavated trench. It has been designed for use with backhoe/loaders and small 360° excavators and the single front strut gives good visibility for the plant operator. One type has a length of 3 m, height 1.5 m and internal width from 0.65–1.05 m.

Trench Struts

Mechanical or hydraulic trench struts have replaced timber for general trench work. These are easily adjusted.

The trench strut has replaced timber struts for general work in trenches as it is easily adjustable and quickly placed. A typical strut is shown below in Figure 4.15.

Several sizes are manufactured and some typical ones are as follows:

Closed (mm)	Extended (mm)
305	457
457	711
686	1105
1041	1702

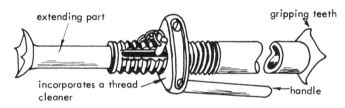

Figure 4.15   Trench strut

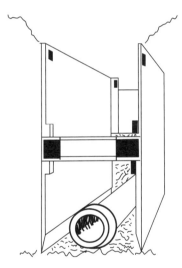

Figure 4.16(a)   Drag-box

4.10.3   Installing Trench Support

As work should never be carried out ahead of supporting the face, a common procedure is for timber poling to be lowered into the end of the excavation and strutted by trench struts, thus providing two 'H' frames.

This can be done with some difficulty by workers on the surface, but more often than not, individuals will stand inside the excavation to erect this first preliminary timbering.

However, the first person into the excavation is at considerable risk and if the task is to erect support for the protection of pipelayers then it follows that this person too requires some form of protection whilst working. A number of devices have been produced to provide safe working areas for persons erecting trench support systems and a few of these are described. An old device was an open fronted steel frame and mesh box, which was lowered into an excavation, allowing a person to work from a safe place. This was supplanted by using adjustable mechanical steel jacks. These formed 'H' frames which together with walings, protected an individual. They were rather cumbersome and have been replaced by adjustable hydraulic aluminium jacks (*see* Figure 4.17). A more recent development is the protective hole-box (*see* Figure 4.16(*b*)).

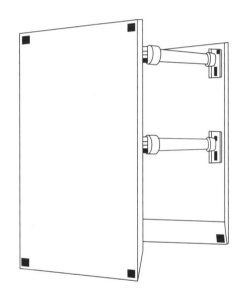

Figure 4.16(b)   Protective hole-box

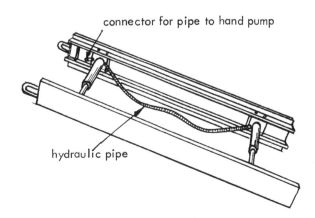

Figure 4.17   Adjustable trench jacks: hydraulic type

Protective Hole-box

The box may be lowered into an excavation and allows maintenance and repair work to be carried out in complete safety. Its length is 1.5 m, height 2.5 m and internal width 0.84–4.44 m. The weight of the unit is 1.3 tonnes (*see* Figure 4.16(b)).

Adjustable Hydraulic Aluminium Jacks

These are lighter (27 kg) than the older steel mechanical jacks and are readily handled and placed by one person.

The hydraulic system operates at approximately 2.5 N/mm^2. Special auxiliary equipment is needed such as a pump complete with hose, valves and gauges; hooks to remove the jack from the trench; fluid release tools for disconnecting the pump hose and releasing the fluid prior to removing the jack from the trench (*see* Figure 4.17).

The jacks are manufactured in several sizes, giving widths from 430 mm to 3 m. A typical size would be 600 mm extending to 1.066 m with a vertical length of 2.130 m.

4.10.4 Special Techniques used when Pipelaying in Wet Ground: Methods of Dealing with Sub-soil Water

(a) Sub-drain in Bottom of Trench (Figure 4.18)

When there is water in the trench bottom, a 100 mm diameter agricultural pipe set below the floor of the trench can often keep the trench dry for pipelaying. The sub-drain would lead to a sump from which the water may be pumped up to an interceptor box before being discharged at an approved disposal point. The interceptor box contains a filter medium and allows sand, etc., to settle out of the water.

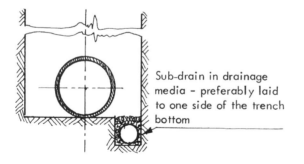

Figure 4.18 Trench sub-drain

(b) Muddy Situations (Figure 4.19)

Where there is too much water/mud for the trench bottom to be kept clear in the above manner, a 'coffin' may be used to keep the laying end clear for pipes to be laid, together with an expandable stopper or 'badger' acting as a readily movable pipe seal.

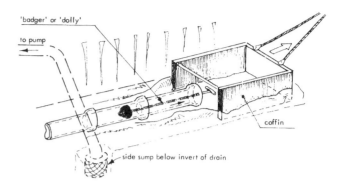

Figure 4.19 Use of coffin and badger in wet trenches

(c) De-watering (One Method of Well Pointing Waterlogged Excavations – Figure 4.20a)

1 Boreholes are sunk at regular intervals on each side of the intended line of trench and about 0.5–1.0 m clear of the excavation.

2 For a deep trench, one set of boreholes may be sunk to about half the depth-to-invert in order to lower the water table sufficiently to start excavation. Subsequently, a second set of boreholes would then be needed, sunk to

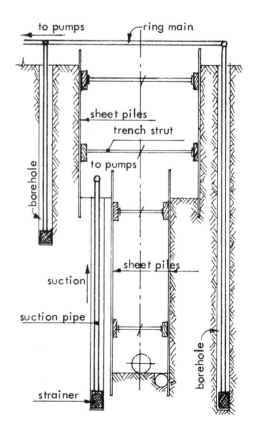

Figure 4.20(a) Well pointing

about 0.5 m below invert level, to allow the lower stages of excavation to be carried out.

3 Vertical pipes (about 50 mm diameter) with strainers on the lower end are inserted in the boreholes and connected to a common suction 'main' or ring main, leading to the pumps which are run continuously. The discharge may have to be piped away for a considerable distance, to ensure that the water does not flow back into the workings.

4 It is essential to provide duplicate pumps and motors, permanently connected into the system. Change over can then quickly be effected and, in extreme conditions, both sets of pumping machinery can be used simultaneously.

5 It will probably be necessary to sheet pile the trench in this very unstable ground.

(d) Electro-osmosis

If two electrodes are placed in the ground and a current is applied, water will flow from one to the other, at which point it can be collected and pumped out to the surface. The method can only be used in wide excavations, wide cuttings, and embankments.

(e) Fixing Ground Water by Chemical Injection

Cement, bentonite, silica, aluminates, resins, etc., make a 'gel' which consolidates the ground through which work is to be carried out – this is used to form a shield to prevent water draining (Figure 4.20b).

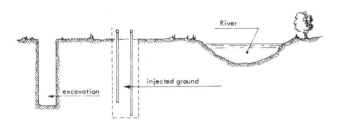

Figure 4.20(b) Chemical injection to protect adjacent trench

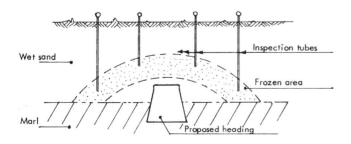

Figure 4.20(c) Freezing of ground water to protect heading excavation

(f) Fixing Ground Water by Freezing

This is a similar idea to that of chemical injection. N.B.: Care must be taken not to freeze water mains or sewers in adjoining ground.

4.10.5 Drainage Excavation in Tunnels and Shafts

Heading can be used in situations where the use of an open trench is not possible due to site restrictions, such as:

(a) Excessive depth or, particularly, a narrow site.
(b) Adjacent buildings, the foundations of which could be undermined.
(c) Where vibration from excavators, etc., could cause damage to old buildings.
(d) On grounds of economy, particularly where the sewer is extremely deep.

Procedure

1 Shafts are sunk to sewer invert level, at suitable intervals, say every 60 m (with, possibly, an intermediate air shaft). The shaft must be large enough to allow easy removal of skips of excavated material.

2 Shafts must be timbered (*see* Figure 4.22) if not of the precast ring type.

(a) Kentledge method (precast ring type – Figure 4.21a)

1 Dig a shallow hole and set up a cutting ring, to correct line and level. Great care and accuracy is needed if the shaft is to be sunk truly vertically.

2 Build the rings on the cutting base ring to a height of about 1.5 m above ground level.

3 Backfill the annular space, round the outside of the cutting ring, with soft clay which must be well rammed.

4 Check the line and level of the rings.

5 Place timber or steel joists on top of the rings and load with pig iron. This will cause the cutting edge to sink into the ground.

6 A suitable lubricant is often of value in speeding up the process.

7 Dig out equally all round and under the cutting edge and allow the rings to settle.

8 Fix additional rings as necessary.

9 Continue excavating and sinking, taking great care to keep the cutting edge level.

(b) Segmental shaft (precast ring type – Figure 4.21b)

1 Excavate a shallow hole and build up four segmental rings.

2 Fill the external annular ring with weak concrete.

3 Starting at one side, excavate slightly more than one ring depth and start to build up an extra ring on the

underside by bolting the segments to the completed ring above.

4 Continue until the new ring is complete.

5 Repeat the process until the required depth is reached, grouting up every two rings.

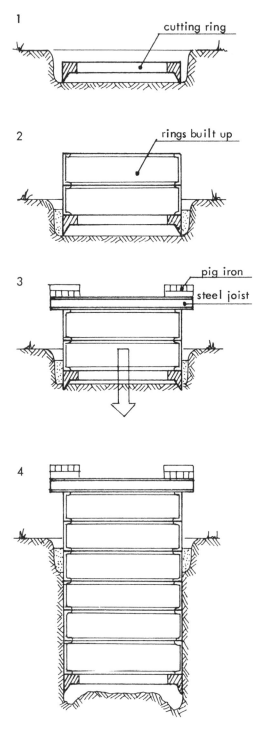

Figure 4.21(a)   Kentledge method

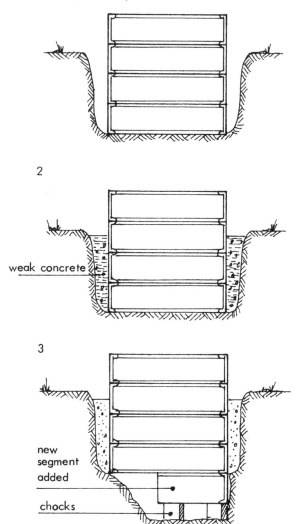

Figure 4.21(b)   Segmental ring method

4.10.6   Trenchless Technology

1   'No Dig' Techniques

Many techniques for the trenchless installation of drainage pipes have been developed in recent years. There are now numerous methods, some of the most common being micro-tunnelling, auger boring, impact moling, pipebursting, directional drilling, and pipe ramming.

Micro tunnelling – this method in its most sophisticated form uses a remotely controlled cutting machine in conjunction with pipe jacking, to install pipes. It can be used in most ground conditions by selecting the appropriate cutting head and is capable of very accurate installation of pipes of up to 900 mm diameter. Fluid jet drilling is one form of microtunnelling, using high pressure water jets in a cutting mole.

The principal machines used are the Bentonite Slurry Shield machine (BSSM) and the Mechanical Spoil Removal machine (MSRM). The BSSM reinforces the side of the bore hole as work progresses and surplus material is removed as a slurry to the surface. The MSRM removes the soil to the surface and both types will drive lengths of 100–120 m depending on the arching and friction characteristics of the ground, the weight, strength and material of the pipe, pipe diameter and jacking force available from the machine. With steerable heads an accuracy of ±30 mm on line and level is possible when using a laser beam.

Auger boring – this can be achieved in most soils up to 75–100 m, and is an established method, normally using steel pipes up to 150 mm diameter. The auger is not steerable and has an accuracy of around 1.5–2% of drive length (*see* Figure 4.22).

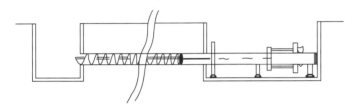

Figure 4.22   Thrust borer with simultaneous pipe jacking (soil removed by auger)

Impact moling – this technique involves driving the mole through the ground by compressed air, establishing a duct in which various services may be placed. A common method extensively used by utilities in most soils, for small diameter mains and services when crossing roads. Not steerable having an accuracy of around 1%, with length of drive up to 70 m.

Pipe bursting – this can be undertaken in most soils and is controlled by a winch cable within an existing pipe. Extensively used for water mains and sewers (often for sewer pipes of 400 mm replacing those of 280 mm) and by gas utilities for upsize pipe replacements.

Directional drilling – used in most soils for long crossings under rivers, etc. Is steerable using up to 1000 mm steel pipes (*see* Figure 4.23).

Pipe ramming – can be used in most soils and is used mostly for short crossings (up to 65 m) beneath roads, etc. Not steerable and may be used with steel pipes up to 1400 mm in diameter.

When 'no dig' techniques are used on sewage work, relatively small driving and exit shafts, which later become manholes, are constructed and the tunnelling is carried out between these shafts. The pipes used in trenchless systems need to have specially designed flush joints and stronger, hence thicker, walls in order to withstand the jacking stresses.

Microtunnelling machines are expensive to hire so the other cheaper systems such as impact moling are still used for such operations as short road crossings. Often an

Figure 4.23   Steerable boring system

oversize duct, within which another pipe is subsequently laid, is driven through the ground.

Decaying pipelines in city areas are often replaced by a new pipe of slightly smaller diameter which is forced through the original pipeline, e.g. 125 mm diameter is forced through a 150 mm diameter pipe.

2   Renovating Existing Pipelines

This has been done for many years by spray linings of cement mortar within iron pipes, or by more recently developed resin applications.

Slip lining is another technique – this method requires a special plastic pipe (known as PE pipe) which is inserted into a defective pipe. The consequent reduction in pipe capacity is the main disadvantage of this process. A later development allows the PE pipe to be reduced in diameter temporarily, but once installed within the defective pipe, the PE pipe is expanded until it becomes a tight inner fit, thus minimizing the reduction of flow capacity.

Another form of renovation consists of inserting a soft lining, this being in the form of a resin-impregnated felt or woven fabric 'sock' which is 'fitted' inside the defective pipe. The insertion is often made by compressed air blowing the sock along the inside of the pipe and forcing it against the existing one. Heat or ultra-violet light is then used to harden the sock and seal it against the inside of the defective pipe.

4.10.7   A Check List for Safe Excavations
Before beginning work
1 Consult with the Safety Officer if particular problems are foreseen.
2 Site security – if an area where vandalism, etc., is known to be prevalent, consult with the police for regular inspection of roadwork signs and lamps, etc.

3 Appoint a 'competent person' and provide registers, abstracts of Acts, etc.

4 Decide where site huts, stores, plant compound, etc., are to be located and provide adequate manoeuvring space.

5 Ensure an adequate supply of signs, lamps, barriers, ladders, trench support materials, etc.

6 Position of spoil heaps – not less than 1 m from the edge of the excavation.

7 Query necessity for bridges and gangways.

8 Location of all public services – expose and mark.

When work is in progress

1 Are operatives using protective clothing and equipment?

2 Are all working faces secure; where used is timbering adequate and secure; are wedges tight and condition of timber satisfactory?

3 Signs of peeling on unsheeted face? Soil seeping through sheeted faces? Are proper pumps installed and is pumping drawing soil from behind timber?

4 If ladders are necessary are they of suitable length, strength and secured?

5 Are spoil heaps far enough back and are they being worked upon whilst men are in the excavation below?

6 Pipes, bricks, tools, etc., should be away from the edge of the trench or positioned so they cannot fall in.

7 Passing traffic should be well clear of trench and stop boards provided for dumpers when tipping.

8 Are operatives following a safe procedure both when erecting timbering and withdrawing timber prior to backfill?

9 Should regular testing for harmful gases be carried out?

10 Are gangways with toeboards and guardrails provided where necessary?

11 Is the work fenced off and are warning notices posted during day-time? Is the work properly guarded and lit at night?

12 Finally, have all the legal requirements been complied with, both in site work and entries in registers recording examinations?

4.11 Types of Pipes and Joints

Pipes are regarded as

(*a*) rigid, if made of cast iron, concrete, asbestos cement, clay or similar materials; and

(*b*) flexible, if made of plastic materials, pitch fibre, steel, etc.

Joints are considered

(*a*) rigid, if made solid by caulking or by welding or bolting the flanges of metal pipes; and

(*b*) flexible, if joints are made with deformable rings or gaskets held between pipe spigots and sockets, sleeves or collars.

Current trends in pipelaying are towards butt jointed plastic pipes with a simple sleeve seal.

Fin drains, now commonly used as an alternative to certain types of pipe, are described in Section 4.19.3.

4.11.1 Spigot and Socket, Salt Glazed Clay Pipes

For many years this was the standard type of drainage pipe and required a cement/sand mortared joint. This was made after caulking the joint with a length of tarred yarn wrapped round the spigot and caulked into the socket, as shown in Figure 4.24(*a*). This type of pipe and joint is suitable for foul water drainage systems.

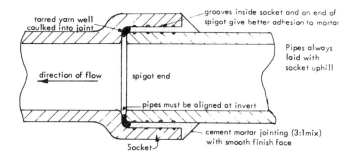

Figure 4.24(a)   Salt glazed spigot and socket pipe

4.11.2 Spigot and Socket, Vitrified Clay Pipes and Hepseal or Similar Type of Joint

These pipes (Figure 4.24*b*) are self-aligning, flexibly joined and self-sealing. They are a modern alternative to the traditional spigot and socket glazed ware (*see* Section 4.11.1) and are easier and quicker to lay. They are suitable for foul drainage and are available in 150, 225, 300, 375 and 450 mm diameters.

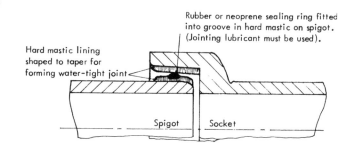

Figure 4.24(b)   Vitrified clay spigot and socket pipe

4.11.3 Double Spigot Vitrified Clay Pipes with Flexible Couplings (Hepsleeve or Similar)

This provides a butt joint with a polypropylene coupling sleeve which encloses the pipe ends as shown (Figure 4.24*c*). It is acceptable for either storm or foul water drainage and is available in 100 and 150 mm diameters.

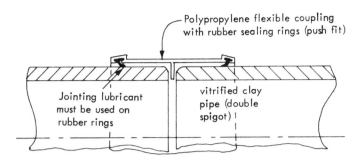

Figure 4.24(c) Vitrified clay double spigot pipe and flexible coupling

N.B.: This is simple pipe (same both ends) and is easily manufactured. The range of pipe sizes is likely to be increased as appropriate sealed joints are developed.

4.11.4 Spigot and Socket, Spun or Cast Concrete Pipes (Flexibly Jointed)

These have joints similar to the Hepseal type of joint, since they are sealed with a rubber ring and are therefore flexible (*see* Figure 4.24d).

Concrete pipes are normally manufactured from Ordinary Portland Cement, but can be made from sulphate-resisting cement. Flexibly jointed concrete pipes are available in four different strengths, i.e. Standard, Light, Medium and Heavy (*see* BS 5911). Larger sizes are manufactured by a spinning technique whereas smaller sizes are cast. Usual sizes are from 225–900 mm diameter.

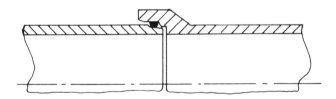

Figure 4.24(d) Concrete spigot and socket pipe

4.11.5 Asbestos Cement Pipes

These are made from a mixture of Portland Cement, asbestos fibre with or without filler, and are suitable for domestic sewage, most trade effluents and surface water drainage.

Although possessing good resistance to corrosion they are not in general use for highway drainage. Pipe sizes range from 100–600 mm diameter and in length 1–5 m.

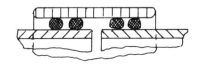

Figure 4.25 Asbestos cement pipe and coupling

4.11.6 uPVC (Unplasticized Polyvinyl Chloride)

These pipes are flexible, are usually up to 6 m long in 110, 160 and 200 mm diameters and are very light to handle. The pipes have either a tight push fit into the coupling joint (Figure 4.26a) or have a push fit into a flexible coupling joint (Figure 4.26b).

Care must be taken when pushing a pipe into the joint sleeve to use a tool which will not mark the face of the pipes (e.g. crowbars are not suitable).

These pipes have an additional advantage in that they can be used as an inner sleeve to an old and leaking length of clayware pipe, thus making a reasonably cheap but effective repair. uPVC pipes have largely replaced pitch fibre pipes in highway work.

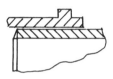

Figure 4.26(a) uPVC pipe – push fit coupling

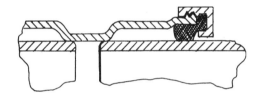

Figure 4.26(b) uPVC pipe – flexible coupling

4.11.7 Cast or Spun Iron Pipes

These are very strong and are to be found in places where the drain is either very shallow or above ground. They have sealed joints (Figure 4.27).

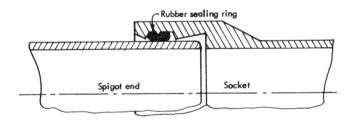

Figure 4.27 Cast or spun iron pipe

4.11.8 OG (Figure 4.28a) and Spigot and Socket (Figure 4.28b) Dense Concrete Pipes

These may be used where a sealed joint is not essential.

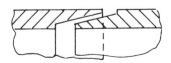

Figure 4.28(a) Dense concrete ogee pipe

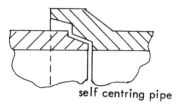

Figure 4.28(b) Dense concrete spigot and socket pipe

4.11.9 Porous Concrete Pipes with Ogee Joints (Figure 4.28c)

These are in common use for subsoil drainage and have the lower half of the barrel lined to make them impervious. Hence, care must be taken when laying the pipes that the impervious portion forms the lower half of the pipeline (see Figure 4.28c).

Figure 4.28(c) Porous concrete pipe

4.11.10 Vitrified Clay Subsoil Pipes with Dry Push fit Flexible Polyethylene Sleeve Joints (Figure 4.29a)

These are suitable for use as filter drains, or as combined filter drains and surface water drains. They are available from 100–300 mm in diameter and range from 1–1.6 m in length.

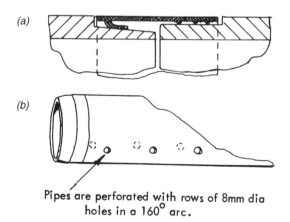

Figure 4.29(a) and (b) Vitrified clay sub-soil pipe and sleeve joint

4.11.11 Agricultural Pipes

Plain earthenware spigot ended pipes used for land drains, unjointed and with butt joints, through which water enters the line of drain. These are often laid by the mole draining method. This is essentially a deep ploughing process cleaving through the ground with the pipes being fed immediately to the rear of the plough blade, the ground closing over the pipeline as the machine proceeds.

For further information on pipes, manufacturers' manuals should be consulted.

4.12 Protection of Pipes

Where the trench bottom is of unstable ground, it has been the practice to excavate an additional depth of about 150 mm and provide a concrete bed on which to lay the pipes. Again, collar/socket holes are made in this concrete bed, so that the pipe barrel is supported along its length by the concrete.

Additional support for the pipes can be provided in the form of a haunch to the pipe or, alternatively, by a complete concrete surround. Owing to the practical difficulties in forming the concrete haunch and, particularly, the surround, it is more usual to fill the full width of the trench to a height of 150 mm above the pipe barrel.

Care should be taken when using a surround, or more so if backfilling a trench with lean concrete, that it does not form a barrier to the movement of sub-soil water and thereby artificially raise the water table.

Concrete bed and haunch   Concrete surround

Figure 4.30 Concrete bed and haunch and concrete surround to pipe

More recent practice tends to use a granular fill in place of the concrete bed and haunch, as this has the merit of providing not only support for the pipes, but it also facilitates drainage of the trench bottom and has been developed particularly since the introduction of flexible pipe joints.

Where the pipes need to be further protected from loads passing over the pipeline, a concrete arch or saddle may be constructed over the pipe. This is reinforced and should be at least 100 mm thick or one quarter of the pipe diameter, whichever is the greater thickness.

4.13 Bedding Construction and Bedding Load Factor

Special design tables should be used when planning pipelines. The construction details are based on a 'bedding load factor'. This is the ratio of the design load on the pipeline to the crushing strength of the pipes.

The load on the pipes consists of two parts: the fill load (weight of backfill material in the trench and over the pipe) and the traffic wheel load (acting on the surface of the fill).

Many factors are taken into account in the design tables for pipelines, but they are beyond the scope of activity of roadworkers.

The roadworker needs to know the designer's actual requirements in order to lay and construct the pipeline to the standards and specification of the engineer.

The designs are based on five Bedding Load Factors which are:

    3.4 (the greatest/strongest pipe protection, i.e.
    2.6 which enables the pipeline to withstand
    1.9 a load of 3.4 times the crushing
    1.5 strength of the pipes themselves)
and 1.1 (the least pipe protection).

For a Bedding Load Factor of 3.4 the pipeline has reinforced concrete protection (*a*) as a bed and cradle for the pipes, or (*b*) as a protective arch over the pipeline (with granular bedding under the pipe) *see* Figure 4.31.

For a Bedding Load Factor of 2.6, the pipeline has plain concrete protection:

    (*a*) as a cradle for the pipes or
    (*b*) as a protective arch over the pipeline, *see* Figure 4.32.

With a Bedding Factor Load of 1.9 selected fill materials are required, but there is no concrete cradle or arch (*see* Figure 4.33).

When the Bedding Load Factor is only 1.5 or 1.1, the pipeline is unprotected and this construction is limited to work in fine grained soils in dry conditions. This factor relates to types of pipe other than vitrified clay, *see* Figure 4.34.

4.14 Rules for Good Pipelaying Practice

Before leaving the subject of pipelaying, it is important that the rules for good practice in pipelaying should be stated.

These are:

(*a*) Close supervision of trench excavation should be maintained, to avoid hand trimming or overdigging.

(*b*) The trench bottom must be firm – to provide a good bed for the pipes.

(*c*) Lay the pipes accurately to line and level.

(*d*) Ensure good jointing of the pipes.

(*e*) Backfill carefully to avoid dislodging or damaging the pipes.

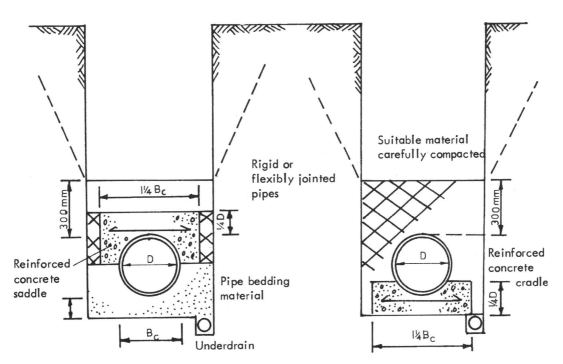

Figure 4.31(a) Bedding factor 3.4 (reinforced concrete saddle)

Figure 4.31(b) Bedding factor 3.4 (reinforced concrete cradle)

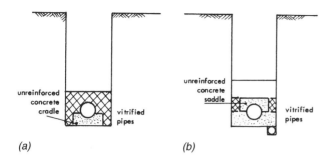

Figure 4.32(a) and (b)   Bedding factor 2.6

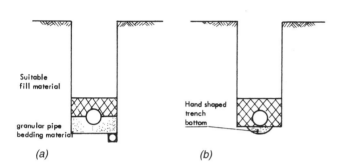

Figure 4.33(a) and (b)   Bedding factor 1.9

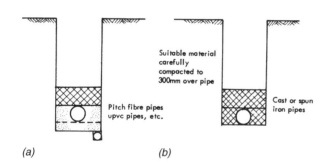

Figure 4.34(a) and (b)   Bedding factor 1.1

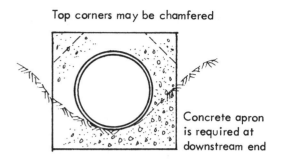

Figure 4.35   Simple concrete culvert

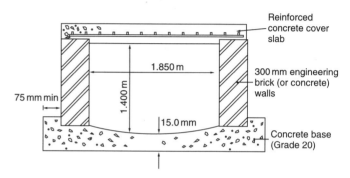

Figure 4.36   Typical brick and concrete culvert

4.15   Culverts

A culvert is defined as 'a structure, in the form of large pipes or an enclosed channel, for conveying water below road level'.

The most simple form of culvert consists of a length of concrete tubes (usually to conduct a watercourse through a road embankment). These tubes may either be surrounded with concrete (as shown in Figure 4.35) or bedded on granular fill.

As an alternative, a flat top culvert may be constructed *in situ* on the lines of that shown in Figure 4.36, which gives typical dimensions.

Larger culverts are often required in order to cope with stream flows at times of heavy rain or melting snow. These larger culverts usually consist of steel tubes which are ribbed for strength and are sectional, each 'ring' of culvert being composed of up to four pieces. Often, they are put together above ground and the completed steel 'tunnel' is lowered by a crane into its position on a prepared bed of granular material in the stream bed. By preparing the culvert above ground, it may not be necessary to divert the stream flow, the stream bed being dug out and filled to the required invert level with granular material on which the steel sectional culvert is then placed. The backfill material is then compacted around it and built up to the required road level.

The ends of the steel culvert are the cut at an angle to approximate to the embankment side slope (Figure 4.37). These ends may be 'capped' with a concrete 'headwall' to

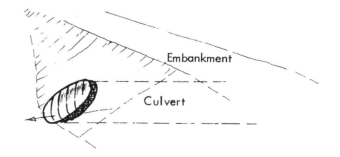

Figure 4.37   Steel tube culvert

improve the finished appearance. The steel sections may be coated with bitumen or galvanized on both inside and outside surfaces.

The medium diameter culvert may be in one piece 'pipes' using a rubber sealing ring and split flanged collar for jointing (*see* Figure 4.38).

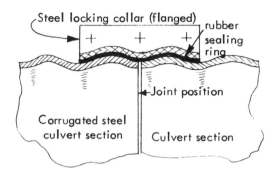

Figure 4.38   Steel tube culvert

Standard culverts or box culverts consist of precast concrete units, being rectangular in section (*see* Figure 4.39(*a*)). These have rebated joints which are sealed by using cement mortar or preformed mastic jointing strips (*see* Figure 4.39(*b*)).

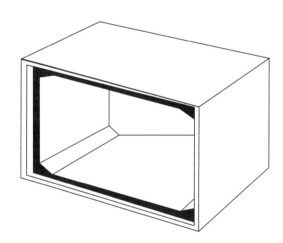

Figure 4.39(a)   Box culvert

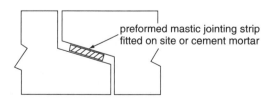

Figure 4.39(b)   Box joint

The culvert units can be designed to suit site requirements and to meet external loading conditions. The units are usually placed on a bed of Type A granular material conforming to the Water Services Association Specifications. End sections can be splayed at an angle or finished flush with headwalls (*see* Figure 4.55).

4.16   Manholes/Access Chambers

The object of constructing a manhole is to provide access to a sewer or drain.

Manholes are built:

(*a*) at junctions between sewers.
(*b*) at a change of gradient.
(*c*) at a change of pipe size.
(*d*) on a long straight length to provide access at such intervals that the lengths between the manholes can be rodded if necessary.

4.16.1   Manhole Design Factors (see Figure 4.40)

The main factors are:

(*a*) The manhole must be fitted with a strong cover and frame (cast iron).
(*b*) The access shaft should be at least 550 mm wide and preferably 600 mm wide.
(*c*) Step irons should be provided every 300 mm (approximately).
(*d*) There should be adequate headroom for workers to work safely in the manhole.
(*e*) The benching must have a crossfall, but should not be too steep for workers to stand on it.
(*f*) The manhole should be watertight.

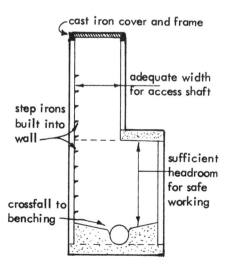

Figure 4.40   Typical manhole

There are many variations on the basic manhole structure, but all have the same objective of providing access to the pipeline. Some of the more common types are as shown.

(a) Inspection Chamber (Figure 4.41)

These are normally shallow, rectangular, with a cover and frame which are the size of the chamber.

Workers operate from above ground when rodding a drain from an inspection chamber.

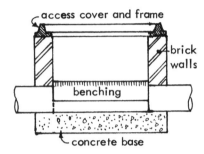

Figure 4.41 Inspection chamber

(b) Silt Trap or Catchpit (Figure 4.42)

This is a chamber built into a line of pipes with the base of the chamber below the line of the drain.

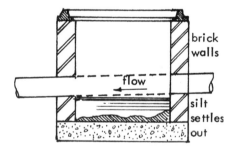

Figure 4.42 Catchpit

The ground/storm water flowing through the chamber reduces its velocity of flow on entering the enlarged area of the chamber and much of the grit and suspended solids in the water settles out into the bottom of the catchpit from where this sludge can be cleared out periodically, via the access cover.

(c) Disconnecting Chamber (Figure 4.43)

This is similar to a catchpit, but is built at the junction between drains. It connects a subsidiary drain (such as a collector drain) to a main drain without a junction, thus allowing silt to settle out.

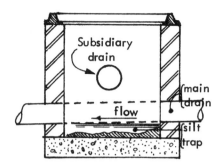

Figure 4.43 Disconnecting chamber

(d) Flushing Manhole (Figure 4.44)

This is a small manhole constructed at the upper end of a length (or several lengths) of pipeline into which a quantity of clean water can be discharged to flush that pipeline. It is only necessary on the 'top' length of a drainage system, where the flow in the drain/sewer is small and there may be a tendency for silt, etc., to settle out and block the drain.

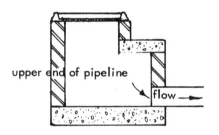

Figure 4.44 Flushing manhole

(e) Side Entry Manhole (Figure 4.45)

This is, essentially, a manhole to which access is provided via a separate opening and connecting shaft which leads into the side of the manhole. It may be convenient to provide access at the footway to a manhole under a carriageway, to avoid traffic disturbance if it is necessary to enter the manhole.

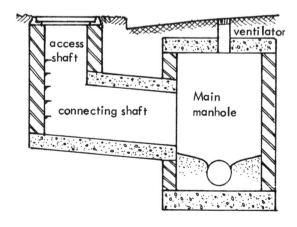

Figure 4.45 Side entry manhole

(f) Rodding Eye (Figure 4.46)

This is provided at the uphill end of a sewer where the depth is not great and where there is no need to provide a manhole, which would be much more expensive.

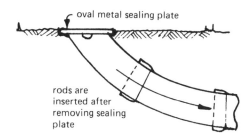

Figure 4.46 Rodding eye

(g) Television Surveys

The use of specialist closed circuit television equipment for sewer surveying is now widespread. Current UK requirements under the Confined Space Regulations 1997 expect risks to be reduced by workers not having to enter the sewer at all. Television survey lends itself to meeting this requirement.

The camera is either winched through the sewer, or driven through, mounted on a self-traction device. Whilst the latter method eliminates the need for line threading, tractor surveys in silted sewers, due to loss of traction, may result in abandonment and loss of equipment.

Television survey equipment should preferably include pictorial representation, supported by computer data storage systems. This has the following advantages over the less sophisticated equipment:

1 surveys can be readily updated;
2 summarizing information on the ranking of sewers for cost-effective remedial action;
3 surveys can be reproduced in a variety of different data formats;
4 historical information can be evaluated;
5 upgraded information can be linked to existing records;
6 progressive deterioration can be monitored.

4.16.2 Manhole Construction Materials

There are three main methods/materials used in the construction of manholes.

(a) Brick walls on concrete base and with concrete cover slab built *in situ*, rectangular in shape (Figure 4.47a).
(b) Precast concrete rings which build up into a circular section manhole and which are set on a concrete base (Figure 4.47b).
(c) Plastic (glass fibre) prefabricated manholes also circular in section with each ring flanged and bolted together.

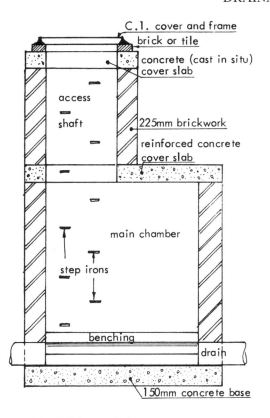

Figure 4.47(a) Brick manhole

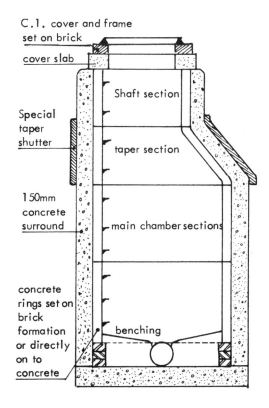

Figure 4.47(b) Precast concrete manhole

Both the concrete and plastic circular manholes require a surround of concrete for protection. The brick manhole walls are built thick enough to provide the necessary strength.

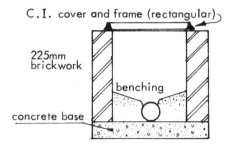

Figure 4.48(a)   Brick-walled inspection chamber

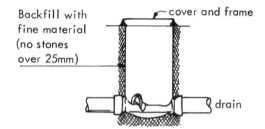

Figure 4.48(b)   GRP inspection chamber

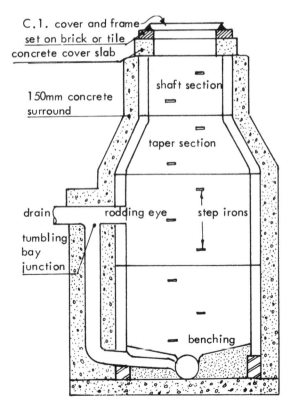

Figure 4.48(c)   Concrete manhole with back drop

There are many variations in detail in the construction of manholes, but typical detailed sectional drawings of each type are shown for comparison between the different types of construction. It is absolutely necessary to ensure that a manhole is watertight, since having piped water to this point, it is essential to prevent it seeping out into the road structure. Inspection chambers in brick and glass reinforced plastic (GRP) are also shown for comparison.

4.17   Gullies and Grips

4.17.1   Gullies

Gullies are fitted below the road channel to collect surface water from the carriageway and trap any silt carried by the water. By connecting the gully to a surface water sewer or open watercourse by means of a length of pipework (usually 150 mm in diameter) the surface water is conducted away from the road and led to a river or stream via the ditch or sewerage system.

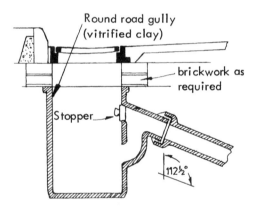

Figure 4.49   Road gully set in position with grid and drain connection

The positions of and number of gullies in a carriageway depend on the area which is being drained, the gradient of the road and the type of surface. They should be frequent enough to prevent water from standing or crossing the carriageway and thus causing an unacceptable hazard to traffic.

On steep slopes gullies usually have a specially shaped gully grid and frame to assist in leading the fast moving water into the gully (Figure 4.50).

Gullies may be of:

glazed stoneware (vitrified clay)
concrete
glass fibre or
plastic.

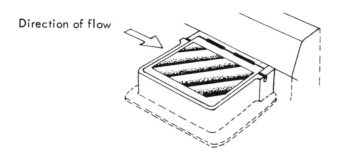

Direction of flow

Figure 4.50 Gully grid for use on steep slope

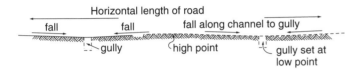

Figure 4.51 Layout of gullies and channel on horizontal road

The stoneware gully is heavy and easily broken or damaged.

The concrete gully is very heavy and difficult to handle.

The plastic/glass fibre gully is light, easily handled, not easily damaged, but subject to distortion.

Gullies should be set on a concrete bed and surrounded with 150 mm of concrete. When fixing a plastic/GRP gully, it is essential that it is weighted and held down when being concreted into position, otherwise it will float on the wet concrete and move.

Gully Connections

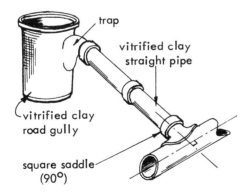

Figure 4.52(a) Gully connection to drain by saddle: vitrified clay gully (also see Section 4.8.5)

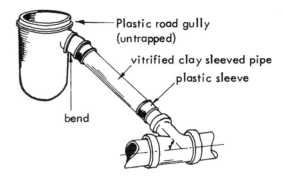

Figure 4.52(b) Gully connection to drain by junction: plastic gully

Saddle Connections

If no provision has been made when laying the main storm sewer for the connection of gullies, then the connection is usually made by a saddle.

Figure 4.53(a) Oblique saddle fitted to drain

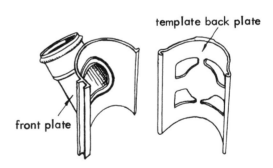

Figure 4.53(b) Saddle for plastic pipe

This entails making a hole in one of the pipes of the sewer (in the upper half, but to one side). The making of the hole, shaped to the saddle fitting, requires considerable care and skill when working with glazed stoneware, vitrified or concrete pipes.

For saddle connections to plastic pipes the job is easier, a template being provided and the pipe being cut with a sharp knife without fear of cracking it away from the hole.

4.17.2 Grips

In rural areas, where many minor roads are not kerbed, the drainage of surface water from the channel is usually by means of grips.

These are shallow channels cut across the verge to conduct the surface water clear of the carriageway and to a convenient roadside ditch.

Where the road is kerbed, there are special kerb fittings which provide an outlet for surface water, whilst also continuing the kerb line. These are usually piped to a nearby ditch or stream (Figure 4.54*a*). Where an obstruction prevents a gully being normally placed in line with the channel, a combination kerb grid can be fitted (Figure 4.54*b*).

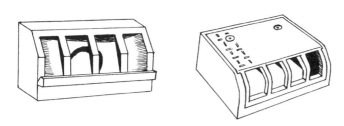

Figure 4.54(a) and (b) Side entry kerb grips

4.18 Headwalls and Aprons

Where a drainage pipe discharges to a ditch or other natural watercourse, it is usually necessary to provide a headwall and apron.

The headwall protects the bank of the ditch from erosion and from falling into the path of the water from the pipe. Sidewalls are often required in addition to the headwall.

The apron provides a strengthening of the ground at the point of issue of the water from the pipe and prevents excessive scouring of the ditch invert. Headwalls are constructed in brick, stone, or concrete and the aprons are almost always of concrete.

Typical details of headwalls and aprons are shown in Figure 4.55. A metal grill should be fitted over the end of the pipe and attached to the headwall, if there is a risk of children entering the sewer.

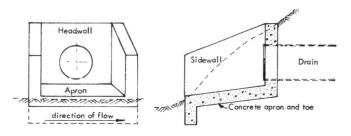

Figure 4.55(a) and (b) Concrete headwall, sidewalls and apron

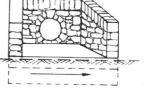

Figure 4.55(c)   Stone headwall

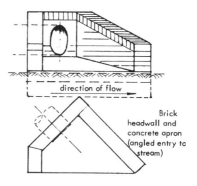

Figure 4.55(d)   Brick headwall and concrete apron (angled entry to stream)

4.19 Other Drainage Equipment and Techniques

There are many ways of dealing with drainage problems and this section deals with a number of fittings, methods and drainage items which the roadworker is liable to encounter.

4.19.1 French Drain

This consists of a fairly shallow trench, excavated to a fall and with a length of pipe (e.g. earthenware with open joints, or porous concrete pipes) laid in the bottom of the trench which is backfilled with material (media) through which water can percolate. The French drain would be connected at the lower end to an inspection chamber or manhole (Figure 4.56*a*).

An alternative method is to use a perforated polythene sheet to envelop the pipe. The trench is then backfilled with the soil which had previously been dug out so that the polythene forms a double sheet up to the surface as in Figure 4.56*b*).

Alternatively, a geotextile type of French drain may be used (Figure 4.56*c*).

These drains are often used for surface water drainage from verges, central reserves and, particularly, in cuttings on embankments to prevent surface water from running off the slopes either on to the carriageway or causing erosion of top soil from the slopes.

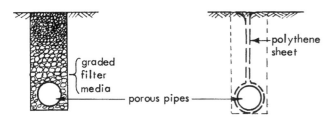

Figure 4.56(a)   *Figure 4.56(b)*

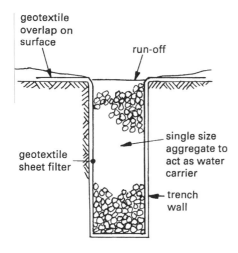

Figure 4.56(c)

4.19.2 Land Drains

These are of open butt-jointed earthenware agricultural pipes, normally laid in herringbone pattern at about 300–600 mm depth to drain an area (e.g. a field) and discharging to a convenient ditch or stream. Blockage by silting is common as there is no provision for filtering the soil around the pipes.

4.19.3 Fin Drains

A fin drain can be used as an alternative to a granular fill type of drain and is particularly useful for lowering the water table, draining wet ground or verges.

It consists of a high crush-strength porous flexible 'sandwich', about 25 mm thick. The walls of the sandwich are of a geotextile material, bonded on one side to a polyethylene backing sheet; the other wall is also of the geotextile type, but without the backing sheet. These walls are joined together by polyethylene injection-moulded spacers, as shown in Figure 4.57.

Surface water running off an impervious area can filter into the fin drain fitted at its edge; ground water filters through the sidewall fabric and runs away through the core of the drain to outlet pipe spigots fixed to the end or side.

Laying of a fin drains requires only a very narrow trench, excavated to falls as with other types of drain, into which the fin drain is laid out against one sidewall of the trench. The excavated material is then used as backfill, leaving little surplus soil to be carted away.

Some of the advantages of fin drains over piped drains are:

1 Excavated width, and therefore volume, is reduced.
2 Excavated soil is used for backfilling, no imported material being needed.
3 Excavation, placing of the fin drain and backfilling can be carried out as a continuous operation, at a rate of up to 15 m per minute.
4 Connectors linking lengths of fin drain are cheaper than manhole construction.
5 The trench does not require support, and is only open for a very brief period.

Fin drains are available in rolls, width 300 mm to 900 mm, length up to 120 mm. They are easily laid by hand or machine (with a specially adapted 'boot' fitted to the excavator. The weight of the 'sandwich' is about 2.75 kg/m².

4.19.4 Soakaway

This is defined as 'a means of disposing of surface water so that it will percolate into the subsoil'.

Soakaways are effective only in permeable soils and should be built to be at least 2 m square and 1.5 to 2.0 m deep, to give reasonable capacity to 'store' water which accumulates during a short heavy storm.

There are two main types:

(*a*) The hole in the ground filled with suitable media (e.g. broken brick, reject gravel, etc.).
(*b*) An underground tank which has honeycombed brick (or concrete) structural walls set on a concrete base. In this type a cover slab is required, usually with access facilities, so that the soakaway may be cleaned at intervals.

4.19.5 Syphons

In drainage, these are usually of the inverted type and are constructed to allow a line of sewer to cross a watercourse (or other pipeline, etc.) at the same level as the sewer without interrupting the flow in that watercourse. Manholes are required either side of the syphon.

4.19.6 Sand Drain

This is a vertical borehole filled with permeable material to facilitate the escape of water. Sand drains may be used to drain surface water which collects on an impervious surface, such as rock, by boring through the rock stratum to a fault or pervious layer.

The diameter is usually 50–100 mm. The hole is then filled with pervious material (e.g. sand). If possible, the borehole will provide access to a pervious layer under the rock or clay, so that natural drainage may take place.

Surface water will filter through the sand drain into the pervious stratum or under-drain (if required). Vertical sand drains are normally provided in quantity in order to drain, say, a carriageway.

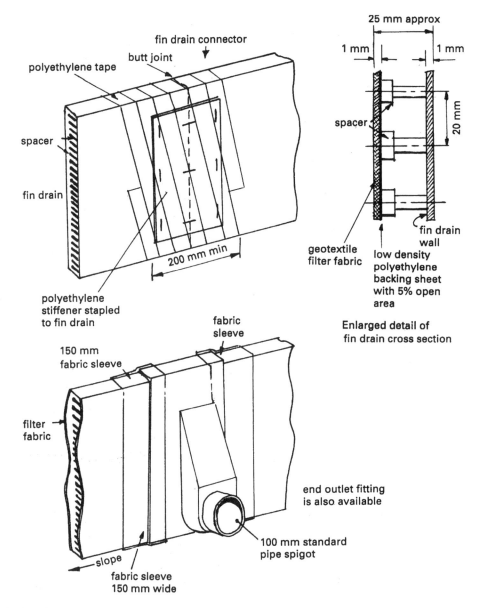

Figure 4.57   Fin drain with butt joint and outlet pipe

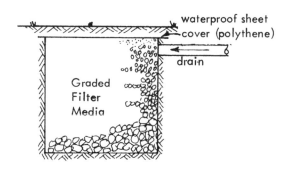

Figure 4.58(a)   Soakaway: media filled type

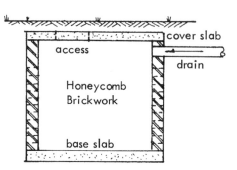

Figure 4.58(b)   Soakaway: underground tank type

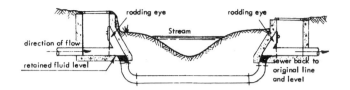

Figure 4.59 Inverted syphon

Sand drains are also used to speed up the consolidation of a newly constructed embankment. The method is to bore vertical holes through the first embankment layer and down through compressible ground until firm soil is reached beneath.

Holes are 500–750 mm diameter and 9–18 m deep. This process is not economic if the depth is less than 4.5 m. Holes are filled with clean coarse sand after which the construction layer is covered with 1 m of the same coarse sand and construction continued as normal, as shown in Figure 4.60.

Methods of Drilling Holes for Sand Drains

The most widely used method is that of the driven mandrel. This consists of a hollow tube with a flat on the bottom end; the mandrel is driven vertically into the ground by a pile driver. The tube is then filled with coarse sand, which is left in the ground as the mandrel is withdrawn. Other methods include using a rotary drill and bucket, rotary drill and water jet or a jetted mandrel.

4.19.7 Interceptor Tank

When surface water from the area of a petrol station is to be discharged into a public sewer, an interceptor tank, as shown in Figure 4.61, must be provided on the outfall drain from the petrol station.

4.20 Water and Air Testing of Drains (and sewers) in Accordance with British Standard BS 8005 Part I 1987 – Guide to New Sewerage Construction

The need for testing pipelines for watertightness is universally acknowledged. The current British Standard Code of Practice provides for both water and air tests for this purpose.

4.20.1 General Testing

All lengths of sewer and drain up to at least 0.76 m in diameter and all manholes and inspection chambers should be tested for watertightness.

Water or air tests should be applied after laying and before backfilling or placing bedding or surround concrete and will reveal the occurrence of cracked or porous pipes and faulty joints. Checks after backfilling may reveal faults in the bedding or support of the pipe, inadequacies in design or accidental damage during, or subsequent to, backfilling and show whether or not the finished work meets the specified requirements regarding watertightness.

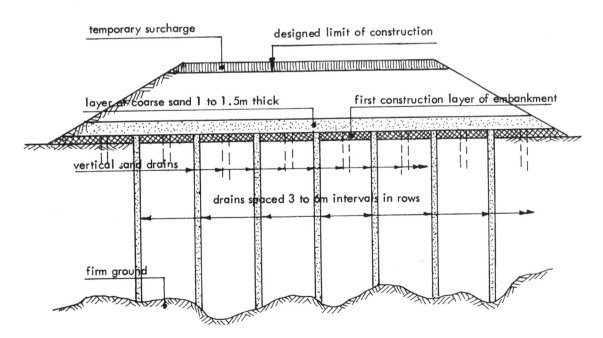

Figure 4.60 Sand drained embankment

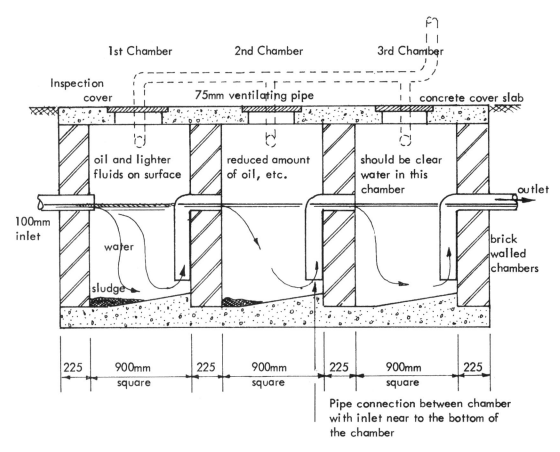

Figure 4.61 Section through interceptor tank

Occasionally a smoke test using smoke cartridges is used, but owing to the possibility of building up high uncontrollable pressures, this method is not recommended in the Code of Practice.

Wherever possible, testing should be carried out from manhole to manhole. Short branch drains connected to a main sewer between manholes should be tested as one system with the main sewer. Long branches, and manholes, should be separately tested.

4.20.2 Water Test

Drains and sewers should generally be subjected to an internal pressure test of 1.2 m head of water above the crown of the pipes at the high end but not more than 6 m at the low end. Steeply graded sewers should be tested in stages if the above maximum head would be exceeded were the whole section tested at once.

Before testing begins, a visual inspection of the pipeline should be made to ensure that there are no obvious faults. The test should be carried out by inserting suitably strutted plugs in the low end of the sewer and in connections, if necessary, and by filling the system with water. For small pipes a knuckle bend may be temporarily jointed in at the top end and a sufficient length of vertical pipe

jointed to it so as to provide the required test head, but for both large and small pipes a hose pipe may preferably be connected to a plug with pressure gauge or stand-pipe. Fall of the test water level may be due, amongst other things, to one or more of the following causes:

1 Absorption by pipes or joints
2 Excessive sweating of pipes or joints
3 Leakage from defective pipes or joints or plugs
4 Trapped air.

Some pipes absorb more water or trap more air at the joints than others. Allowance should be made for this by adding water to maintain the test head for appropriate periods. While the aim should be to commence the test proper as soon as possible, preferably within one hour, the appropriate period will best be determined by conferring with the pipe manufacturers.

The loss of water over a period of 30 minutes should be measured by adding water from a measuring vessel at regular intervals of 10 minutes and noting the quantity required to maintain the original water level. For the purposes of this test, the average quantity added for pipes up to 450 mm diameter should not exceed 1 litre per hour per linear metre of nominal

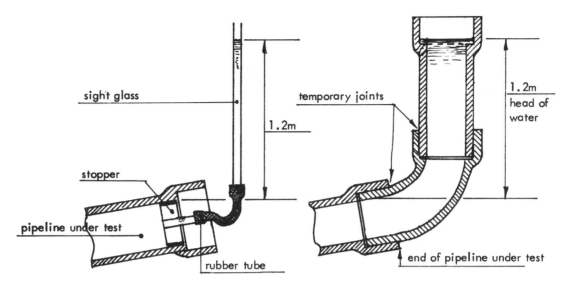

Figure 4.62 Water test

internal diameter (measured in metres). For example, a 300 mm diameter pipeline, 100 m long is allowed a loss rate of $1 \times 100 \times 0.3 = 30$ litres/hour but for larger pipes expert advice should be obtained.

Any leakage, including excessive sweating which causes a drop in the test water level, will be visible, and the defective part of the work should be removed and made good.

4.20.3 Air Test

It is often more convenient to test sewers by means of internal air pressure. However, while an excessive drop in pressure when employing the air test may indicate a defective line, the location of the leakage may be difficult to detect and the leakage rate cannot be measured. Failure to pass this test is not necessarily conclusive and when failure does occur, a water test as described above should be made and the leakage rate determined before a decision as to acceptance or rejection is made.

The length of pipe under air test should be effectively plugged as previously described and air pumped in by suitable means (e.g. a hand pump) until a pressure of 100 mm of water is indicated in a glass U-tube connected to the system. The air pressure should not fall to less than 75 mm during a period of five minutes without further pumping after allowing a suitable time for stabilization of the air temperature.

In order to make the test as valid as possible, the following points must be taken into account:

(*a*) The pipeline must be protected from change of temperature (e.g. effects of sunshine, showers, etc. during the test).

(*b*) The pipes may be slightly permeable to air.

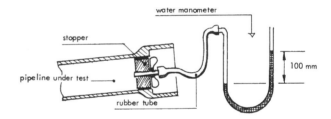

Figure 4.63 Air test

(*c*) The stoppers may be faulty – they should be checked for embedded grit or pitting of the rubber.

(*d*) Rubber tubing used for connecting plugs to the pump or U-tube must be in good condition.

(*e*) Each end of the pipeline must be firmly anchored to prevent any movement during the test.

4.21 Clearing of Blocked Drains

This is dealt with in Chapter 14, Highway Maintenance.

Revision Questions

1 What would be the typical self cleansing velocity for a 150 mm diameter pipe?
2 What sort of protection can be offered to operatives working in deep trench excavations?
3 What is a 'Grip'?
4 What type of tests might you undertake on a new sewer to test the integrity?

Note: See page 305 for answers to these revision questions

5.1   The Purpose of the Road

The main purpose of the road structure is to provide a means of reducing the stress or pressure due to a wheel load to a value which the ground under that structure can support.

A vehicle standing on a road structure exerts a direct load (static stress) on the small area of contact between its tyres and the road surface. When the vehicle is moving, there is additional dynamic stressing due to the up and down movements of the vehicle, caused by slight unevenness of the surface, gusting wind, etc. This has the effect of 'hammering' the surface as the vehicle travels along the road.

The intensity of the static and dynamic stress is greatest at the surface of the road and spreads in a pyramidal shape throughout the depth of the structure. As the spread of the load increases, so the stress is reduced, until at the formation level, the stress is low enough for the natural ground to support it without distortion or damage.

For simplicity, the pyramidal spread of the load can be considered to be at 45° to the horizontal and this gives an approximately correct stressing figure. In reality, the spread is slightly greater in the upper layers of the road structure.

Flexible construction carriageways have been designed to last 20 years, allowance being made in the design for traffic growth year by year (2% growth per year is the usual allowance). However, a road can only achieve its designed life period and carrying capacity if the actual job of construction is properly done, all the materials are of the required standard, to the design specification and always correctly used.

The *Design Manual for Roads and Bridges* (Highway Agency) is based on a 40 year life period before major reconstruction is needed, subject to periodic repair and maintenance.

Adequate drainage of the area is also of the greatest importance in providing a sound formation on which to build the road structure. Waterlogged ground allows excessive movement of the structure leading to early damage and destruction.

Flexible construction is so called because it allows a small amount of vertical movement of the road structure under load.

5.2   The Structure of the Pavement

This usually consists of four layers of road construction material, these being built up on the formation as shown in Figure 5.1.

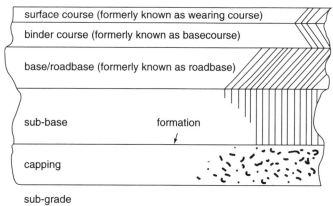

Figure 5.1   Section through pavement

By definition, formation is the surface of the ground in its final shape after completion of the earthworks and of consolidation, compaction, or stabilization *in situ*.

The four layers of the road structure have been re-named since 1 January 2004 and are:

1 The sub-base, which (a) assists in load-spreading, (b) assists in subsoil drainage (if a free drainage material is used), and (c) acts as a temporary road for construction traffic.

2 The base/roadbase (formerly roadbase), which is the main load-spreading layer of the structure.

3 The binder course (formerly basecourse), which supports the wearing course and also assists in protecting the road.

4 The surface course (formerly wearing course), which (a) provides a skid-resistant surface, (b) waterproofs the pavement and (c) withstands the direct loading of the traffic.

It is customary to shape the formation with a fall to the side, or camber, in order to assist drainage of the surface. If this is not done, an overfill of 150 mm is needed to protect the formation prior to final shaping.

The term 'sub-grade' is used to define the natural foundation or fill which directly receives the loads from the pavement. Hence, the top surface of the sub-grade is the formation.

On new construction, a capping layer is used to protect the sub-grade from damage.

5.3 The Load-Carrying Requirements of a Flexible Pavement

It is necessary to know something of the way in which traffic loads are carried and spread through the various layers of the road structure.

5.3.1 Load Distribution (Figure 5.2)

Tests at the Transport Research Laboratory have confirmed that load spreading occurs, but show that the angle of distribution is not constant as suggested by Figure 5.2.

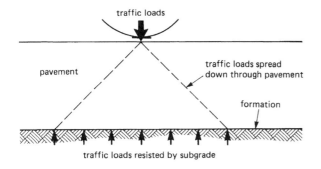

Figure 5.2 Distribution of load through the road structure

As the stresses (i.e. loads per unit area) from the vehicles travelling the road are greater nearer the surface, stronger materials are needed in the surfacing than in the lower layers.

In addition there are lateral deflective forces caused by the pounding effects of heavy traffic. This has led to the development of upper and lower roadbases, the lower being of stronger material. *See* Figure 5.7 and design example Table 5.3.

Another factor of great importance is the surface profile. An uneven surface will not only be unsuitable for the safe passage of traffic, but will also cause greater, and variable stresses in the pavement, leading to 'fatigue' of the structure and shortening its life.

These two factors have led to the development of layered construction, the lower layers of which are thicker and of cheaper materials, in order to provide the necessary spread of the load. Each layer must be shaped and

compacted as accurately as possible, the surface layer thus being shaped into an accurate and even surface.

The Specification for Highway Works (Clause 702) recognizes the need for surface tolerances of the layers to be more accurate the nearer they are to the surface (Figure 5.3).

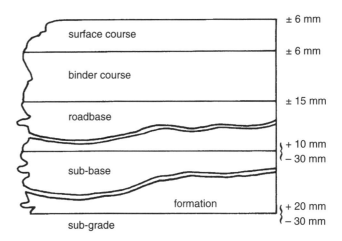

Figure 5.3 Tolerances of surface levels of pavement courses and formation

5.3.2 Structural Design

Information for pavement design for roads in the United Kingdom is provided by the Highway Agency and its research establishment, the Transport Research Laboratory.

Road Note 29, a TRL report published in 1970, as a 'Guide to the structural design of pavements for new roads' was replaced in 1984 by LR1132 'The Structural Design of Bituminous Roads'. This was supplemented by Departmental Standards issued by the DfT.

The basis of the design method is the monitoring of performance of pavement materials of differing thicknesses, under varying ground and traffic conditions. For the performance to be properly monitored, it is essential to have a procedure based on common factors.

For instance, if, due to unexpectedly rapid growth of traffic, a road fails long before the end of its designated life, this failure cannot be properly evaluated if traffic estimations at the time of the design were not recorded. These factors are:

1 The number of commercial vehicles expected to use the road initially.
2 The estimated annual rate of commercial traffic growth.
3 The designed life period of the road.
4 The type of soil on which the road structure is built.
5 The level of the natural water table in relation to the proposed formation level.
6 The type of road structure which will be used in building the pavement.

Figure 5.4 shows these as an outline procedure.

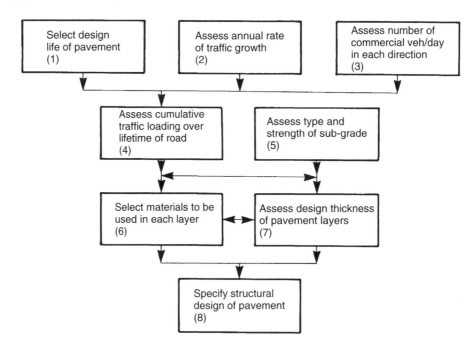

Figure 5.4   General objectives in structural pavement design

5.4   Overall Concept of Pavement Design

5.4.1   Design Life of the Pavement

The current Highway Agency standards recommend a design life of 40 years for flexible (fully bituminous roadbase) and composite (bituminous and cement bound roadbase or fully cement bound roadbase) construction. A first stage, based upon predicted traffic flows over 20 years, is followed by an overlay or partial reconstruction as traffic loadings justify (*see* Section 5.4.7).

5.4.2   Assessment of Annual Rate of Traffic Growth

This normally varies from 2% to 6%, based on national census data, but it can vary considerably for an individual road, for instance where new development takes place.

5.4.3   Assessment of Number of Commercial Vehicles in Each Direction

The damaging effect on the road structure from vehicles depends upon the axle weight, but is not directly proportional to weight: the damaging effect of an axle load of 8200 kg is 100 times that of an axle load of 2700 kg; the damaging effect of an axle load of 11 800 kg is more than four times that of an axle load of 8200 kg. Thus a slight increase in axle weight can result in a greatly increased effect on the road structure.

This means that, for example, the drive to a private house that is strong enough to carry a private car daily may be severely damaged by a coal lorry or furniture van driven onto it only once.

Hence, in designing a road, the private vehicle traffic is ignored, and only the number of commercial vehicles (of unladen weight more than 1500 kg) likely to use it is estimated. This number is determined from traffic assessment studies or from example values as shown in Table 5.1.

Type of road	Estimated traffic flow (c/v per day in each direction) at the time of construction
1 Minor residential roads	10
2 Through roads and roads carrying regular bus routes	75
3 Major through roads carrying regular bus routes	175
4 Main through roads	350

Table 5.1 Example of commercial vehicle (c/v) flows used when accurate assessments are not available for residential and associated developments

5.4.4 Assessment of Cumulative Traffic Loading Over Lifetime of Road

Having established

1 the design life;
2 the traffic growth;
3 commercial vehicle numbers;

this assessment is converted into a factor known as standard axles (i.e. an axle with an 80 kN total force). Over the lifetime of the pavement the number of standard axles is measured in millions.

Figure 5.5 shows an example of how one stage is carried out, i.e. collation of millions of commercial vehicles, over a design life with a growth rate.

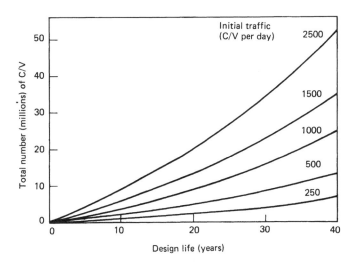

Figure 5.5 Relation between total number of commercial vehicles and design life – growth rate 3%

5.4.5 Assessment of Type and Strength of Sub-grade

From site investigations, core samples, etc., the type of soil can be determined and its CBR value (*see* Chapter 10) established either by test or by use of existing values such as those shown in Table 5.2.

It is the CBR value which determines the type and thickness of the sub-base and capping layer.

5.4.6 Selection of Materials and Thickness to be used in each Pavement Layer

These are designed by reference to the *Design Manual for Roads and Bridges* (volume 7). Table 5.3 and Figures 5.6 and 5.7 show examples for a flexible pavement.

5.4.7 General Observations

One of the major problems with pavement design is related to increased traffic damaging the structure to the extent

Type of soil	CBR (%)	
	Well drained water table at least 600 mm below formation	Poorly drained
Heavy clay	2 2 2.5 3	1 1.5 2 2
Silty clay	5	3
Sandy clay	6 7	4 5
Silt	2	1
Sand (poorly graded)	20	10
Sand (well graded)	40	15
Well graded sandy gravel	60	20

Table 5.2 Estimated laboratory CBR values for British soils compacted at the natural moisture content

---

Design Example

Design traffic: 75 million standard axles
Roadbase: Hot dense macadam (HDM)
Design thickness: 320 mm (combined thickness of layers)

Options:
(a) 45 mm hot rolled asphalt surface course*
    55 mm hot dense macadam binder course
    220 mm roadbase HDM

(b) 45 mm hot rolled asphalt surface course*
    275 mm HDM roadbase

(c) 50 mm HRA surface course
    60 mm HDM binder course
    240 mm HDM roadbase

---

*In (a) and (b), the contractor may choose to lay 50 mm HRA surface course.

Table 5.3 Pavement design example

that major reconstruction is required. Several motorways have required reconstruction work well within the original design life of 20 years, having already met the total traffic load originally predicted. For example, parts of the M4 reached the designed traffic loads within 15 years, resulting in the need for major reconstruction or at least, in some areas, overlaying the existing surface with another surface course.

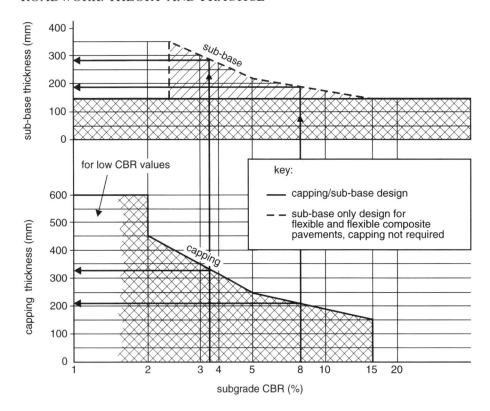

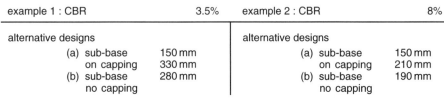

example 1 : CBR   3.5%   example 2 : CBR   8%

alternative designs			alternative designs		
(a) sub-base	150 mm		(a) sub-base	150 mm	
on capping	330 mm		on capping	210 mm	
(b) sub-base	280 mm		(b) sub-base	190 mm	
no capping			no capping		

Figure 5.6 Capping and sub-base thickness design

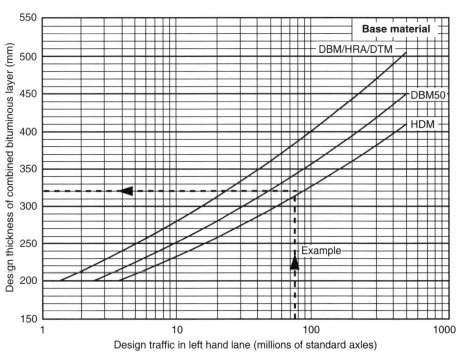

Figure 5.7 Design thicknesses for flexible pavements

This explanation is of little consolation to the travelling public who see our roads apparently constantly under repair. The Highway Agency is aware of the situation and constantly reviews the traffic projections (as mentioned earlier). As the Highway Authority for these roads, they have to provide the funds, and therefore to achieve a balance between the need and the resources available.

It is extremely rare for a flexible pavement to withstand traffic and weather for the initial 20 years design life without some form of surface treatment being required. This may be to make good abrasion by traffic, loss of skid resistance, minor deformation or reduced quality of running surface. An overlay can satisfy all these conditions and make a contribution towards strengthening to meet future traffic loads.

5.5 The Sub-base

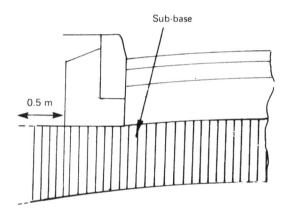

Figure 5.8

The functions of the sub-base are to:

(*a*) assist in load spreading as part of the structural design;
(*b*) make up the pavement to 450 mm thickness on frost susceptible soils;
(*c*) provide some protection to the sub-grade as soon as it is exposed;
(*d*) provide a platform on which to lay the base/roadbase, and support for the kerb, bed and backing.

The sub-base should be laid as soon as possible after final stripping to formation level. This is done in order to prevent deterioration of the formation due either to:

(*a*) rain, which causes the exposed ground to become soggy and may even cause erosion; or
(*b*) sunshine, which can dry out the surface and cause cracking in the sub-grade.

If the sub-grade is not going to have the sub-base laid over it immediately, it should be protected from the weather as described in Section 3.2.4. Both rain and sun on the exposed sub-grade result in a loss of bearing strength of the ground.

In pavement construction, it is good practice to extend the sub-base beyond the kerb line and as far as the side drains. This helps to remove surface water during construction and allows it to run off well clear of the main road structure, and also to support the kerb race (Figure 5.8).

5.6 The Base(roadbase) (Figure 5.9)

Because the roadbase is the main load-spreading layer, the materials used must be carefully chosen. On very minor works, local materials are often used, economically and successfully, both for roadbases and for sub-bases.

The main materials have different binding characteristics, as follows:

1	bound by natural interlock	– as with dry bound macadam;
2	water bound	– as with wet mix;
3	cement bound	– as with wet-lean concrete and other cement-bound granular materials
4	bituminous bound	– as with dense macadam and hot rolled asphalt

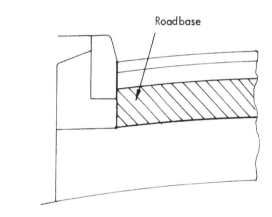

Figure 5.9

5.7 Sub-base and Base(roadbase) Materials

The Specification for Highway Works groups these materials as follows:

Unbound materials
Clause 803   Granular sub-base, Type 1
Clause 804   Granular sub-base, Type 2

N.B.: All clause references relate to the Specification for Highway Works.

Bituminous-bound materials
Clause 903   Dense macadam roadbase
Clause 904   Rolled asphalt roadbase

Concrete and cement-bound materials
Clause 1030   Wet-lean concrete
Clause 1036   Cement-bound granular material category
              1 & 1a (CBM1 & 1a)
Clause 1037   Cement-bound granular material category
              2 & 2a (CBM2 & 2a)
Clause 1038   Cement-bound granular material category
              3, 4 & 5 (CBM3, 4 & 5)

5.8   Unbound Sub-base and Base(roadbase) Materials

5.8.1   Type 1 (Clause 803) and Type 2 (Clause 804) Unbound Mixtures for Sub-base

Type 1 unbound mixture for sub-base must be made from crushed rock, crushed slag, crushed concrete, re-cycled aggregates or well burnt non-plastic shale. It may contain up to 10% by mass of natural sand. Table 5.4(*a*) gives the grading.

Type 2 unbound mixture for sub-base is made from natural sands, gravels, crushed rock, crushed slag, crushed concrete, re-cycled aggregates or well-burnt non-plastic shale. Where permitted it may contain more than 50% asphalt arisings.

As can be seen from the table, Type 1 has a greater proportion of coarse aggregate; consequently it is far stronger due to its interlocking qualities. Type 2 is a much smaller sized material.

Sieve size (mm)	Proportion passing sieve	
	Type 1 (%)	Type 2 (%)
63	100	100
31.5	75–99	75–99
16	43–81	50–90
8	23–66	30–75
4	12–53	15–60
2	6–42	—
1	3–32	0–35
0.063	0–9	0–9

Table 5.4(a)   *Type 1 and Type 2 unbound materials (BS EN 13285)*

5.8.2   Type 3 (Clause 805) Open-graded Unbound Mixture for Sub-base

Type 3 is made from crushed rock, crushed blast furnace slag or crushed concrete. The grading requirements are shown in Table 5.4(*b*).

Sieve size (mm)	Proportion passing sieve
80	100
40	80–99
20	50–78
10	31–60
4	18–46
2	10–35
1	6–26
0.500	0–20
0.063	0–5

Table 5.4(b)   *Type 3 Open graded unbound mixture for sub-base*

5.8.3   Slag Bound Material (Clause 805) – Formerly Known as Wet Mix

Slag bound material (SBM) is a 'plant manufactured material' which is granular aggregate bound by granulated blast furnace slag (GBS). The SBM contains aggregate with usually between 10% and 25% by mass of GBS, with an activator to increase the rate of curing. The activator is usually lime and together with water will form after mixing, a 'damp' mix, rather than a 'wet' mix, with the result that segregation during transport and laying is minimized and at the same time the material is more easily compacted. SBM should be laid only during the months of March to September and should be placed and spread evenly in such a manner as to prevent segregation and drying. It should be laid and compacted in layers which enable the specified thickness, surface level, regularity requirements and compaction to be achieved. Compaction should be carried out immediately after the material has been spread and should normally be completed within 6 hours of the addition of the GBS and the activator.

5.8.4   Other Unbound Material

These materials are used on roads carrying light traffic, car parks, hard shoulders, etc., and these are not covered by the specifications. They are specified to local agreement between the supplier and purchaser and include the following:

Quarry Waste or Scalpings

The stone should be naturally graded, not over 100 mm maximum size and not include more binding material than is required to help hold the stone together.

Hardcore

This should consist of reasonably clean broken bricks, stone, concrete, etc. not over 150 mm in size. Unbroken bricks do not interlock but tend to ride on one another.

Care should be taken with hardcore from demolished building sites, as with mechanical loading much debris, i.e. timber, metal or plaster, may be picked up and make the material unsuitable.

Hoggin (or similar self-binding gravel)

The clay content should not be more than sufficient to bind the material together. A careful check should be made to ensure that there are no lumps of clay.

Clinker

This should be hard and well burnt, and graded down from 50 mm. Care should be taken to differentiate between clinker and ash; the latter is not suitable.

Shale

Shale is usually obtained from colliery tips and has been burnt as a result of spontaneous combustion. It should be used only from sources known to be low in sulphur, reliable and stable, as exposure can in some instances result in breakdown of the material.

Where provision of a definite thickness of frost-resistant material is appropriate and the sub-base forms part of it, the material should be examined in accordance with the Road Research Laboratory Frost Heave Test if there is any doubt about its suitability.

PFA: Cement-stabilized

Extensive laboratory and field tests have been carried out into the use of cement-stabilized pulverized fuel ash.

Tests have confirmed that suitable and economic proportions for a cement-stabilized PFA mixture are 10% cement and 90% PFA. Moisture content is important and must be not less than the optimum moisture content of the PFA.

Seven day compressive strength tests at approximately 90% compaction frequently show figures in excess of 2750 kN/m². The age-hardening characteristics of PFA are particularly useful in this instance, and contribute to a considerable increase in the bearing strength when the material is used as roadbase.

Cement-stabilized PFA can overcome the frost heave problem. Properly mixed and compacted, it shows considerable resistance to softening during the freeze–thaw cycle.

Dry Bound Macadam

In this method of construction 50 or 37.5 mm single size crushed rock or slag is spread in layers 75–100 mm thick and rolled.

A 25 mm thickness of 4.7 mm down similar material is then spread on the layer and vibrated into the coarse aggregate, the process being repeated until no more fines can be worked in. Any excess fine material must be removed by sweeping and the layer rolled with an 8 tonne minimum weight roller. Additional layers are then constructed to make up the total thickness of roadbase required.

5.9 Spreading, Levelling and Compacting Unbound Materials

All plant-mixed materials should be protected from the weather; drying or wetting changes the moisture content and may cause separation within the material. The material should be spread evenly; on large-scale works it must be laid by a paving machine or by spreader box.

On smaller scale works, a loading shovel, bulldozer or grader can be used.

The levels can be controlled by:

(*a*) pegs and line
(*b*) sight rails and travelling rods or profiles
(*c*) a guide wire which can be followed by the machine operator, using manual controls, or by the use of an automatic levelling device.

The Specification requires the material to be laid in layers, each of which should be not less than 110 mm compacted thickness, with no layer being more than 225 mm, unless otherwise specified. If a thickness of more than 225 mm is required, it must be laid in two or more layers.

The surface tolerance allowed is from +10 mm to −30 mm, as shown in Figure 5.3.

The thickness of material which can be adequately compacted depends upon the type of compaction plant being used. Examples of the relationship between the plant and compaction thickness are given in Table 5.5.

5.10 Bituminous Bases(roadbases)

5.10.1 Dense Base Macadam (Clause 903)

The need for strong but truly flexible bases that will not crack has led to the use of these dense bituminous materials.

The main requirements in the composition of the dense macadam for use in bases are that the materials have a fines content (aggregate passing 3.35 mm sieve) of 32–46% and are made with high-viscosity binders, i.e. 50 pen, 125 pen or 190 pen bitumen. The more viscous of the alternate binders, namely 50 or 125 pen bitumen, are the appropriate ones to use on roads designed for more than 2.5 million standard axles, while the less viscous 190 pen bitumen is suitable for roads carrying less traffic. The term 'pen' is an abbreviation for 'penetration grade' and is a measure of the hardness of a bitumin binder. It

Type of compaction plant	Category	Number of passes for layers not exceeding the following compacted thicknesses:		
		110 mm	150 mm	225 mm
Smooth-wheeled roller (or vibratory roller operating without vibration)	Mass per metre width of roll: over 2700 kg up to 5400 kg over 5400 kg	16 8	unsuitable 16	unsuitable unsuitable
Pneumatic-tyred roller	Mass per wheel: over 4000 kg up to 6000 kg over 6000 kg up to 8000 kg over 8000 kg up to 12 000 kg over 12 000 kg	12 12 10 8	unsuitable unsuitable 16 12	unsuitable unsuitable unsuitable unsuitable
Vibrating roller	Mass per metre width of vibrating roll: over 700 kg up to 1300 kg over 1300 kg up to 1800 kg over 1800 kg up to 2300 kg over 2300 kg up to 2900 kg over 2900 kg up to 3600 kg over 3600 kg up to 4300 kg over 4300 kg up to 5000 kg over 5000 kg	16 6 4 3 3 2 2 2	unsuitable 16 6 5 5 4 4 3	unsuitable unsuitable 10 9 8 7 6 5
Vibrating-plate compactor	Mass per square metre of base plate: over 1400 kg/m^2 up to 1800 kg/m^2 over 1800 kg/m^2 up to 2100 kg/m^2 over 2100 kg/m^2	8 5 3	unsuitable 8 6	unsuitable unsuitable 10
Vibro-tamper	Mass: over 50 kg up to 65 kg over 65 kg up to 75 kg over 75 kg	4 3 2	8 6 4	unsuitable 10 8
Power rammer	Mass: 100 kg up to 500 kg over 500 kg	5 5	8 8	unsuitable 12

Table 5.5 Examples of compaction requirements for granular materials

is obtained from the standard penetration test which is described in Section 10.20.

5.10.2 *Rolled Asphalt to BS 594 (Clause 904)*

Rolled asphalt it the oldest established bituminous material used for roadbase construction and has load-spreading properties superior to those of other flexible roadbases.

The composition of rolled asphalt for roadbase construction is a mixture to BS 594, containing 65% of coarse aggregate.

5.10.3 *Characteristics of Dense Bituminous and Hot Rolled Asphalt Bases(Roadbases)*

The use of bituminous roadbase materials results in many advantages from both the design and constructional standpoints, including the following.

Load-Spreading Properties

Rolled asphalt and dense coated bitumen roadbases have better load-spreading properties than those of materials hitherto normally used for roadbase construction. This feature was recognized in Road Note 29 which permitted a reduction in the thickness of surfacing required for all types of roadbase material over a range of standard axles where these materials were used.

Speed of Construction

The materials can be quickly and accurately laid by machine and are strong as soon as they are cool. Due to the dense nature of the material, the application of the first layer of the base early in construction affords protection to the sub-base and formation in bad weather. Another layer of a hot laid wearing course may then be superimposed or, if desired, the material may be used by traffic (when it has cooled), a feature particularly important in the reconstruction of existing roads.

Imperviousness and Frost Resistance

The compacted material provides excellent weather protection and resistance to water passing through it, and is not itself affected by frost.

5.11 Concrete and Cement-bound Materials

The grades of concrete covered in the Specification (Clause 1001) range from those suitable for unreinforced and reinforced slabs to Cement Bound Granular Material Category 4.

5.11.1 Concrete

Concrete for rigid pavements and wet lean concrete for rigid or composite pavements are discussed in Chapter 6.

5.11.2 Cement-bound Materials

Soil cement, cement-bound granular material and lean concrete are all categories of the group known as cement-bound material (CBM), categories 1, 1a, 2, 2a, 3, 4 and 5 have differing aggregate size and crushing strengths.

Cement-bound materials are mixtures of raw material and cement which have a moisture content compatible with compaction by rolling, and capable of meeting the requirements for surface level, regularity and finish.

CBM1 and 1a are the categories with the finest and weakest of these materials and should have a minimum average crushing strength of 4.5 N/mm^2 after 7 days. CBM2 and 2a are both coarser and stronger, being of 40 mm nominal size, with a strength of 7 N/mm^2.

CBM3 and 4 may be either 40 mm or 20 mm nominal size and must be made of natural aggregate, slag or re-cycled coarse aggregate with a strength of 10 N/mm^2 for CBM3 or 15 N/mm^2 for CBM4.

CBM5 is the strongest category designed for high speed roads and has 40/20 mm aggregate and a strength of 20 N/mm^2.

The Specification covers requirements for batching, mixing, laying, compaction and curing. The mix is determined from site trials and can include PFA (pulverized fuel ash) or GGBS (ground granular blast furnace slag), used in combination with Ordinary Portland Cement (OPC).

(a) Mixing and Batching

CBM1 and 2 can be mixed in place or mixed in a plant and can be batched either by weight or by volume.

CBM3, 4 and 5 must be mixed in a plant and batched by mass.

The mix-in-place method of construction is used where a previously prepared layer is stabilized *in situ* with cement. The prepared layer can consist of either imported or as-found material which should be spread and compacted to a uniform density such that, after stabilization, the requirements for surface level and regularity will be met.

If it is necessary to add water, this can be done either as part of the mixing operation, or after the material which is to be stabilized has been shaped and compacted prior to the addition of the cement.

The cement should be spread by means of a spreader which is fitted with a device to ensure uniform rate of spread over the whole area.

The mixer should be equipped with a device for controlling the depth of processing. Mixing should continue until the required depth and uniformity of processing has been obtained.

The mix-in-plant method of construction requires the material, cement and water to be mixed in a central plant and then transported to the point of laying and spreading.

The mixer should be capable of mixing evenly and have an output sufficient to meet the demands of the spreading and compacting operations.

(b) Transporting

Mixed material must be transported as quickly as possible to the point of laying and also be protected from the weather during this operation.

(c) Laying

All cement-bound material must be placed and spread evenly in such a manner as to prevent segregation and drying. Roadbase cement-bound material must be spread by means of either a paving machine or a spreader box approved by the engineer.

Cement-bound material must be spread in one layer so that after compaction the total thickness is as specified. Joints between each day's work, or where work has had to be interrupted, should be vertical.

(d) Compacting

Compaction must be carried out immediately after spreading. All loose segregated or otherwise defective areas must be removed and replaced to the full thickness of the layer.

(e) Curing

Immediately after compaction the material should be cured for a minimum of 7 days. Curing may be by:

1 covering with impermeable sheeting;
2 bituminous spraying; or
3 spraying with a curing compound.

Traffic should not be allowed on the surface until the 7 day strength has been attained.

(f) Surface tolerance

All roadbases should be laid to a surface tolerance of ±15 mm (*see* Figure 5.3).

5.12 Surfacing

Both the surface course and binder course are included in the part of the road structure which is known as surfacing.

These upper layers of the road have to carry the greatest intensity of loadings as they come directly under the wheels of the traffic. In other words, the top layers of the road have the greatest work to do.

In comparatively few cases single course surfacing may be laid. The normal structure, however, requires the two-course structure.

The upper layers of the structure, therefore, have the job of:

1 carrying the direct load from the traffic and spreading it to reduce the stress on the lower parts of the road structure;
2 providing a safe, skid-resistant surface;
3 providing a good and well-shaped running surface;
4 providing good drainage from an impervious surface, thus protecting the road structure from the weather;
5 giving a 'confidence inspiring' surface on which the vehicle driver 'feels' safe.

The stresses of present day traffic and the concept of DBFOs (design, build, finance and operate) has led to the development of new materials. These are dealt with in Chapter 10.

From the 'safe surface' aspect, the resistance to skidding is obviously the most important feature. The camber and degree of super-elevation on corners is also very important.

The skid resistance of surfacing material is defined under one of four headings.

1 rough harsh;
2 rough polished;
3 smooth harsh;
4 smooth polished.

These are usually represented by figures obtained by using the skid resistance measuring equipment. In this case a good resistance gives a reading of PSV 65+ and a poor surface would have a reading of about 40.

Road surfacing material is tested for skid resistance and gives 'polished stone values'. This test is an artificial wear one, normally carried out in the laboratory, though there are portable skid testers for use on the actual road (*see* Chapter 10 Materials and Testing).

The polished stone values (PSV) are such that the higher the reading the greater is the resistance of the material to polishing and having its edge worn smooth and rounded.

5.13 Binder Course Materials

The objects of the binder course (Figure 5.10) are to:

1 distribute the traffic loads over the roadbase, which is usually of somewhat weaker material;
2 provide a good shaped and regular surface on which to lay the relatively thin surface course.

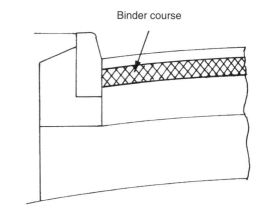

Figure 5.10

In new construction, the thickness of the binder course is usually between 45 mm and 105 mm. Where a binder course is laid as a regulating course, however, to strengthen an existing road structure, the thickness may vary considerably.

The type of material used for a binder course is selected according to the intensity of traffic loading expected, whilst the nominal size of the stone (20, 28 or 40 mm) depends on the thickness of the layer. The thicker the binder course, the larger the stone size.

On new roads being designed to carry over 2.5 million standard axles, a binder course of rolled asphalt or dense-coated macadam should be used. These materials are similar to those for roadbase construction and provide the same benefits, particularly the excellent load-spreading properties and the ability to carry the heaviest traffic as soon as the material is cool.

Materials suitable for use in the binder course include:

20 mm nominal size open graded;
40 mm nominal size dense;
28 mm nominal size dense;
20 mm nominal size dense;
hot rolled asphalt binder course;
40 mm nominal size single course.

Details of the use of these materials is given below:

5.13.1 20 mm size Open Graded Macadam (BS 4987)

The open graded compositions are coarsely graded with little or no fine material. The materials produced can be laid easily at low temperatures, being low viscosity mixes, and become stable and durable under traffic. The average thickness of a compacted course should be 45–75 mm. The use of this material is declining and is restricted to general repair work and to roads carrying less heavy traffic.

5.13.2 Dense (BS 4987)

There are three nominal sizes of dense-coated macadam for the binder course, 40 mm, 28 mm and 20 mm. The three nominal sizes enable the correct type to be used to suit the thickness required. The 40 mm should be laid 95–140 mm thick; the 28 mm, 70–100 mm thick; and the 20 mm, 50–80 mm thick. All the materials have a high fines content. The binders used for 20 and 40 mm material are 190 and 290 pen and for 28 mm are 35, 50, 125 and 190 pen.

All binders set hard as soon as they cool and hence from a construction stand-point the materials have to be treated as 'hot-laid' in order to achieve adequate compaction and strength. The selection of binder viscosity depends on the intensity of traffic, the higher viscosity being appropriate for use under the heavier categories of traffic.

These materials have a number of particularly good features such as flexibility, stability, frost resistance and excellent load-spreading properties.

The use of dense binder course materials can add considerable strength to an existing road, their use is particularly appropriate for strengthening a road which may be required to carry heavy traffic at short notice. From the constructional standpoint the main advantage is that as soon as the material is cooled to atmospheric temperature, traffic can be switched on to it if required. This can be particularly useful in the reconstruction of a city street or congested area, such as a junction or roundabout where obstruction can cause considerable traffic delay.

5.13.3 Rolled Asphalt (BS 594)

This has a coarse aggregate content graded to BS 594, the asphaltic content being petroleum bitumen. It should be laid in 50–75 mm layers and has qualities very similar to the dense binder courses mentioned previously. The thickness of the layer determines the size and grading of the coarse aggregate content of the course. Asphalt regulating courses are binder course mixtures with small sized (10 mm) aggregate to enable the layers to be laid thinly.

A further point to be considered in the selection of the binder course material is the type of surface course with which it is to be covered. If the surface course is to be made with a high viscosity binder necessitating laying while still at a high temperature, the binder course must be made with a binder of similar viscosity to avoid any softening or movement in the binder course due to the laying temperature of the surface course material. Alternatively, there should be an interval of some months between the two operations.

5.13.4 40 mm Single Course (BS 4987)

Where say a 90 mm surfacing is required for new construction or a heavily trafficked road, a two-course job, i.e. binder course and surface course, is to be preferred because it allows more re-shaping to be done, and gives a better riding surface. However, many of the less heavily trafficked roads which require re-shaping or strengthening can be dealt with quite satisfactorily at a lower cost by single course surfacing.

The medium textured material provides for this in a 40 mm down nominal size graded aggregate laid 75–105 mm thick; the durability of a single course job depends upon the surface treatment it receives early in its life.

The best method is to surface dress as soon as possible after laying and preferably before the first winter. Alternatively the materials can be lightly dressed with fine cold asphalt just to fill the voids and surface dressed after not more than twelve months.

5.14 Surface Courses (Figure 5.11)

The purpose of the surface course is to:

1 provide a durable skid-resistant surface;
2 protect the pavement from the effects of the weather;
3 withstand the effects of abrasion and stresses from the traffic;
4 provide a good regular shaped running surface.

A wide variety of bituminous materials is used for surface courses, laid in thicknesses ranging normally from

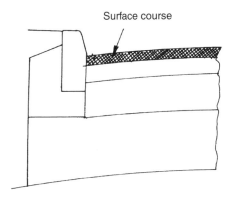

Figure 5.11

20–40 mm. Important points in new construction are the additional strength which the surface course may add to the pavement and the extent to which it forms an impervious layer over the construction.

N.B.: When the surface course does not itself provide an impervious layer (for example, in the case of open-textured material) it is important in new construction, if the lower layers of material and the sub-grade would be weakened by the ingress of water, that the surface course be surface dressed or an impervious membrane incorporated in the construction. The top of the roadbase or the binder course could be surface dressed and outlets provided at this level to drain off any water which may collect. The use of dense-coated macadam or rolled asphalt for the binder course or roadbase meets this requirement as the materials are sufficiently dense to provide adequate protection to the underlying construction.

The type of material for the surface course is selected according to the anticipated traffic intensity and, for new construction, recommendations are given in Section 5.3.2.

For maintenance and resurfacing work, the requirements for strength and waterproofing may not be so important as in new construction, hence a wide variety of materials can be suitable. Dense and close graded bituminous materials often give a very satisfactory performance even on roads carrying heavy traffic, if of adequate strength of structure.

Materials suitable for use as a surface course are:

1 hot rolled asphalt
2 10 mm size close graded
3 14 mm size close graded
4 6 mm size medium graded
5 3 mm size fine graded
6 6 mm size dense
7 10 mm size open graded
8 14 mm size open graded
9 20 and 10 mm size porous

Further information on bituminous materials is provided in Chapter 10 although a brief description of surface courses and their uses follows here.

5.14.1   Hot Rolled Asphalt (BS 594)

This is a strong and durable surface course. It is very dense, made with a high fines and asphaltic cement content with crushed rock, slag or gravel added, to provide additional stability and reduce the cost.

It is normally laid 40 mm thick with 20 mm coated chippings rolled into the surface to provide better skid resistance.

5.14.2   10 and 14 mm size Close Graded (formerly called 'Dense') (BS 4987)

These materials are used in new construction and for resurfacing using crushed rock, slag, gravel or limestone.

Limestone and gravel tend to polish, resulting in loss of skid resistance and are, therefore, not always suitable.

They are often used to provide the road with additional strength, using the more viscous and harder binder. They are closely graded, are dense and virtually impervious, are produced at a very high temperature and are consequently more difficult to lay than open or medium textured materials.

5.14.3   Dense Tar Surfacing (BS 5273)

Dense tar surfacing is generally not used in UK road construction, but sometimes is considered for lightly trafficked roads. This is available in 10 mm or 14 mm nominal sizes with coarse aggregate contents of 35% or 50% and laid to a thickness of 30 mm or 40 mm respectively. If desired, coated chippings may be rolled into the surface of the 35% coarse aggregate content mixtures.

For motorway service areas and similar purposes, where its excellent resistance to softening by oil droppings is especially valuable, the material should be of 14 mm nominal size with coarse aggregate content of 50%, using crushed rock or slag aggregate.

5.14.4   6 mm Size Medium Graded

This material is intended solely for lightly trafficked roads, footways, school playgrounds, patching and other similar situations. The aggregate may be of crushed rock, slag or gravel. Although porous when laid, it has a tendency to close up and become less porous under traffic.

5.14.5   3 mm Size Fine Graded (BS 4987)

This was formerly called fine cold asphalt. The name has been changed to avoid confusion with hot rolled asphalt. It is, in fact, a coated macadam and should never have been called 'asphalt'.

It is not suitable for use as surfacing where there is heavy traffic, as it tends to polish very easily. It is used occasionally with 14 or 10 mm coated chippings, in which case it is laid 25 mm thick to allow for embedment of the chippings.

5.14.6   6 mm Size Dense (BS 4987)

This material is now included in the British Standard, but has been used for many years for resurfacing lightly trafficked roads and for newly constructed playgrounds, footways and similar applications.

5.14.7   10 mm and 14 mm Size Open Graded (BS 4987)

These traditional coated macadam 'carpets' are unable to withstand current heavy traffic stresses and their durability is dependent upon the existing supporting road structure.

As the open grading allows water to permeate, coated grit is often used to assist in sealing the surface, with a further surface treatment within two years.

The 10 mm size is often used for patching and reinstatement repair.

5.14.8 Stone Mastic Asphalt (SMA)

SMA is recommended for use in 'thin surface' course systems (Clause 942). This gives a very durable, hard wearing surface which is not as susceptible to rutting as most other surfaces and also reduces surface noise. The normal layer thickness is between 20 and 40 mm.

5.14.9 Porous (BS 4987)

These surfacings have been developed in an effort to reduce spray from vehicle wheels and also to reduce the risk of aquaplaning in wet weather. They are made in 10 mm and 20 mm sizes with a relatively hard binder and 2% hydrated lime to reduce 'stripping' under the action of water.

5.14.10 Friction Courses

At low speeds, the sharp micro-projections on the aggregate particles provide the major part of the wet skid resistance. This requires a coarse aggregate with a high polished stone value, ideally with sharp micro-textured fine aggregate.

Road stone aggregates of all sizes must have a sufficient number of small angular projections in order to penetrate the thin film of viscous water which is adsorbed on to the surface of the stone, and so make direct contact with the vehicle tyre tread.

A pre-Cambrian gritstone is very suitable for road stone for the high friction dense material, and fines produced by crushing this same stone are better than most sands. It has been found that a smaller quantity of binder is needed than is the case with hot rolled asphalt.

The dense asphalt mix can provide a material which retains the surface macro-texture throughout the life of the surface course. This should eliminate any need to cut grooves in the surface or to surface dress the road.

Certain of the dense macadams incorporate two aggregates having different rates of wear. This produces an open surface macro-texture to provide good surface drainage, but the aggregate mixture is so graded that it is dense (only about 5% voids) and impermeable below the top layer of stone.

5.14.11 'Quiet' Surfacing Material

This is a new material formulated to provide a 'quiet' surface which produces less traffic noise than other materials. The DfT Ten Year Transport Plan mentions that quieter road surfaces 'will be specified in future contracts as a matter of course'.

The aggregate used has smooth, flat surfaces and when rolled produces a surface with few voids, which reduces the amount of spray from tyres in wet conditions. The surfacing contains a network of interconnecting sub-surface voids allowing better drainage.

The road is slippery when first laid, due to a 20–40 mm layer of hard setting polymer-based binder over the aggregate. This binder wears away under traffic movements allowing the surface to increase its grip.

5.15 Preparatory Work Before Laying

The existence of satisfactory sub-soil drainage and, in new construction, a sub-base of adequate strength to bear the equipment which will be used to lay the coated macadam or hot rolled asphalt is assumed. Some form of permanent lateral support should be given to the material, such as previously laid kerbing, kerb bed and backing.

Where the previous course is either of coated macadam, rolled asphalt or of other granular material, it is essential that it is adequately compacted using a roller exerting at least as great a load per unit width of roll as that to be used for rolling the coated macadam or rolled asphalt.

When an existing road is to be surfaced, all weak places should be strengthened, major inequalities of profile remedied and depressions filled with suitable material thoroughly compacted prior to the laying of coated macadam or asphalt surfacing. Any large depressions or irregularities should be taken out using the paver and this can be done by setting up a line to the correct profile/gradient for the paver to work to. Any excess of bitumen on the old surface should be removed. Heated or cold planing is the usual method used.

British Standard 4987 states that the accuracy of the finish of the surface on which the coated macadam is to be laid should be determined in the longitudinal direction by placing a 3 m straight edge at any position on the road parallel to the centre line of the carriageway. The depth of the gap at any place between the points at which the straight edge is in contact with the road should not exceed 25 mm below two course work or 13 mm below surface course work. The transverse profile should conform to a similar standard of accuracy, using a correctly shaped template instead of the straight edge.

According to BS 594, when an existing surface is to be improved either by heating and planing or by the addition of a regulating course, the maximum depth under a 3 m straight edge placed longitudinally, or under a transverse template (shaped to the true shape of the road surface) must not exceed 13 mm for two-course construction work, or 10 mm when the asphalt is to be laid as a single course.

The surface on which the material is to be laid should be free from standing water, loose materials and foreign matter.

Provision should be made for draining by offtakes at the level of the bottom of the new material.

5.15.1 Tack Coats

A 30% or 40% bitumen emulsion is normally used for tack coats.

Where the old surface is very smooth or polished, or has been heat-planed and the resulting surface is too smooth or is deficient in binder, the engineer may decide that it is necessary to use a tack coat to promote adhesion between the old and new material. Unless laid on a new basecourse, a tack coat will normally be required under a fine textured surface course. Tack coat is normally sprayed at a rate of 0.4–0.6 litre/m², and may be covered lightly with 6 mm chippings or fine cold asphalt to avoid being picked up by lorry wheels or the workmen's boots (coverage can be checked easily by spraying the surface using a full drum and measuring the area covered).

Pooling of emulsion should be avoided, and the spray should be far enough ahead of the work to allow the emulsion to break. It is essential that the sprayer is cleaned after use. Bitumen emulsions are described in more detail in Chapter 10.

5.16 Transporting Flexible Road Materials

The main features of the Specification (Clause 901) for bituminous materials are as follows.

Bituminous materials must be transported in clean vehicles and must be covered over when in transit or awaiting tipping. The use of dust, coated dust, oil or water on the interior of the vehicles to facilitate discharge of the mixed materials is permissible but the amount shall be kept to a minimum, and any excess shall be removed by tipping and brushing.

The mixed material must be supplied continuously to the paver and laid without delay. The rate of delivery of material to the paver must be so regulated as to enable the paver to be operated continuously. Wherever practicable, material must be spread, levelled and tamped by approved self-propelled pavers.

5.17 Laying the Surface Material

In the construction of new roads, the laying of the macadam base and surface course is done by a paving machine.

There are two main types of paver:

1 Pneumatic wheeled type (most common);
2 Caterpillar tracked type.

In each of these types, the paver has a large front hopper into which lorries tip directly and from which the road material is fed to the rear of the machine, being, at

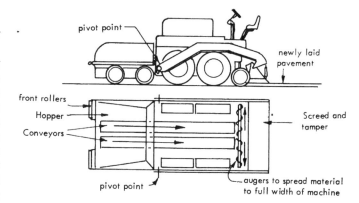

Figure 5.12 Road paving machine

the same time, spread to the pre-set width for laying and in an even thickness layer.

The paver has a vibrating tamping bar to consolidate the road material as it is laid. This tamping can provide as much as 80% of the total compaction of the material, the remaining 20% being obtained by rolling.

The laying widths range from 2.1–4.2 m or wider where twin machines are linked together.

Almost all macadam surfacing materials are delivered hot, in sheeted lorries, directly from the coating plant at the quarry to the site of the roadwork. For instance, hot rolled asphalt should be delivered at a temperature of about 180°C in order to remain sufficiently warm after placing for the binder to be fully effective and full compaction to be attained.

5.18 Specification for Laying Bituminous Materials

On trunk roads and motorways, the Specification for Highway Works has to be used. For other works, BS 594 and BS 4987 are commonly used.

Some of the most important points covered by the specification are as follows:

Laying

Wherever practicable, materials must be spread, levelled and tamped by an approved self-propelled paving machine. As soon as possible after arrival at the site the mixed material shall be supplied continuously to the paver and laid without delay. Delivery of material to the paver must be so regulated as to enable the paver to be operated continuously.

The rate of travel of the paver and its method of operation must be adjusted to ensure an even and uniform flow of material across the full laying width, freedom from dragging or tearing and without segregation of the material.

Hand laying of any bituminous material is permitted only in the following circumstances:

1 For laying regulating courses of irregular shape and varying thickness.
2 In confined spaces where it is impracticable for a paver to operate.
3 For footways.
4 At the approaches to expansion joints at bridges or viaducts.
5 For small areas.

Compaction

Material must be laid and compacted in thicknesses which enable surface level and regularity requirements to be met and adequate compaction to be achieved.

Material must be uniformly compacted as soon as rolling can be effected without causing undue displacement of the mixed material and must be completed while the temperature of the mixed material is greater than the minimum rolling temperature. Rolling is continued until all roller marks have been removed from the surface.

Compaction must be carried out using 8–10 tonnes deadweight smooth wheeled rollers having a width of roll not less than 450 mm, or by vibratory rollers or a combination of these. Surface course and binder course materials are always to be surface finished with a smooth wheeled non-vibrating roller.

The material must be rolled in a longitudinal direction with the driven rolls nearest the paver. The roller should first compact the material adjacent to any joints and then work from the lower to the upper side of the layer overlapping on successive passes by at least half the width of the rear roll.

Rollers must not stand on newly laid material while there is a risk that it will be deformed thereby.

Coated chippings must be applied by means of an approved mechanical chipping spreader distributing evenly. Addition of chippings by hand operation is limited to the following circumstances:

1 To confined spaces where it is impracticable for a chipping spreader to operate.
2 As a temporary expedient when adjustments have to be made to the distribution mechanism.
3 Where hand laying of the surface course is in progress.
4 To correct unevenness in the distribution of chippings.

All chippings must be applied uniformly to the surface and be rolled into the surface course in such a manner that they are effectively held and provide any specified texture depth.

Hand-raking of surface course material which has been laid by a paver and the addition of such material by hand-spreading to the paved area for adjustment of level is permitted only in the following circumstances:

1 At the edges of the layers of material and at gullies and manholes.
2 At the approaches to expansion joints at bridges and viaducts.
3 Where otherwise directed by the engineer.

Where joints between laying widths or transverse joints have to be made in surface courses, the material must be fully compacted and the joint made flush in one or other of the following ways, method 3 always being used for transverse joints:

1 By heating the joint with an approved joint heater at the time when the additional width is being laid but without cutting back or coating with binder.
2 By using two or more pavers operating in echelon or by using a multiple-lane-width paver.
3 By cutting back the exposed joints for a distance equal to the specified layer thickness to a vertical face, discarding all loosened material and coating the vertical face completely with a grade of hot bitumen before the next width is laid.

All joints must be offset at least 300 mm from parallel joints in the layer beneath.

The engineer may require the application of a bituminous spray tack coat complying with Clause 920 to the surface on which the laying is to take place.

Bituminous material must be kept clean and uncontaminated. The only traffic permitted to run on bituminous material which is to be overlaid is that engaged in laying and compacting the next course or, where a binder course is to be blinded or surface dressed, that engaged on such surface treatment.

Upper roadbase material, in pavements without binder course, and binder course material must not remain uncovered by either the surface course or surface treatment, whichever is specified in the Contract, for more than three consecutive days after being laid. The engineer may extend this period by the minimum amount of time necessary if compliance is impracticable because of weather conditions or for any other reason.

Road Rolling

The Department for Transport has published information on 'preferred methods' of compaction of bituminous material, to develop the compaction guidance given in BS 4987.

This suggests rolling patterns which are intended to improve compaction by giving greater attention to matching paver output to roller capacity. Essentially, these patterns consist of a 'rolling length' of staggered pattern and determine the distance by which the roller should advance on to the uncompacted mat at each stage in the rolling, so that the roller approaches as closely as practicable to the paver at each pass.

By this means it is hoped that better compaction can be achieved with consequent longer life of the surface.

In order to establish a suitable pattern, the following information is needed:

1 Time available for compaction
2 Width of mat to be laid
3 Paver speed when laying mat
4 Type, number and dimensions of rollers available
5 Number of passes to be achieved
6 Range of operating speeds of rollers.

Tandem rollers are the most suitable rollers for the 'preferred method'.

See Figure 5.13 for an outline of this process.

5.19   Check List to Use when Laying Road Material

5.19.1   Spreading by Machine

1 Before starting, make sure that the screed is clean.
2 Check the tamper.
3 Heat the screed plate.
4 Do not allow sheets to be removed before time.
5 Have at least two loads on the job before you start.
6 Keep a steady feed on the screed. N.B.: Ensure steady lorry contact to avoid lifting the screed unit. The lorry should be stationary and allow the paver to come up behind it. Do not reverse the lorry directly on to the paver rollers.

7 Material at the side of the hopper should be shovelled in towards the centre to prevent it becoming chilled.

8 When laying is interrupted, empty the machine, roll off, cut the joint.

9 Leave narrow strips at the side a little high to allow for compaction.

10 Leave the material slightly high at the joints and pinch it with the roller as soon as possible.

11 Work strictly within the prescribed temperature ranges when laying and rolling.

5.19.2   Spreading by Hand

1 Tip the load in no more than two heaps on to a clean, hard surface, and turn it over at least once before laying.

2 Protect the heap with tarpaulin sheets.

3 Use heated tools, especially with hot laid materials.

4 Follow up closely with the roller.

5 Watch for segregation, and correct it with fines.

6 Lay one load completely, if possible, before the next load is tipped.

7 Adjust the levels of manhole covers and frames, gully frames, etc., after the binder course and before laying the surface course.

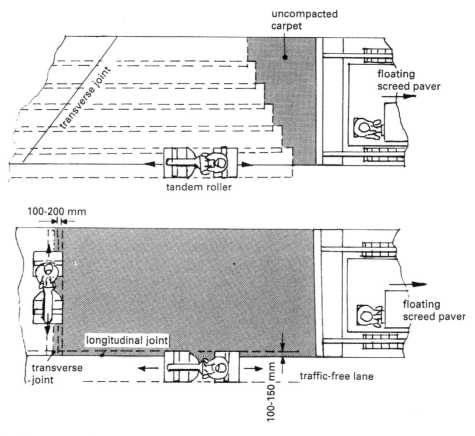

Figure 5.13   Road roller compaction patterns

5.19.3 Rolling

1 Roll with a suitable roller without delay.

2 Use the largest roll nearest to the machine (i.e. rear rollers of three-wheeled smooth roller).

3 Work from the lower edge.

4 Roll the joint first.

5 Roll straight and steady.

6 Stagger the reversing points by 1 m or so.

7 On some materials use a 6 tonne roller behind the machine, followed by a 10 tonne. Always use two rollers if the area to be covered in one day is over 4000 m².

8 Make a minimum of six passes (half a roll overlap) and leave no marks. Do not commence too early, especially with hot materials. Keep off when compaction is achieved.

9 Do not allow traffic on to newly laid material until it has cooled sufficiently to withstand traffic without deformation.

5.20 Laying Thicknesses and Rates of Spread

Bituminous material must be laid at a thickness which allows for mechanical interlock of the stone and for adequate compaction.

Tables 5.6 and 5.7 show recommended laying thicknesses and rates of spread.

5.21 Delivery Checks and Controls

When material is delivered to the site, it is advisable that the following information is recorded in order to ensure good quality control:

1 Location of work and person recording information.

2 Material used.

3 Layer of structure (i.e. surface course, binder course, etc.).

4 Where each load is laid (usually chainage for new works or lamp column numbers for existing roads).

5 Date.

6 Delivery temperature and time of arrival of each load.

7 Laying temperature and time of discharge of each load.

8 Temperature and time of rolling of each load.

9 Registration number of each delivery vehicle.

10 Delivery note number.

11 Remarks. This should include any other relevant information, such as extreme weather conditions, sample taken, plant breakdown, etc.

The engineer may decide that not all this information is required for certain materials.

Material description	BS 4987-1 reference	Nominal size (mm)	Nominal layer thickness (mm)	Minimum thickness at any point (mm)
Fine graded surface course	7.7	3	15 to 25	10
Medium graded surface course[a]	7.6	6	20 to 25	15
Dense surface course[a]	7.5	6	20 to 30	15
Porous asphalt surface course	8.2	10	30 to 35	25
Open graded surface course	7.2	10	30 to 35	25
Close graded surface course[a]	7.4	10	30 to 40	25
Open graded surface course	7.1	14	35 to 55	30
Close graded surface course[a]	7.3	14	40 to 55	35
Porous asphalt surface course	8.1	20	45 to 60	40
Open graded binder course	6.1	20	45 to 75	40
Dense, heavy duty and high modulus binder course[a]	6.5	20	50 to 100	40
Dense, heavy duty and high modulus binder course	6.4	28	70 to 120	55
Dense, heavy duty and high modulus base(roadbase)[a]	5.2	28	70 to 120	55
Single course	6.2	40	75 to 105	65[b]
Dense, heavy duty and high modulus base(roadbase)	5.1	40	100 to 150	90

Note: Thicknesses in excess of those given in the table can provide better compaction if adequate equipment is used but may lead to problems with surface irregularity and level control.

[a]Preferred mixture.
[b]80 mm if used as a single course with no subsequent surface course.

Table 5.6 Specified nominal and minimum layer thicknesses for coated macadam

5.21.1 Check Weights

Regular checks should be made to ensure that the tonnage ordered is as stated on the conveyance note.

5.21.2 Temperature

The temperature at which bituminous materials are mixed, laid and compacted is very important, particularly in the case of dense macadams and asphalt. The specified temperatures are shown in Table 5.8 and these should be recorded on the record sheet.

Average thickness of course (mm)	Approximate rate of spread	
	Open graded and porous macadams (m²/t)	Dense, close graded, medium and fine graded macadams (m²/t)
20	20 to 27	18 to 24
25	17 to 22	15 to 19
30	14 to 17	13 to 15
35	12 to 16	11 to 13
40	12 to 15	10 to 12
45	10 to 13	9 to 11
50	9 to 12	8 to 10
60	8 to 10	7 to 8.5
65	7 to 10	6 to 7.5
75	7 to 8	5 to 6.5
100	4.5 to 6	4 to 5.0

Table 5.7   Approximate rates of spread of coated macadam

Type of mixture including binder type and grade		Minimum temperature of mixture in lorry within 30 min after arrival on site °C	Minimum temperature immediately prior to rolling °C
Bitumen			
Dense, close graded,	290 pen	100	80
medium graded	190 pen[a]	110	85
surface courses			
and fine graded			
Dense and close	125 pen	120	95
graded surface			
course			
Open graded and	290 pen	85	65
single course	190 pen	95	75
Porous asphalt	190 or	110	85
	125 pen		

[a]For slag macadam, temperatures 10°C lower than those recommended may be used.
Note1: See Table 3 for temperatures of dense, heavy duty and high modulus binder course and base (roadbase) macadams.
Note2: Fluxed and deferred set mixtures (see BS 4987-1:2001, Annex A) may be delivered/rolled at lower temperatures than those in this table.

Table 5.8   Recommended delivery and rolling temperatures for coated macadams other than dense, heavy duty and high modulus binder course and base(roadbase) macadams

5.21.3   *Faults in Manufacture, Laying and Compaction*

Due to the speed at which material has to be manufactured, transported, laid and compacted, faults may occur.

Imperfections in the finished surfaces can usually be attributed to one or more of the following causes:

(*a*) Design of specification:
  e.g.   incorrect material specified
       wrong thickness
       inadequate specification

(*b*) Mixing and transporting:
  e.g.   incorrect grading
       incorrect binder
       incorrect binder quantity
       incorrect temperature
       poor quality aggregate (dirty or wet)

(*c*) Laying unit:
  e.g.   unprepared road surface (dirty or uneven)
       no tack coat where one is needed
       poor workmanship
       poor equipment

Table 5.9 illustrates many of the possible causes of faults and how they may be recognized. Some can be corrected as the job progresses, others may require:

further consideration,
analysis by the laboratory,
consultation with the supplier or manufacturer,
amendment to the specification,
closer control on future works,
training of operatives,
machine repair or adjustment, etc.

5.22   Surface Alignment and Regularity

In order to maintain a good ride, it is essential that the road structure layers are laid accurately and to constant thicknesses, as well as to the specified level.

Clause 702 of the Specification provides details of the tolerances which are permitted and these are summarized as follows.

5.22.1   *Horizontal Alignment*

±25 mm from one edge of the pavement.

5.22.2   *Surface Levels*

The level at any point is to be the designed level subject to the tolerances shown in Table 5.10.

5.22.3   *Regularity*

Regularity is measured with a rolling straight edge along a line parallel to the edge of the pavement.

Table 5.11 shows the maximum number of irregularities permitted when the straight edge is set at 4 mm or 7 mm. No irregularity over 10 mm is permitted.

POSSIBLE DEFECTS IN HOT LAID FLEXIBLE PAVEMENTS	DEFECT IN PAVEMENT	Insufficient bitumen	Mixture too coarse	Excess fines in mix	Excess bitumen	Excess moisture in mix	Mix too hot or burned	Mix too cold	Poor spreader operation	Inadequate rolling	Over rolling	Rolling when mix too hot	Rolling when mix too cold	Rollers standing on hot pavement	Rollers too heavy	Paver operating too fast	Traffic allowed on before cool	Mix laid in too thick a course
Some of these can be corrected on site, others at the mixing plant.	Bleeding				×													
	Brown dead appearance	×					×	×										
	Rich fat spots				×													
	Poor surface texture			×	×		×	×	×	×	×	×				×		
	Rough uneven surface			×			×	×	×			×	×	×	×	×		
If there is any doubt about the quality of a load, telephone the Highways Lab. to come and take a sample for testing.	Honeycomb or ravelling		×	×			×	×	×					×				
	Uneven joints			×			×	×	×				×	×				
	Roller marks			×	×		×		×			×	×	×	×			
	Pushing or waves			×	×	×			×			×				×	×	×
	Cracking (fine cracks)	×		×							×	×				×		
	Cracks (large cracks)					×						×				×		
	Aggregate broken by roller		×				×	×				×	×			×		
	Tearing of surface on laying	×	×				×	×	×								×	
	Surface slipping on base				×	×		×				×	×		×		×	

Table 5.9 Faults in flexible carriageway: causes and defects

Table 5.10 Tolerances in surface levels of pavement courses

Road surfaces	
– general	±6 mm
– adjacent to a surface water channel*	+10 – 0 mm
Binder course*	±6 mm
Top surface of base in pavements without binder course*	±8 mm
Base other than above*	±15 mm
Sub-base under concrete pavement surface slabs laid full thickness in one operation by machines with surface compaction	±10 mm
Sub-bases other than above	+10 – 30 mm

*Where a surface water channel is laid before the adjacent road pavement layer, the top of that layer, measured from the top of the adjacent edges of the surface water channel, shall be to the tolerances given in Table 5.10

Table 5.11 Maximum permitted number of surface irregularities

	Surfaces of carriageways, hard strips and hard shoulders				Surfaces of lay-bys, service areas, all bituminous binder courses and upper roadbases in pavements without binder courses			
Irregularity	4 mm		7 mm		4 mm		7 mm	
Length (m)	300	75	300	75	300	75	300	75
Category A* roads	20	9	2	1	40	18	4	2
Category B* roads	40	18	4	2	60	27	6	3

*The category of each section of road is described in the Contract

For short lengths, or where the use of the rolling straight edge is impracticable, a 3 m straight edge is used, and the permitted irregularities in this case are as follows:

The maximum allowable difference between the surface and the underside of the straight edge, when placed parallel with, or at right angles to, the centre line of the road at points decided by the engineer shall be for:

pavement surfaces	3 mm
binder courses	6 mm
upper roadbase in pavements without binder courses	6 mm
sub-bases under concrete pavements	10 mm

5.22.4 Surface Roughness – Texture Depth

In order to achieve a good skid resistance value, it is necessary for the surface course to have a 'roughed' surface. For coated macadams, this can be achieved by the careful choice of aggregate grading and size. On fine graded macadam and hot rolled asphalt this roughing is provided by applying coated chippings.

The 'roughness' is measured and referred to as 'texture depth'. A typical value for a high speed road is 1.5 mm. The measurement of texture depth is done either by the sand patch method or by use of a mini texture meter.

'Sand Patch' Method (using sand or glass beads)

The method is known as 'Sand Patch' because fine sand was used. However, the size of the sand particles needed was only generally available in the UK. For the test to be global, glass beads are now manufactured which usually replace the sand particles. To find the texture depth (TD) the following procedure is used.

1 The surface to be measured must be dry and should first be swept with a soft brush.

2 Fill a cylinder (80 mm high with an internal diameter of 20 mm) with beads and level off with a straight edge.

3 Pour the beads into a heap on the surface to be tested – in windy conditions, a tyre or similar item may be used to protect the pile.

4 Spread the beads over the surface, working the material into a circular patch – using a flat wooden disc of 65 mm diameter (this has a hard rubber disc 1.5 mm thick, stuck to one face and usually a handle on the wooden side). The beads are spread so that the surface depressions are filled to the level of the peaks.

5 Using dividers, measure the radius of the patch formed by the beads and record this to the nearest millimetre.

6 Carry out a number of tests at intervals along the road surface, ideally along a line parallel to the kerb, in order to obtain an average value (see Figure 5.14).

It is necessary to prepare a conversion table, from which to obtain values of the texture depth. This is done by first calculating the volume (V) of the cylinder. The texture

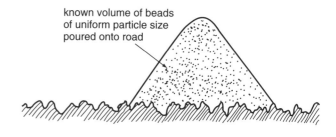

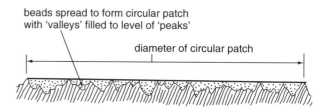

Figure 5.14 Sand patch method of measuring surface texture depth

depth (TD) is then given by TD = V/R where R is the radius of the patch.

A table is then prepared showing TD for values of R in increments of 1 mm from 50 mm to 180 mm. The smaller the radius R of the sand/bead circle, the greater is the surface roughness or texture depth. The test procedure is suitable only for texture depths in excess of 0.25 mm. On smoother surfaces, a smaller cylinder and finer material are required. The sand patch test has severe limitations in use as it is a slow and laborious process and is prone to operator bias in selecting the areas to be checked. Furthermore, it cannot be carried out on a damp or wet carriageway or in windy conditions.

Mini Texture Meter

The Mini Texture Meter, the result of research and development by the TRL, has been found to give reliable results. The unit is basically a hand operated, manually propelled device for determining the texture of road surfacings.

The construction details of this instrument are outlined in Figure 5.15. It can be seen that it consists of a two-wheeled

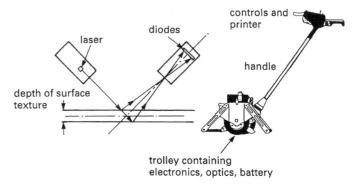

Figure 5.15 Mini Texture Meter (MTM)

trolley, housing the electronics, optics and battery, together with a detachable handle which houses the machine controls and the printer. As this instrument uses a class 3B laser diode which emits invisible radiation, great care must be taken to avoid exposure to the beam.

The unit has a displacement operating range of 20 mm with a surface texture output resolution of 0.01 mm. The results are printed out, giving an average texture reading every 10 m with a summary at the end of 50 m. These readings are described as 'sensor measured texture depth' (SMTD) and this is referred to as the 'SMTD result' for a given road surface.

At present, the MTM unit is specified for use in dry road conditions only. It will work on damp roads, but the results may be inaccurate if there is any water in the depressions of the texture, as the laser beam will be reflected off the water surface in those depressions.

5.22.5 Density

High density is one of the most critical requirements of paving materials. To achieve this, it is important to reduce air voids in the material to a minimum. To obtain the maximum possible density it is necessary to specify

1 The thickness of the layer of road material.
2 The temperature of both delivery and laying.
3 The type of roller to be used.
4 The number of passes to be made by the roller.

Testing for density can be carried out by core sampling or by nuclear density meter (NDM).

Core sampling is the traditional method. The core is cut out of the road structure and tested for density in the laboratory. Disadvantages of this method include slowness of the cutting process, subsequent repairing of the hole, and time taken to perform the test. *See also* Chapter 10, Section 10.17.

The sample core extraction requires a special trailer equipped for this purpose and entails coning off an appreciable area of carriageway in order to undertake the operation.

The advantages of the nuclear density meter are that testing is much more rapid, the apparatus is relatively small and can be carried by hand, and the process does not damage the road structure.

5.23 Colour Finishes of Bituminous Surfacings

There are three basic methods for obtaining a coloured/decorative finish with a bituminous surfacing:

1 The colour can be incorporated into the surface course at the time of manufacture.
2 Suitable decorative chippings can be applied to some types of surfacing during the laying operation.
3 An overall decorative surface treatment can be applied to the surfacing after laying.

5.23.1 Colour in the Surface Course

There are two ways in which colour can be incorporated in a surface course during manufacture: first by the incorporation of a pigment in the macadam/asphalt mix, together with an aggregate of complementary colour; second, simply by manufacturing the mix with an aggregate of the required colour. In the former case, an overall colour effect is immediately obtained, but in the latter case the colour will become apparent only after considerable trafficking has removed the surface binder film; the process will therefore be effective only on well-trafficked roads and even then the colour is likely to take some time to develop.

As far as the pigmented mixture is concerned, red is the only colour regularly available, although green may occasionally be available to order.

Proprietary macadam-type mixes are available in which the normal black bitumen binder has been replaced with a clear resin. By the incorporation of appropriately coloured or decorative aggregate in the mix the manufacturer can provide a variety of coloured finishes.

5.23.2 Chippings Applied to a Surface Course

Bituminous mixes with high mortar contents are particularly suited to carry an application of decorative chippings rolled into the surface at the time of laying. These mixtures include rolled asphalts to BS 594 containing no more than 40% of coarse aggregate and 3 mm size (fine-graded macadam) to BS 4987. Rolled asphalts are suitable for most traffic situations but fine graded macadam is appropriate only for lightly trafficked and pedestrian areas.

To provide a decorative finish to both of these types of surfacing pigmented bitumen-coated or clear resin-coated chippings can be applied during laying. The bitumen or resin coating is to ensure adhesion to the surfacing and it is not recommended that uncoated chippings be used as there will be little adhesion. However in the case of 3 mm fine graded macadam laid on areas subjected to little trafficking (e.g. private drives and footways), a light scattering of uncoated chippings, e.g. white spar, can give a very attractive finish if uniformly applied and well embedded to ensure their retention for as long as possible.

Decorative chippings cannot be successfully rolled into the surface of more stony bituminous mixes such as coated macadams or high stone-content rolled asphalts. With these types of surfacing, surface treatments as described below will need to be considered.

5.23.3 Surface Treatment

There are two surface treatments which can be applied to bituminous surface courses to provide a decorative finish. These are surface dressing and pigmented bituminous slurry coats. Surface dressing consists of a bitumen, bitumen emulsion or tar spray applied to the surface course following which an overall application of natural stone chippings is applied and well rolled. The finish obtained is relatively deep-textured and will be the colour of the chippings. The other treatment requires the use of a fine aggregate mixture of the surfacing by brush or squeegee to a thickness of approximately 3 mm. A fine-textured finish is achieved and a number of colours can be supplied.

Neither of these two surface treatments provides any significant degree of structural strength and it is therefore essential that they are applied only to an adequate construction. The slurry material is normally used only on untrafficked or lightly trafficked surfaces unless the particular material supplied is suitable for heavy-duty use.

A wide range of colours is available in naturally occurring aggregates from the browns and buffs of gravels through the whites and creams of limestones and spars to the darker greys of some basalts and the pinks, reds and greens of some granites.

5.24   Conservation and Use of Reclaimed Materials

5.24.1   Reclaimed Materials

Reclaimed materials are increasingly being used in earthworks and pavement construction and subject to environmental controls are used in new construction, highway improvements and maintenance of all types of road, including motorways and trunk roads.

Reclaimed materials can already be on site, such as roads under construction, or off site as with residues from industrial processes, mining and demolition. All these sources reduce the demand on natural aggregates.

The Specification for Highway Works (MCHW 1) permits a wide range and choice of reclaimed materials including road planings, crushed concrete and mineral by-products such as Slags and Pulverized Fuel Ash (PFA). Table 5.12 gives details of selected secondary aggregates.

5.24.2   Use of Reclaimed Materials

Materials indicated as complying with MCHW 1 for a particular application may not necessarily comply with all the requirements of the series listed, only particular clauses. For example, in the 600 series, Unburnt Colliery Spoil can satisfy the specification as a general fill, but is excluded as a selected fill. *See* Table 5.12.

Application & series Material	Embankment and fill 600	Capping 600	Unbound sub-base 800	Bitumen bound layers 900	Cement bound sub-base 1000	Cement bound roadbase 1000	PQ Concrete 1000
Crushed Concrete	A	B	A	C	A	A	A
Bituminous Planings	B	B	C	A	B	C	C
Demolition Wastes	B	B	C	C	B	C	C
Blastfurnace Slag	A	B	A	A	A	A	A
Steel Slag	C	C	A	A	B	C	C
Burnt Colliery Spoil	A	B	A	C	B	C	C
Unburnt Colliery Spoil	B	C	C	C	B	C	C
Spent Oil Shale	B	B	A	C	B	C	C
P F A	A	A	C	C	B	A	A
F B A	B	B	C	C	B	C	C
China Clay Waste	B	B	B	B	B	B	B
Slate Waste	B	B	B	C	B	B	B

A Specific Provision
B General Provision – permitted if the material complies with the Specification requirements, but not named within the Specification
C Not Permitted

Table 5.12   Specification for Highway Works (MCHW 1) application of secondary aggregates

5.24.3 Definition of Terms Used in Table 5.12

Crushed concrete

Selected granular fill specified by name. It is however excluded for use over buried steel structures.

Bituminous planings

These can be acceptable for direct addition to new mixes or used as granular fill, once granulated or crushed.

Demolition wastes

Crushed demolition debris can be used as a bulk fill providing there are no contaminants in the material that may be an environmental hazard, e.g., water soluble salts.

Blast Furnance Slag

A by-product from the production of iron, resulting from the fusion of fluxing stone (fluorspar) with coke, ash and the siliceous and aluminous residues remaining after the reduction and separation of iron from the ore. BFS is not subject to the Waste Licensing Regulations (1994).

Steel Slag

By-product of the manufacture of steel from pig iron. There are two types, basic oxygen slag (BOS) and electric arc furnace slag (EAS). EAS normally has lower free lime and magnesia contents than BOS, as a result of the process and is more easily weathered.

Burnt Colliery Spoil

Also known as burnt limestone. This is the residue following ignition of coal mine spoil heaps which results in partial or complete combustion of coal particles in the spoil and consists of calcined rocks.

Unburnt Colliery Spoil

Also known as minestone, a by-product of mining and is derived from the rocks which lie above, below and within the coal seams. These rocks consist mainly of siltstones and mudstones and in some areas, sandstones and limestones.

Spent Oil Shale

The residue of shale, mined in the Lothian region of Scotland, after heating to drive off volatile hydrocarbons. Similar to well burnt colliery shale.

Pulverized Fuel Ash

PFA or Fly Ash is extracted by electrostatic precipitation from the flue gases of modern coal burning power stations and is similar in fines to cement.

Furnace Bottom Ash

FBA is the coarser fraction of ash produced in coal burning power stations resulting from the fusion of PFA particles, which fall to the bottom of the furnace. It varies in size from fine sand to coarse gravel and has a porous structure.

China Clay Waste

A by-product from the extraction of china clay from decomposed granite consisting largely of two distinct materials; 'Stent' which is 'waste' rock and 'Tip sand' which is predominantly quartz with some mica. The product tends to be a consistent form.

Slate Waste

By-product of slate quarries primarily producing roofing slates. The waste represents some 70–90% of gross quarried volume.

5.25 Recycling the Existing Road Surface

Recycling the existing road surface has emerged as a good cost effective method in many recent trials. Two key issues are being addressed, the quality of the material to be recycled and the type of binder added to make the rejuvenated asphalt mix work. Two techniques may be used – either the *ex situ* or the *in situ* method.

Ex situ Method – planings are transported to a crushing and mixing plant where the stone is screened to a designed grading and mixed with cement and bitumen 'foamed' through the injection of air and water. This is then transported back and laid as needed.

In situ Method – the top layers of worn out carriageway are pulverized to a depth of up to and around 500 mm with rotovating plant, which simultaneously adds water to bring the material to its optimum moisture content. Calculated quantities of a bituminous or hyraulic binder such as cement, or a combination of both, are then mixed in before compaction.

The *ex situ* method is proving to be more popular as it allows multilayered construction and greater control of the material.

Revision Questions

1 What are the 4 construction layers in a flexible pavement and what functions do they perform?
2 Using a vibrating roller of between 1800 kg and 2300 kg, how many passes would have to be made over granular material with a compacted thickness of 110 mm?
3 What would be the minimum average crushing strength of CBM 1 and 1A Materials?
4 What could be the possible cause of 'BLEEDING' in a flexible pavement?

Note: See page 305 for answers to these revision questions

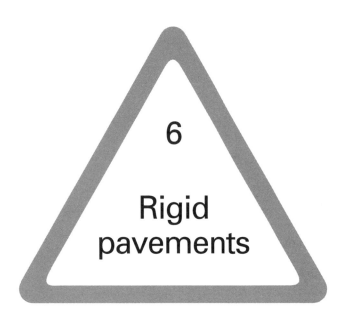

6.1 Introduction

Current practice in carriageway construction favours flexible pavements and many concrete roads have been overlaid with flexible binder and surface courses. The concrete road once overlaid, is known as 'composite' construction. It must be noted that these overlays need to be 100 mm thick in order to prevent reflection cracking in the surfacing, due to expansion and contraction of the concrete slabs.

The construction of rigid pavements is a mainly specialist job, often requiring complex and expensive laying machinery.

In this chapter, the main principles involved in this work are covered in a generalized manner. Some detail of small-scale concreting work, which a roadworker might be required to undertake, is given at the end of the chapter.

Details of the design and maintenance of rigid (concrete) pavements are contained in the *Design Manual for Roads and Bridges*, Volume 7, produced by the Highways Agency and the *Specification for Highway Works*, which is published by HMSO.

In this type of construction, the main structure of the pavement is a concrete slab which, in relation to the flexible pavement, is the equivalent of the surface course, binder course, roadbase and sub-base combined. It is termed 'rigid' because this concrete slab does not deflect within itself under traffic load, and it is designed to last 40 years before needing major reconstruction.

Since the concrete slab provides both the surface course and the main structural strength of the rigid road, it must be constructed of high strength and high quality concrete and great attention must be paid to the surface

finish in order to provide a good running surface with good skid resistance under all weather conditions.

Concrete expands and contracts as the air temperature rises and falls, so provision must be made to accommodate this variation, often referred to as 'thermal expansion and contraction', in the length of the road slab. Hence, transverse joints are provided at regular intervals.

Transverse joints can be of three types, contraction, expansion or warping joints. Longitudinal joints are needed for wide slabs.

As the purpose of these joints is to prevent the pavement from cracking under stress, the joint is designed to cater for the anticipated type of movement.

During recent years, bituminous overlays have been included in the original design, with the concrete slabs requiring fewer joints, due to the protecting surface.

6.2 The Rigid Road

The concrete carriageway slab may be constructed in one of the following ways:

Unreinforced concrete (URC)
Jointed reinforced concrete (JRC)
Continuously reinforced concrete pavement (CRCP)

It is essential that a separation membrane be provided between the road slab and the sub-base.

In the case of unreinforced and jointed reinforced slabs, this membrane must be of impermeable plastic, 125 µm thick, laid flat and without creases, with a minimum of 300 mm overlap at all joints in the plastic.

For continuously reinforced slabs, a bituminous spray is used as the waterproof membrane, and this is applied prior to concreting.

This separation/waterproof membrane serves two purposes: firstly, it prevents the wet concrete from losing some of its moisture content into the sub-base below, so that it can undergo the complete chemical reaction and achieve its full strength when set and cured. Secondly, it allows the slab, when set, to move reasonably freely over the sub-base to relieve thermal stresses.

6.2.1 Road Slab Design

The *Design Manual*, Volume 7, contains information about the required thicknesses of rigid pavements, based on the expected number of standard axles (from 1 to 400 million) designed to use the road.

The minimum thickness is 150 mm for CRCP (continuously reinforced concrete pavement) and up to 330 mm in the case of URC (unreinforced concrete).

The pavement quality concrete replaces the surface course, binder course and base of flexible and flexible composite roads. The sub-base and capping layer are designed in a similar manner to those for flexible construction.

6.2.2 Sub-base Materials

Generally, materials for the sub-base should be hard, durable, chemically inert and suitably graded, should be capable of being compacted to a high density and not be susceptible to appreciable frost heave. Granular materials such as cement-bound material or wet lean concrete are suitable.

Where a capping layer is needed, either cement-bound material CBM3 or wet lean concrete must be used, with a thickness of 150 mm.

6.3 Concrete Quality

Increasing volume and weight of traffic have caused designers to increase the strength of pavement surface concrete to provide greater durability in order to offset the resulting wear. Changes in cement manufacture have produced an ordinary Portland cement that hardens more rapidly than earlier types. There is thus a tendency to reduce the cement content of the slab mix.

6.3.1 The Concrete Mix

Normally, Ordinary Portland or Portland Blast Furnace cement is used, Grade 40 being required for surface concrete mixes. Lower-strength mixes are adequate for concrete roadbases (in two-layer slab construction).

Grade 40 mix requires a minimum of 320 kg of cement in each fully compacted cubic metre of concrete; for durability of this concrete, a water/cement ratio of below 0.45 is necessary.

The coarse and fine aggregate must be carefully selected, either from natural material (to BS 882) or from crushed air-cooled blast furnace slag (to BS 1047). Additionally, pulverized fuel ash may be used as part of the aggregate.

Once the proportions of the mix have been determined, they must remain unchanged throughout the construction of the road, unless the mix is changed with the agreement of the engineer in charge.

The workability of the concrete at the point of placing must be sufficient for the concrete to be fully compacted and finished without undue flow. The workability is determined by the Compacting Factor test, or the Vebe test. If the concrete is of grade C20 or below, the slump test may be used to check the workability.

Concrete for the top 50 mm of the road slab should be air-entrained (*see* Section 10.27). As almost all road slabs are reinforced, it is convenient to lay the lower part of the slab, up to the level of the reinforcement, using ordinary concrete and then to add a top layer of air-entrained concrete over the reinforcement. Alternatively, air-entrained concrete may be used throughout the slab.

Another point in favour of using the same mix throughout the depth of the slab is that it halves the number of test specimens required to check the concrete strength.

6.3.2 Strength

The strength of the pavement-quality concrete is measured by the crushing strength of 150 mm concrete cubes. These are made and cured in pairs from concrete delivered to the paving plant. At least one pair of cubes must be made for every 600 m^2 of concrete slab laid, with a minimum of 6 pairs being made each day, for each different mix.

Generally, one cube from each pair is tested at 7 days, the other being tested at 28 days.

For a grade C40 concrete, a 7-day strength of 31 N/mm^2 should be attained when ordinary Portland cement is used.

6.3.3 Workability

The Compacting Factor is a suitable workability test for any of the stiff mixes used when paving by machine, because it can be carried out alongside the paver. Workability should be constant, and can be obtained by noting the power input to the mixer. If necessary, plasticizing or retarding admixtures may be used to suit local or weather conditions.

Approximate values for the compacting factor are as follows:

0.80 for single layer construction
0.80–0.83 for top layer in two-layer construction
0.77–0.80 for bottom layer

The designed Compacting Factor for the mix must be maintained within 3%.

Low workabilities are needed to ensure that inserted dowel bars are retained in position. Higher workabilities are necessary to allow texturing and finishing to be completed satisfactorily within the time available. In practice, a compromise workability between these is used.

6.3.4 Air Content

The total quantity of air in air-entrained concrete must be determined at the point of delivery to the paving plant by a pressure-type air meter.

Where aggregate of 20 mm nominal size is being used in the mix, the air content should be 5% by volume. With 40 mm stone, the air content must be at least 4%.

The air-entraining agent is added to the mixture, care being taken to ensure uniform distribution of the agent throughout the batch during mixing.

The main reason that the use of air-entrainment admixture is preferred as a method of improving workability is that it increases the frost-resistance of the concrete. This is because the air-entrained concrete mix has a lower water content.

6.4 Reinforcement

Reinforcing steel may be of prefabricated sheets or hot rolled steel bars (of either Grade 250 or Grade 460) or cold worked steel bars.

Reinforcing steel must be free from oil, dirt, loose rust and scale.

When prefabricated steel sheet reinforcement is used, the sheets must overlap by more than one complete mesh or be butt welded.

Where steel bars or mesh are used, if the reinforcement is positioned before concreting, it must be fixed on supports or preformed spacers (see Figure 8.45) and retained in position at the required depth below the finished surface as follows:

(a) 60 ± 10 mm cover for slabs less than 270 mm thick
(b) 70 ± 10 mm cover for slabs 270 mm thick or more.

In two-layer construction, when sheet steel fabric is used, it may be laid on the surface of the bottom layer of concrete.

Generally, transverse reinforcement must terminate 125 mm from the slab edges, with a tolerance of ± 25 mm. The applies also to longitudinal joints where tie bars are used. The reinforcement must terminate 300 ± 50 mm from transverse joints.

Some road slabs are constructed of continuously reinforced concrete (i.e. no joints). In this case, the steel reinforcement must be of Grade 460 deformed bars assembled on site. This reinforcement is positioned at mid-slab depth, terminating 125 ± 25 mm from the slab edges and longitudinal joints.

6.5 Surface Finish

The surface of the slab, after final regulation, is usually brush-textured in a direction at right angles to the longitudinal axis of the carriageway.

This surface texturing must be applied evenly across the slab by a stiff brush not less than 450 mm wide.

The minimum texture depth should be 0.75 mm when measured between 24 hours and 7 days after the construction of the slab. This figure may be reduced to 0.65 mm when taken at least 6 weeks before the road is to be opened to traffic.

There will be a considerable reduction of texture depth during the early stages of the road being used by traffic, due to the removal of the weaker surface concrete mortar. To ensure that the minimum texture depth prior to opening the road to public traffic is achieved, it may be necessary to obtain a texture depth of 1.0 mm initially.

If the texture depth is as great as 1.3 mm after use by construction traffic, it will tend to produce excessive tyre/road noise within a vehicle.

6.6 Curing

Curing is essential to provide adequate protection from evaporation and against heat loss or gain by radiation, and so to allow the concrete to attain its designed strength.

Without adequate curing, the concrete strength could be only half the strength of the corresponding cubes cured in water in the laboratory.

The best form of curing is to keep the concrete constantly damp. This can be achieved by

(a) covering the concrete with plastic sheeting
(b) spraying on plastic material which hardens into a plastic sheet and which can be peeled off later
(c) spraying on resin-based aluminized curing compound, which contains sufficient finely dispersed flake aluminium to produce a complete coverage of the surface with a metallic finish.

These all protect the surface of the slab, the underside being protected by the plastic separation membrane, which provides a barrier to ensure that essential moisture is not partially lost into the sub-base material.

6.7 Joints in Pavement Slabs

The rigid pavement is divided into individual panels by joints in both the longitudinal and transverse directions, except in the case of continuously reinforced concrete pavements which have only longitudinal joints, if they are 6 m width or over.

In the normal jointed slab road there is a plane of weakness at each joint, caused by, in many cases, crack inducers fixed to the sub-base at the joint position, together with a surface groove, needed to accommodate the joint sealant. This sealant has to seal the joint to prevent surface water, dirt, etc., from getting into the joint and damaging it.

6.7.1 Transverse Joints (Figure 6.1)

Transverse joints must be formed at right angles to the longitudinal axis of the carriageway, except at junctions and roundabouts.

The spacing of these joints (i.e. the distance between them) depends upon the weight of longitudinal reinforcement used.

When concreting of the carriageway is carried out during the summer months, it is not necessary to provide expansion joints. At other times of the year, every third joint must be an expansion one, the remainder being contraction joints.

To prevent differential vertical movement between adjacent slabs, dowel bars are provided, set at the mid-depth of the slab and parallel to the longitudinal axis of the road. One end of the dowel bar is de-bonded, so that it does not stick to the concrete of one slab; the other end is cast into the concrete of the adjacent slab.

Transverse joints on each side of a longitudinal joint must be in line with each other.

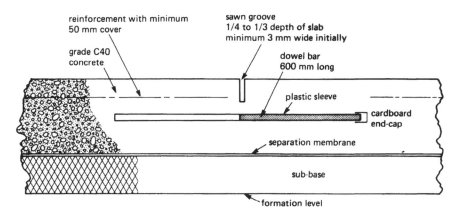

Figure 6.1(a)   Sawn groove transverse joint (contraction)

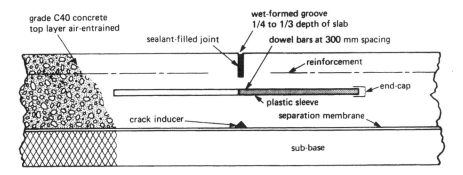

Figure 6.1(b)   Wet formed groove transverse joint (contraction)

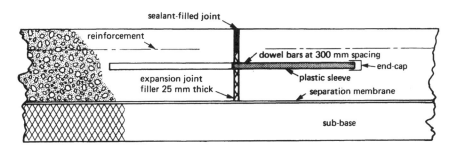

Figure 6.1(c)   Transverse joint (expansion)

The joints must have a sealing groove which may be:

(*a*) a sawn joint groove, sawn as soon as possible after the concrete has hardened sufficiently to enable a sharp-edged groove to be produced without damage to the concrete. Grooves must be between 1/4 and 1/3 of the slab depth and at least 3 mm wide;

(*b*) a wet-formed joint groove, formed in the plastic concrete by vibrating a metal plate or joint former into the concrete, prior to the final regulation and finishing process. If the grooves are over 15 mm wide, some of the disturbed concrete may have to be removed.

In addition, dowel bars must be provided. Bottom crack inducers are not required where the joints are sawn. They may also be omitted from the wet-formed joints in the summer period.

Expansion joints must, in addition to dowel bars and joint grooves, have a joint filler board, 25 mm thick, made of a firm compressible material, sufficiently rigid to resist deformation during the passage of the concrete paving plant. The depth of the filler board must be the slab depth less the depth of the groove. Holes for dowel bars must be accurately bored or punched out.

6.7.2 Longitudinal Joints (Figure 6.2)

Where a concrete carriageway is more than 4.2 m wide, it is necessary to provide one or more longitudinal joints, these being arranged to coincide with the lane divisions of the carriageway. Longitudinal differential vertical movement between the slabs is prevented by fixing tie-bars of mild steel, usually 12 mm diameter. The tie-bars prevent the joint from opening by more than a fraction of a millimetre, so maintaining the interlocking of aggregate particles on the two sides of the crack.

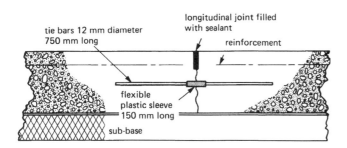

Figure 6.2 Longitudinal joint detail

The tie-bars are fixed at mid-depth of the slab and, in a continuous laying process, a longitudinal joint groove is formed in the surface of the concrete at the time of laying.

It is usual for the tie-bars to be positioned before the concrete is laid, supported by a metal cradle, in order to ensure that they remain in the correct position and parallel to one another, at right angles to the longitudinal axis of the road. The spacing between bars is usually 600 mm.

Wet-formed longitudinal grooves are made in the concrete ahead of the finishing beam, a groove former being inserted from a special dispenser. The concrete displaced by this operation is recompacted by a vibrating compactor at least 300 mm wide and set centrally over the groove. Alternatively, the groove may be sawn.

6.7.3 Dowel Bars (Figure 6.3)

It has been found that after dowel bars have been inserted from the top surface of a slab, full compaction has not been achieved around them, the concrete being more porous and susceptible to frost damage than the remainder of the slab. Hence, when dowel bars and tie-bars are to be inserted vertically into fresh concrete, this is now only permitted into the bottom layer of two-course construction.

It may be preferable to use dowel bar assemblies or cradles, with the bars being positioned prior to concreting.

For expansion joints, dowel bars are of Grade 250 steel, 25 mm diameter and 600 mm long for slabs under 240 mm thick, and 300 mm apart. For thicker slabs the bars are 32 mm diameter.

When used in contraction joints, the dowel bars are 20 mm diameter, 400 mm long for slabs under 240 mm thick

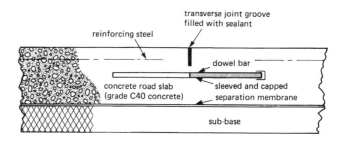

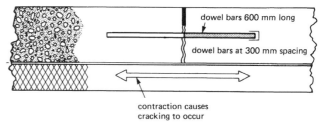

Figure 6.3 Dowel bars in transverse joints

and 25 mm diameter for thicker slabs. They are positioned 300 mm apart.

The dowel bars are positioned at mid-slab depth ±20 mm.

Dowel bars must be covered with a thin plastic sheath as follows:

(a) for expansion joints, for half the length plus 50 mm from one end;
(b) for contraction joints, for at least 2/3 of the length.

At expansion joints, a closely fitting cap 100 mm long of waterproof cardboard or synthetic material is placed over the sheathed end, leaving a space for expansion of about 25 mm between the end of the cap and the end of the dowel bar.

Great care must be taken to ensure that the dowel bars are in a straight row and parallel to each other longitudinally. Any misalignment would result in very high stresses being set up as the adjacent slabs expand and contract due to temperature changes, causing severe cracking to occur at or near to the edges of the slab.

6.7.4 Joint Grooves

Joint grooves are formed in the surface of the slab both transversely and longitudinally. They may be formed in the plastic concrete at the time of laying, or, alternatively, they may be sawn in the surface after the concrete has set. Sawn grooves are preferred as they avoid disturbance of the surface of the plastic concrete, unlike wet forming of the grooves. The time of the sawing of the grooves is critical; if sawn too soon, aggregate particles will be plucked out; if too late, the concrete will already have cracked, not necessarily at the desired position.

Narrow crack-inducing grooves should be sawn first, being widened for sealing later.

The machine which forms the grooves at the time of laying must include a horizontal vibrating plate at least 300 mm wide, across the line of the joint, to ensure that the concrete displaced from the groove is fully re-compacted into place. This must be followed by screed-ing to obtain a satisfactory surface finish ready for texturing.

For grooves over 13 mm wide, the forming mechanism must remove from the slab the volume of concrete displaced by the grooving process.

Preformed strips are often used to seal longitudinal joints when pavements are being constructed in 2 or 3 lane widths in one operation, these being inserted ahead of the finishing screed, from a special dispenser.

6.7.5 Sealing of Joint Grooves

Temporary fillers and formers are left in position in the grooves until they are prepared for permanent sealing. The grooves must be cleaned by grit-blasting to remove any dirt or loose material immediately prior to the appli-cation of the permanent seal.

The sealants used are of three types: hot poured (bituminous), cold poured (polymer resin with a curing agent) and preformed elastomeric compression seals.

With hot poured sealants, care must be taken not to overheat the material and also to complete the pouring within the specified time for that sealant.

Two-component cold poured sealants must be thoroughly mixed and applied to the groove as soon as possible after mixing, either by using a caulking gun or by pouring into the joint.

Preformed elastic compression seals are compressed to at least 70% of their uncompressed width and insertion into the joint is assisted by the use of a lubricant adhesive. This type of material must be in one piece for at least one full lane width when used in transverse joints. In longitu-dinal joints, these seals must be in continuous lengths where possible, any joints being butted together and fixed with additional adhesive.

6.8 Treatment of Manholes and Gullies

Manhole covers, gullies and their frames must be isolated from the main pavement slab and be housed in 'localized' slabs, which must be larger than the manhole shaft, gully grid or gully. This allows these items to be removed, repaired or replaced without affecting the main concrete road slab.

Recesses for manholes and gullies are formed by casting the main slab against formwork boxes placed and fixed accurately and with truly vertical sides.

When the formwork boxes are removed subsequently, for the concrete to be placed and compacted around the manhole cover or gully grid, preformed joint filler 25 mm thick is fixed to the edges of the main slabs with an allowance at the top for a sealant to be applied later.

Where possible, the positions of the manholes and gullies in the concrete pavement are adjusted so that they are either adjacent to the edge of the slab or to a joint, or that they lie within the middle third of the slab.

6.9 Machine Laid Concrete

There are two main types of machine used for laying concrete pavements:

(a) Slip-form pavers, which may be either conforming plate or oscillating beam models.
(b) Concreting trains.

With the slip-form paver, the objective is to lay an uninterrupted flow of concrete without the need for prefabricated and pre-set joint assemblies, which would prevent the slip-form machine from performing in this most economical way.

Ways have now been found of:

1 inserting longitudinal tie bars through the conform-ing plate using a high speed punching mechanism;
2 injecting dowel bars into the concrete as it flows from the front of the machine at predetermined and static positions in the slab.

This is done by inserting the bars through a battery of guide tubes which terminate at a point behind the leading edge of the conforming plate. Dowel bars are loaded into storage magazines in between deliveries of concrete and are pushed mechanically from the guide tubes at a rate corresponding to the speed of the paver. At the point where the dowel bars are inserted a plane of weakness (i.e. reduced cross-sectional area) is formed by cutting a slot using a cutting plate. Crack inducers are not normally used in this situation.

Longitudinal joints are formed and filled with plastic sealing strip as the machine proceeds.

6.9.1 Conforming Plate Slip-form Pavers

This type of machine is built round a pair of parallel side forms, which are linked together by a horizontal top plate. These shape the concrete slab as the machine moves forward, the concrete being fed to the front of the conforming plate via a hopper.

Figure 6.4 shows the principles of concrete placing by conforming plate paver (diagrammatic layout). Greater detail of the machine is shown in Figure 6.5.

Ready-mixed concrete is fed into the hopper at the front of the machines, by lorries. These lorries can approach by reversing along the prepared sub-base.

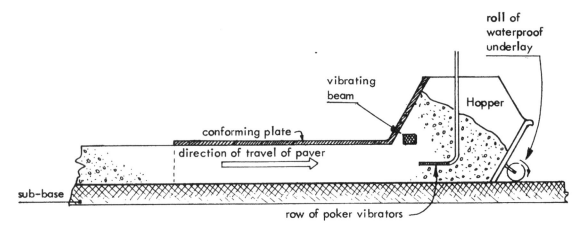

Figure 6.4   Diagrammatic section of conforming plate paver

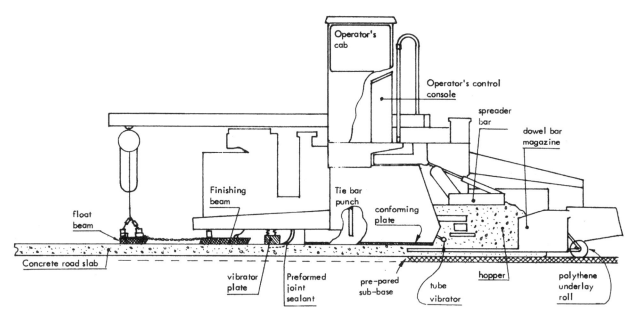

Figure 6.5   Conforming plate slip-form paver

The waterproof underlay is laid from rolls immediately under the front of the hopper.

As the machine moves forward, the weight of the concrete in the hopper forces the material under the conforming plate. The compaction of this concrete is assisted by a vibrating beam across the hopper and a row of poker vibrators. In the larger slip-form pavers, a spreader is fitted to assist in distribution of the concrete across the full width of the hopper. The machines can readily be adjusted to lay various widths of pavement up to a maximum of 13 m.

Dowel bar insertion is precisely controlled by a plunger mechanism linked to the forward movement of the machine, the plunger ejecting a dowel bar from each feed tube (from the magazine) at the same instant. The feed tubes are 300 mm apart across the machine (*see* Figure 6.6).

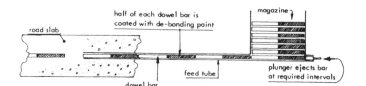

Figure 6.6   Dowel bar feed system: conforming plate paver

6.9.2   Oscillating Beam Slip-form Pavers

In this type of machine, the side forms give the shape to the edge of the road slab and the oscillating beams compact and give the shape to the top of the road slab. Initial compaction may be assisted by vibrators in the hopper.

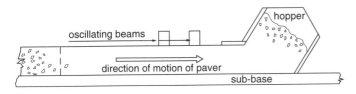

Figure 6.7(a)   Sectional elevation: oscillating beam paver

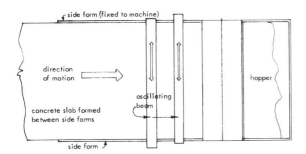

Figure 6.7(b)   Plan view of oscillating beam paver

This is shown diagrammatically in Figure 6.7. The method of propulsion of the paver is omitted for clarity.

Longitudinal joints are cut mechanically as the machine proceeds, usually centrally in the slab being laid, the slot being filled by a temporary plastic insert fed from a drum on the paver into the slot as it is cut, immediately behind the conforming plate or oscillating beam.

6.9.3   Control System

Normally a slip-form paver has to have two tensioned guide lines or wires, one on each side of the slab being laid, at a constant height above and parallel to one edge of the true pavement surface. Additionally, one of the wires must be at a constant horizontal distance from the slab edge.

The wires are supported by stakes which must not be more than 8 m apart.

The paver has an electronic sensing system which picks up signals from contact with the guide wire and these signals initiate alterations at the controls, either raising or lowering the level of the slab surface, or causing the machine to veer to the left or right (*see* Figure 6.8).

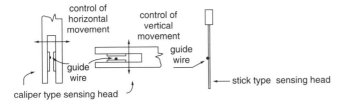

Figure 6.8 Electronic sensing devices for controlling direction and level

The accuracy of fixing of the guide wires is, therefore, of the utmost importance if the slab is to be laid to the correct line and level.

6.9.4   Laying Reinforced Concrete by Slip-form Paver

Where steel sheet reinforcing fabric is to be laid in the concrete slab, it is preferable to use two pavers of the oscillating beam type.

The leading paver lays the lower portion of the slab, which must be a stiff enough mix to support the reinforcing fabric which is then laid on it.

The second paver then lays the upper part of the slab, using air-entrained concrete, and completes the finishing processes.

For this concrete construction, it is necessary to feed the concrete to the hopper by side entry.

6.10   Finishing Processes (Figure 6.9)

These operations are carried out by machines separate from, but often attached to and drawn by, the slip-form paver. The processes are:

1 Transverse joint forming.
2 Re-compaction and screeding of the transverse joint.
3 Surface texturing, by wire brushing.
4 Application of curing agent, usually a sprayed plastic membrane.

6.11   Concreting Trains (Figure 6.10)

Concreting trains consist of a series of machines travelling in sequence along prepared side forms. These side forms may be either rails, which are firmly fixed to the concreting side forms or independently supported, or concrete 'banquettes' which will probably be used later as the channel at each side of the carriageway.

For the use of a concreting train to be considered, the job must be extensive, as, in addition to the train itself, large-scale concrete batching plants will be required, and will have to be set up at suitable points along the route of the road to supply the large quantities of concrete that will be needed.

Ahead of the concreting, the rails to support the train have to be fixed, probably over a length of several kilometres; the underlay has to be laid, and crack inducers fixed at the planned transverse joint positions. Longitudinal crack inducers may also be needed, together with tie bars, made up into prefabricating units on support cradles, all needing to be fixed in place before concreting can start.

The total length of the train itself may be up to 600 m.

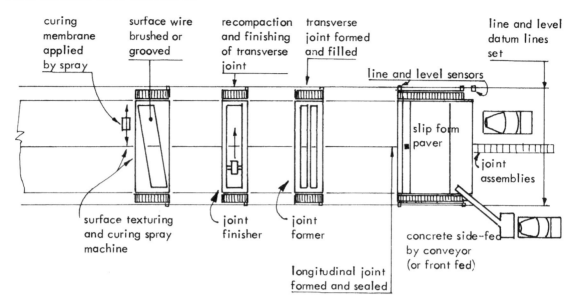

Figure 6.9 Slip-form paving

As an example of the scale of operation, in the construction of the M6 from Preston northwards, a distance of 20 km, a concreting train was used. Concreting of the 11 m wide carriageways was carried out at the rate of 0.8 km per day at peak output. To concrete at this rate required eight 10 m³ capacity mixer trucks and a concreting plant capable of producing 8000 tonnes daily.

This type of machine-laid concrete road construction is called 'fixed-form' paving, to distinguish it from the slip-form method.

6.12 Comparison of Slip-form and Concreting Train Methods of Laying a Rigid Pavement

1 Concreting train: this consists of a series of machines travelling in sequence along prepared side forms (which previously have been fixed to a true line and level).
2 Slip-form paver: this is a machine which does almost all of the operations involved in the concrete road construction and which travels on its own tracks on the sub-base.

6.12.1 Concreting Train

Each machine does one or two operations in the concrete road construction.

The units of the train in sequence are:

1 First spreading and compacting unit.
2 Dowel bar insertion unit (vibratory).
3 Second spreading unit and compacting unit.
4 Transverse joint forming unit.

5 Diagonal beam final shaping unit.
6 Surface texture unit.
7 Curing membrane spray unit.
8 Mobile cover unit.

The train is for use only for very large-scale jobs.

The supply of concrete is by lorry to the side of the spreading units.

The underlay is often laid well ahead of the train.

Tie bars (for longitudinal joints) are also positioned ahead of the train.

6.12.2 Slip-form Paver

As it proceeds, it forms the road slab by means of either a conforming plate and side forms or by vibrating beams and side forms. These side forms are part of the machine, so that, as it proceeds, it takes its side forms with it, leaving the freshly laid slab exposed on top and sides.

Its line and level are determined by electronic sensing systems which 'follow' a wire, which has been accurately positioned beside the road slab which the machine is to lay.

Concrete supply by lorry is usually to the front of the machine (or can be to the side).

Polythene underlay dispensed from rolls under the hopper as the machine advances.

Dowel bars fed out from magazines loaded from the front.

6.13 Concreting by Hand

Many small-scale concreting jobs cannot economically be tackled with machine-laid concrete, and it is here that the roadworker needs to develop his knowledge and hand

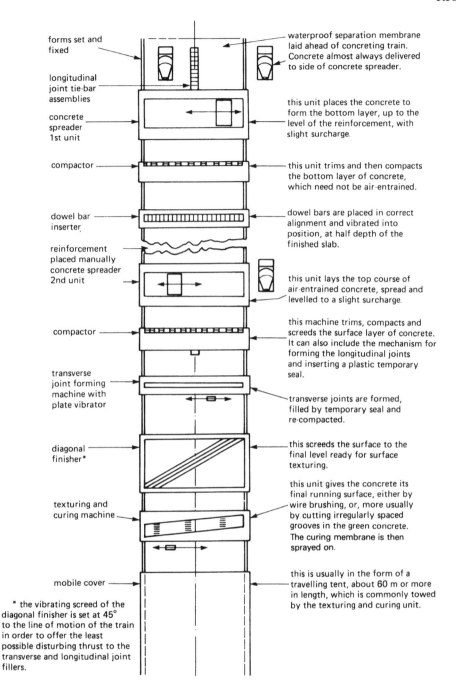

Forms set and fixed

longitudinal joint tie-bar assemblies

concrete spreader 1st unit

compactor

dowel bar inserter

reinforcement placed manually concrete spreader 2nd unit

compactor

transverse joint forming machine with plate vibrator

diagonal finisher*

texturing and curing machine

mobile cover

waterproof separation membrane laid ahead of concreting train. Concrete almost always delivered to side of concrete spreader.

this unit places the concrete to form the bottom layer, up to the level of the reinforcement, with slight surcharge.

this unit trims and then compacts the bottom layer of concrete, which need not be air-entrained.

dowel bars are placed in correct alignment and vibrated into position, at half depth of the finished slab.

this unit lays the top course of air-entrained concrete, spread and levelled to a slight surcharge.

this machine trims, compacts and screeds the surface layer of concrete. It can also include the mechanism for forming the longitudinal joints and inserting a plastic temporary seal.

transverse joints are formed, filled by temporary seal and re-compacted.

this screeds the surface to the final level ready for surface texturing.

this unit gives the concrete its final running surface, either by wire brushing, or, more usually by cutting irregularly spaced grooves in the green concrete. The curing membrane is then sprayed on.

this is usually in the form of a travelling tent, about 60 m or more in length, which is commonly towed by the texturing and curing unit.

* the vibrating screed of the diagonal finisher is set at 45° to the line of motion of the train in order to offer the least possible disturbing thrust to the transverse and longitudinal joint fillers.

Figure 6.10   Typical concreting train for fixed-form paving

skills in order to produce a satisfactory and durable concrete roadway or footway.

Hand-laid concrete does, however, need the use of some special equipment, particularly formwork, tamper bars, etc.

For concrete pavements, the preparation of the sub-base to true and accurate level, well compacted, is essential if the work above it is to be satisfactory.

Formwork can only be set accurately on a sound base. It must be firmly fixed, to contain the liquid concrete and to withstand the load from the tamper bar, particularly when it is a vibratory one.

Traditionally, the alternate bay method of construction was used, but this needed large quantities of formwork which had to remain in place for some considerable time, due to the 'two-stage' concreting.

It is more usual nowadays to concrete in a continuous slab, forming the transverse joints by grooving (pushing a timber strip into the surface of the freshly laid concrete to a depth of about 25 mm and leaving it in position until the concrete has set) and inserting fibrous expansion jointing material at every third joint probably choosing this point at which to stop concreting each day.

6.14   Alternate Bay Construction Procedure

In this method, the sub-base is first prepared and side forms fixed.

The length to be concreted is then divided into bays by fixing transverse end stops (often of timber) in such a way that they face into bays 1, 3, 5, 7, etc. (as shown in Figure 6.11). Waterproof underlay is also laid in these bays. The odd numbered bays are concreted first: hence the term 'alternate bay construction'.

As soon as convenient after this concreting, say after 24 or 48 hours, the end stops are carefully removed and the even-numbered bays (2, 4, 6, 8, etc.) are prepared for concreting. The already concreted odd-number bays act as the end stops for the even-numbered bays.

If the joint is to be a normal construction joint, then concrete can be placed immediately against the existing concrete. Where expansion joints are to be provided, say between bays 2 and 3, 5 and 6, etc., dowel bars will already have been fixed, carefully aligned, into the end stop and will have been concreted into the odd-numbered bay. The exposed half of each dowel bar is either sleeved or coated with debonding paint and the expansion joint (usually fibrous material) placed in position prior to starting concreting. Waterproof underlay is laid in each of the even numbered bays and concrete can then be placed.

Joints are completed by inserting plastic filler strip or wooden laths into the top 25 mm of the concrete at each joint.

Normally tamping, screeding and surface finishing is done from the side forms. The concrete must, of course, be cured after placing.

6.15   Formwork or Shuttering

Normally steel road forms are used. These are 3 m long and of a height to provide the necessary slab thickness, say 150 mm or 200 mm. They are set to line and level and must be secured by at least 3 pins per 3 m length, having one pin fixed at each side of every joint. Form sections must be joined tightly together by a locked joint free from play.

Care must be taken to provide an accurate and continuous top edge for the tamper to run on. Care must also be taken to keep the front face of the forms vertical.

When fixed, they must be cleaned and oiled.

The road form is usually fixed on the sub-base and adjusted finally by driving the locking wedges into the slot

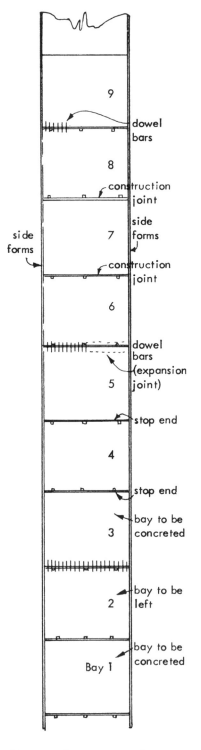

Figure 6.11   Alternate bay construction

in front of, and behind, the pin. The form may be moved forward by driving the front wedge or pulled back by driving the rear wedge (*see* Figures 6.12 and 6.13).

The pins must be firmly driven to withstand the lateral thrust on the forms from the freshly placed liquid

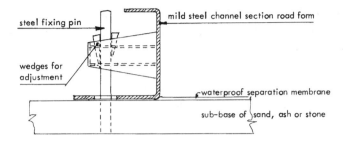

Figure 6.12 Sectional elevation: steel road form

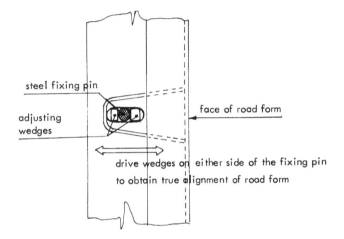

Figure 6.13 Plan: steel road form

concrete and to withstand the load from the vibrating tamper bar. The top face of the forms provides a datum level for the slab surface.

6.16 Transverse Joints

6.16.1 *Construction Joint (Figure 6.14)*

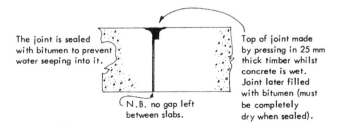

Figure 6.14 Construction joint (dowel bars not shown)

6.16.2 *Expansion Joint (Figure 6.15)*

Expansion joints should be provided at every third joint. The purpose is to allow longitudinal movement of the slab, to accommodate thermal expansion and contraction.

The dowel bar is used to keep the two adjacent slabs level during this movement. Great care must be taken to keep the dowel bars correctly aligned longitudinally and parallel to each other.

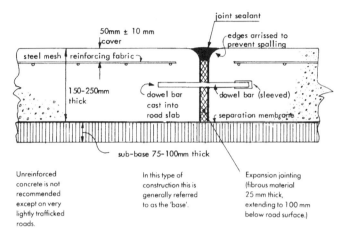

Figure 6.15 Expansion joint

6.16.3 *Dummy Construction Joint (Figure 6.16)*

This type of joint is made by forcing a timber strip (the length = width of slab) into the surface of the slab to a depth of 25–50 mm, arrissing edges. (The edges are arrissed or rounded, using a special trowel, to prevent a sharp edge being formed, with increased risk of spalling.) Then remove timber and fill with waterproof sealant after concrete has set.

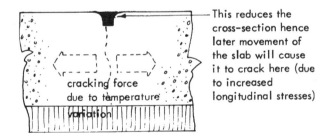

Figure 6.16 Dummy construction joint

6.17 Longitudinal Joints

Where the full width of the carriageway is not concreted in one piece, longitudinal joints must be made to allow lateral expansion/contraction but not to permit any vertical misalignment.

Slab 1 is cast first, with a 'former' fixed on to the 'inside' face of the shuttering to form the joint groove (*see* Figures 6.17 and 6.18).

The shuttering needs to be of split type with grooves to accommodate the tie-bars cut out of the top edge of the lower part of the formwork.

Additional support must be provided to keep the tie-bars in alignment until the concrete slab 1 has hardened.

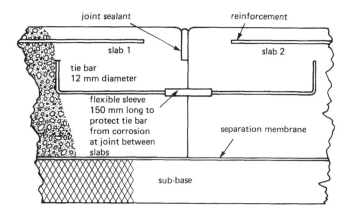

Figure 6.17 Longitudinal joint: joint groove formed in slab 1

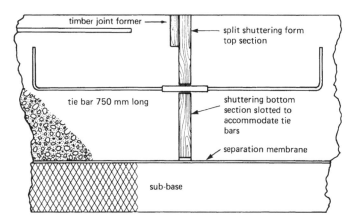

Figure 6.18 Formwork for longitudinal joint

6.18 Release Agents

To prevent damage to the concrete, the surface of the form must be coated with a release agent before the concrete is placed. There are several suitable release agents of which the main ones are as follows:

6.18.1 Neat Oil with Surfactant

Suitable for all types of formwork. Staining of concrete may occur if the oil is over-applied. The oil film can be affected by heavy rain.

6.18.2 Chemical Release Agent

Suitable for all types of formwork, it is based on light volatile oils which dry on the surface of the form to leave a thin rain-resistant coating.

6.18.3 Mould Cream Emulsion (Oil Phased)

Ideal for use on timber forms. Not suitable for steel. Has a limited storage life.

6.18.4 Neat Oil (Without Surfactant)

Must not be used where the appearance of the concrete is important. Inexpensive.

6.18.5 Water-phased Emulsion (Oil in Water)

Again must not be used where appearance of concrete is important. Cheap and easy to apply.

6.18.6 Wax

Recommended for forms/moulds made of concrete. Difficult to apply thinly and evenly.

6.19 Striking of Formwork

Always take great care when striking formwork to avoid damage to arrises and projections. Before removing soffit forms and props, the concrete surface should be exposed carefully to check that it has hardened sufficiently for the forms or props to be removed safely.

It is not possible to specify precise periods which should elapse before shuttering/formwork is struck (i.e. taken down/removed). This period will depend upon the concrete used, the weather, exposure of the site, method of curing, whether any subsequent treatment is to be given to the concrete, etc.

There are certain minimum periods which should elapse before the formwork is struck. Typical examples are:

Formwork location	Surface temperature of concrete	
	16°C	7°C
Vertical formwork (columns, walls, large beams)	9 hours	12 hours
Slab soffits (props left under)	4 days	7 days
Beam soffits (props left under)	8 days	14 days
Props to slabs	11 days	14 days
Props to beams	15 days	21 days

6.20 Tamping

The tamper bar (Figure 6.19) is often of timber (about 200 mm × 75 mm) and of length = (road width) + 2 × (road form top flange width).

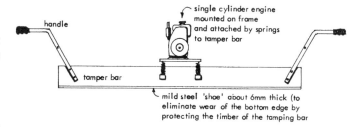

Figure 6.19 Typical tamper bar (vibrating)

When tamping, a small excess of concrete should be kept against the front face of the bar to provide adequate material for filling any voids (Figure 6.20). It may be necessary to use a shaped tamper bar. This is shaped to the profile of the carriageway, thus building in camber, etc. (Figure 6.21).

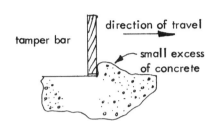

Figure 6.20

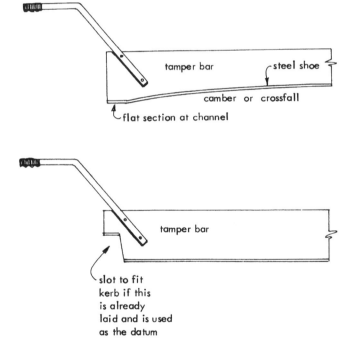

Figure 6.21 Details of tamper bar shoes

The concrete is vibrated using the formwork as the datum for the surface. To improve skid resistance, it is normally necessary to follow-up with a hand tamper to produce small transverse 'ridges' in the surface. This also assists by providing small channels for the run-off of surface water.

6.21 Surface Finish and Texture

This is normally carried out using a hand tamper bar, which is slightly longer than the width of the concrete slab. It is used by an operative at each end tamping the freshly laid concrete in a transverse direction. For a footway, the degree of surface texture achieved by this method may be adequate.

For a road slab, a suitable wire brush can be drawn by hand across the surface of the wet concrete, usually after it has had some time to start the setting process, as the mix used for hand laying would be wetter than that used in mechanized concreting and would be too wet to texture when first laid.

6.22 Curing of Hand Laid Concrete

It is just as important to cure carefully concrete which has been laid by hand as it is in the case of machine-laid slabs.

Hand spray units are available to apply plastic setting agents, or polythene sheeting may be laid on the surface of the slab.

6.23 Warping Joints

In *unreinforced* concrete road slabs, bending stresses are caused by temperature differences between the top and bottom of the slab. In hot weather, the slab becomes quite hot to touch whilst the underside remains cool. In frosty weather, the surface may be colder than the underside of the slab.

These temperature variations between top and bottom of the slab are referred to as vertical temperature gradients and the bending moments or stresses caused by this are thought to be greater than those due to longitudinal contractions of the slab.

In order to relieve these stresses, transverse *warping joints* are provided. These joints are made at regular intervals with the objects of:

(*a*) providing a discontinuity in the slab to release any build-up of warping moments/stresses;
(*b*) acting as a load transferring device;
(*c*) being sealed against water and grit.

The slab discontinuity is by means of a transverse construction joint with a crack inducer at the bottom and a sealing groove cut into the top of the slab. When the slab cracks, the load transfer is provided by interlock of the aggregate at the crack, and the crack in the concrete is prevented from opening too far by a system of longitudinal tie-bars. Sealing of the joint is by filling the groove with a suitable conventional flexible sealant.

In slabs of 200 mm thickness and over, these joints are provided at 60 m intervals. For slabs of less than 200 mm thick the joints should be at 40 m intervals. The spacing of the longitudinally placed tie-bars is determined by the thickness of the slab as follows:

Slab thickness (mm)	Tie-bar spacing (12 mm diameter)
300	180
240	240
180	360

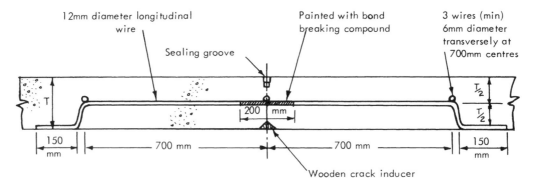

Figure 6.22   Typical warping joint

The 12 mm diameter tie-bars are specified to be 1.4 m long with the centre 200 mm de-bonded. They are kept together by transverse wires of 6 mm diameter, at least three transverse wires being required.

Every third tie-bar is made longer and bent so that 1.4 m effective length of tie-bar lies at half slab depth with the assembly supported on the feet of these longer tie-bars which rest on the base of the slab as shown in Figure 6.22.

6.24   Sampling and Testing

Details of sampling and testing of concrete are given in Chapter 10 Materials and Testing. These include:

The slump test
The compacting factor test
Cube crushing tests

6.25   Fast-track Concrete Paving

The essential feature of this process of rapid concrete pavement construction is precise control of the concrete mix production. The aim is to produce a very rapid 'age-hardening' concrete mix which can be laid rapidly and, after a few hours' curing, can be opened to traffic – thus overcoming one of the main problems of concrete pavements, the length of time normally required before traffic can run on the new surface.

For this fast curing, it is necessary to use a concrete mix which is designed to achieve, in a few hours, the strength which has traditionally been required at 28 days (typically, 2.75 N/mm²). This can be done by careful control of the water–cement ratio, cement characteristics and content, and may be aided by the addition of the normal admixtures as required. Either ordinary Portland cement or rapid-hardening Portland cement is used. For the curing period of only a few, possibly about six, hours from laying, coated hessian insulation blankets are most suitable.

Careful organization and control of the laying, curing and groove cutting are essential features of the process. All labour must be adequately trained to progress the concreting at the required speed.

Fast-track concrete paving can be applied to:

Complete carriageway construction.
Partial replacement by an 'inlay' of one or more lanes.
Strengthening of an existing bituminous or concrete pavement by means of a concrete overlay.
Airfield runway construction/reconstruction.
Maintenance/reconstruction of a carriageway on the basis of closing the road, laying the new concrete and reopening to full traffic use within two days.

There are obvious advantages to this speed of operation, including:

1 Reduced contract period – hence reduced cost.
2 Minimizing the use of expensive concrete paving plant – again reducing costs.
3 Reducing traffic delays and their costs.
4 Reducing the quantity of equipment, such as side forms, etc., as these can be struck and reset within the day's work.

6.26   Reinstatement

This chapter has concentrated upon general rigid construction activities.

The requirements of the New Roads and Streetworks Act 1991 for the reinstatement of openings in highways are covered in Chapter 15.

Revision Questions

1   Where should you position reinforcement steel bars on when constructing reinforced concrete slabs?
2   What is the purpose of an expansion joint and why?
3   Name some release agents for preventing formwork sticking to concrete?
4   What sort of tests might you use to test the suitability or strength of concrete?

Note: See page 305 for answers to these revision questions

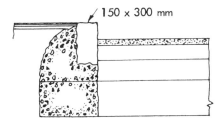

7

Kerbing, footways & paved areas

7.1 The Purpose of Kerbing

In the past, many rural roads were constructed without kerbs but with the heavier flow of traffic using all types of road, it has become increasingly necessary to use kerbs. The main purpose in using kerbing and other edge detail is to strengthen the sides of a carriageway and to prevent lateral spread of the pavement which would cause a deterioration of the road structure. In addition to this main purpose, kerbs also:

(*a*) form a wall which causes surface water to run along the channel and into the gullies. This surface water runs off the road to the channel due to the camber of the road.

(*b*) act as an elevated boundary line to footways, so keeping pedestrians above the road channel level. The surface of the footway should fall to the kerb line, thus causing surface water from the footway to run off into the road channel.

(*c*) act as a demarcation line between road and footway, thus being a safety feature to assist drivers.

(*d*) assist in preventing vehicles from mounting the footway.

(*e*) prevent encroachment of vegetation on to the road.

7.2 Kerbing Materials and Other Types of Edge Detail

Kerbs may be formed from precast concrete, natural stone, and *in situ* asphalt or concrete. Due to the very high cost of cutting and dressing the natural stone kerb,

it is rarely used. The precast concrete kerb is by far the most common form of edge detail but other materials such as extruded asphalt and concrete are increasingly used.

7.2.1 Natural Stone Kerbs

These are usually quarried and dressed from granite or sandstone. Kerbs are supplied in random lengths which vary from 500 mm to 2.00 m having widths of 100, 125 and 150 mm, together with depths of 200–300 mm in 25 mm steps. The ends are chisel dressed to be square and the top face is fair picked. If sandstone, then the face may be sawn and so will be smoothly finished. *See* Figure 7.1, which shows a kerb laid in advance of flexible pavement.

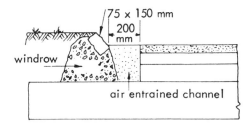

Figure 7.1 *Granite kerb in heavily trafficked urban road*

Perhaps setts should be mentioned here as these are used as a form of kerbing when bedded in concrete. A sett is usually a rectangular block of stone 125–225 mm in depth and approximately 100 mm in width and length. *See* Figure 7.2 which shows setts laid in advance of flexible pavement.

Figure 7.2 *Splayed kerb of granite setts laid on windrow*

7.2.2 Raised Precast Concrete Kerbs (BS 7263 Part 1 2001)

Most present-day kerbs are hydraulically pressed but some special kerbs may be manufactured by vibrating the concrete. A hydraulically pressed kerb has a pimpled finish. The vibrated kerb has a smooth finish. However, the pressed kerb is considered to be a better product as it is stronger and has a surface more resistant to abrasion.

Kerbs may be classified as straight, radiused or special. The section or profile of a kerb or edge detail can be vertical, half-batter, splayed or curved.

Straight Kerbs

Straight kerbs are standardized in length within the range of 450 mm to 915 mm, the 915 mm kerb being the norm:

(a) Vertical kerbs

These are normally used only where protection of the footway is of primary importance such as in city streets.

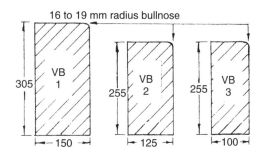

Figure 7.3 Vertical bullnose kerb sections

(b) Half-batter kerbs

The half-batter kerb is in very common use where the footway adjoins the carriageway. Whilst it gives some protection against a vehicle leaving the road, it also allows close rolling of the carriageway without damaging the kerb face.

A common non-standard kerb in use is one having the same profile as HB2, but being only 210 mm instead of 255 mm in depth. The kerb shown as HB3 is used in situations where depth in construction is not available, i.e. bridge decking.

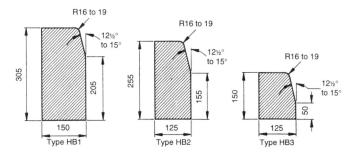

Figure 7.4 Half-batter kerb sections

(c) Splayed kerb

The splayed kerb allows vehicles to leave the carriageway in an emergency. It is, therefore, used where raised edge detail is required but where there is no adjacent footway.

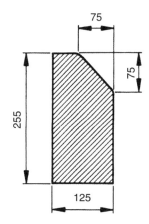

45° splayed kerb: type SP

Figure 7.5 Splayed kerb section N.B.: These kerbs may be obtained with depths of either 210 mm or 305 mm (non-standard)

(d) Barrier kerb

This is sometimes provided to confine vehicles to the carriageway at particularly hazardous points such as high embankments, sharp curves, etc. The main requirement of a barrier kerb is to slow a vehicle but still allow it to be controlled, preventing it from entering other traffic lanes and meeting other hazards.

Figure 7.6 Barrier kerb

Radius Kerbs

These are manufactured to the same profiles as straight kerbs. Radiused kerbs are manufactured to suit certain designed internal and external curves. The standard length of a radius kerb is 915 min, but it can be supplied with a length of 610 mm. The radius should be marked on one of the unexposed faces of each kerb and may be of 1.0 and 2.0 (external only), and 3.0, 4.5, 6.0, 7.5, 9.0, 10.5 and 12.0 m (internal and external).

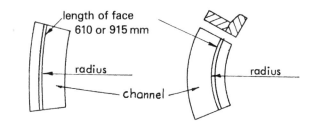

Figure 7.7 (left)   Internal kerb and channel
Figure 7.8 (right) External kerb and channel

Special Kerbs

This category covers droppers, quadrants, and footpath edgings.

(a) Transition Kerbs (Droppers)

This is a kerb which has a standard section tapering off to a horizontal top face for carriageway crossings. It allows the lowering of the main kerb line to assist a vehicle to mount the kerb safely and to cross the footway.
(DL1/DR1 or DL2/DR2 for BN and SP kerbs) (*see* BS 7263 Part 1)

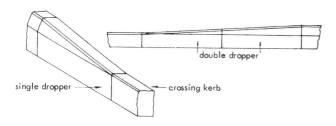

Figure 7.9   Droppers and crossing kerbs

Droppers are available in single and double patterns, the whole of the drop being formed in one 915 mm kerb for the single pattern and taking two kerbs for the double pattern.

(b) Quadrants

Quads are quarter circles with half-batter profile and are supplied in 125, 150 and 255 mm depths. The radius may be 305 or 455 mm. Corners (internal and external angles, *see* Figure 7.11) together with offlets (*see* Figure 7.12) are specified in BS 7263: Part 1. They are a common item and are reproduced in several forms.

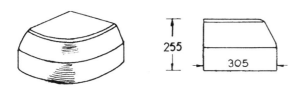

Figure 7.10   Standard concrete quadrants

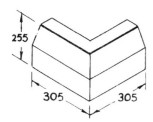

Figure 7.11   Corner or external angle

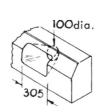

Figure 7.12   Offlet or weir kerb

(c) Edgings

These may be used for backing footways and have profiles which are square, half round, chamfered or bullnosed. They are 915 mm in length and have depths of 150, 200 and 250 mm. All are 50 mm in thickness.

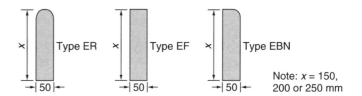

Figure 7.13   Standard sections of concrete edging

7.2.3   Channels and Special Kerbs

A plain flush kerb is used where there is no footway and where control of surface water is not required. A weakness in using the flush kerb is that surface water may reach the sub-grade down the joint on either side of the kerb. Serrated kerbs (*see* Figure 7.15) give an audible warning to light vehicles in danger of leaving the carriageway under conditions of poor visibility.

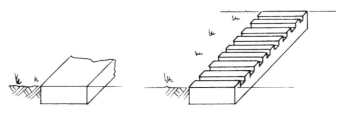

Figure 7.14
Channel block: plain

Figure 7.15   Channel block: serrated

A channel is used in conjunction with a raised kerb mainly, to enable an artificial longitudinal fall to be introduced where no natural fall exists in the carriageway. Otherwise, there is not a good case for a precast concrete channel to be placed in front of the kerb face, as additional points of weakness are introduced at the longitudinal joint between the carriageway and the channel. Also, the channel block is often damaged by a roller during compaction of the binder course.

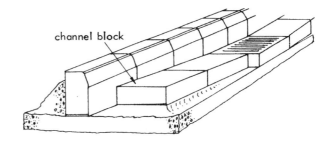

Figure 7.16   Use of channel blocks

As with raised kerbs the channels are 915 mm in length and can be radiused in the same manner.

Special precast concrete channels are manufactured such as the dished channel which allows vehicles to leave the carriageway but also provides positive control of surface water. Another type incorporates a concealed duct section within the channel which removes all water from the location (*see* Figure 7.18(*a*)), but maintenance is difficult because of blockages.

Precast combined drainage and kerb blocks are now available and are manufactured to BS 7263. They may be either 'one piece' or two-part blocks, split horizontally, so

that the lower half can be laid first, as an open channel, the upper portion being placed over it and being so shaped that the two parts interlock. General details of the cross-section of these blocks are shown in Figure 7.18(*b*).

The water inlet slots may be either continuous, or discrete apertures, with entry areas of not less than 0.0075 m^2 (measured in the vertical plane) and not less than 0.015 m^2 in area per metre length.

Discharge openings are either circular, of at least 225 mm diameter, or elliptical (minimum area 0.05 m^2).

Vertical joint faces between the blocks must be sealed with mastic sealant applied with a trowel.

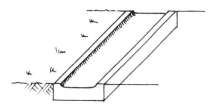

Figure 7.17   Channel block: open type

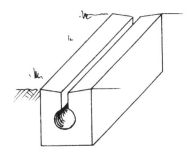

Figure 7.18(a)   Channel block: concealed duct type

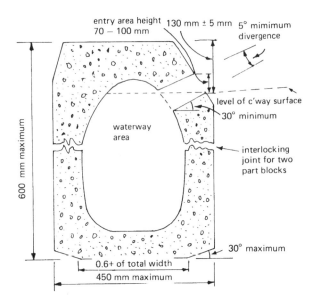

Figure 7.18(b)   Cross-section of precast combined drainage and kerb block

7.2.4   Precast Kerb Channel Combination Units

A vertical joint between kerb and channel can be avoided by combining them into one profile. These are widely used in North America for residential and urban roads in the form of precast units, which are usually 450–600 mm wide and 250 mm high, with a kerb height of 100 mm.

The 'continental type' kerb (Figure 7.20) combines the advantages of a raised mountable kerb and a dished channel in one unit.

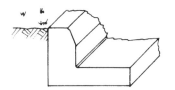

Figure 7.19   Precast kerb/channel unit

Figure 7.20   Precast kerb/channel unit

7.2.5   Cast in situ

Alternatively, kerbs may be constructed *in situ*, normally ahead of paving operations, either between forms or by kerb extruders.

Different types of machine are available for extruding *in situ* concrete raised kerbs. They are often used at the backs of hard shoulders and have withstood overriding without deterioration. As they are more difficult to alter in line or level than a precast or stone kerb, their use has not been advocated in ordinary roads. Some typical sections are shown in Figures 7.21 and 7.22.

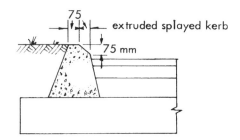

Figure 7.21   In situ *kerbs*

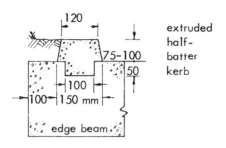

Figure 7.22 In situ *kerbs*

When extruding, dimensions may be varied to increase the channel width to say 2 m to provide a parking lane. An integral concrete kerb of this type is formed by the slip-form paver conforming plates being shaped at the sides to produce a kerb upstand (*see* Chapter 6).

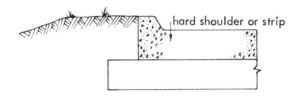

Figure 7.23 Extruded kerb and channel unit

7.2.6 Asphalt Extruded Kerbs

This method is cheap for long uninterrupted lengths and satisfactory if no subsequent adjustment of line or level is needed. When extruded on to the carriageway a cationic bitumen emulsion tack coat is first applied.

7.2.7 Steel Kerbs

Under conditions of heavy traffic in towns, failure of precast concrete kerbs is most common at radii points. In these conditions, replacement by natural stone or steel faced kerbs is the most satisfactory solution. Although expensive, steel faced kerbs give the maximum resistance to abrasive action by traffic.

7.2.8 Hard Shoulders and Hard Strips

Hard shoulders are normally 3.3 m wide and are provided on motorways and occasionally on trunk roads as a refuge for stationary vehicles.

As an edge detail, hard shoulders should:

(*a*) be strong enough to take heavy stationary loads without damage;
(*b*) allow surface water to flow clear of the carriageway and offer no barrier to drainage from the carriageway base and sub-base;
(*c*) have adequate structural integrity and skid resistance to enable it to be used as a temporary running surface during roadworks;
(*d*) be free of manhole covers or access chambers.

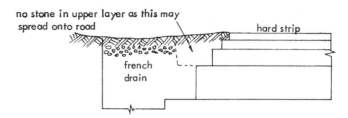

Figure 7.24 Detail of hard strip

Important new roads built in rural areas often have a hard strip 1 m in width on each side. This strip provides a partial refuge for stationary vehicles and the design requirements are the same as those of hard shoulders.

7.2.9 Edge Markings

These may be lines (continuous or broken) and/or reflective studs provided to supplement the effect of other edge details or to define the edge of the carriageway in their absence.

Lines are spray plastic, paint or thermo-plastic and preferably reflective. Studs are made of plastic or metal.

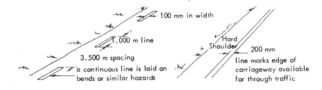

Figures 7.25 (left) and 7.26 (right)   Marking of hard shoulder

7.2.10 Flush Verge

Often adopted on rural roads to reduce initial construction costs. Usually in the case of flexible construction, the layers of the pavement are stepped, the grass verge being extended over the lower layers. The edge detailing of a hard shoulder or strip is either flush verge or a raised kerb.

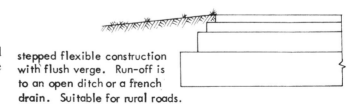

Figure 7.27 Stepped flexible structure with verge: rural situation

7.3 Properties of Natural Stone and Precast Concrete Kerbs

As these are the most common forms of kerbing to be found on normal roads it may be useful to compare their properties.

7.3.1 Natural Stone Kerbs

These are quarried and have:

(*a*) good resistance to impact and abrasion;
(*b*) a very long life (durability) when properly laid;
(*c*) little need of maintenance when properly laid;
(*d*) great resistance to weather effects.

But, natural stone kerbs:

(*e*) are very expensive;
(*f*) are difficult to lay (due to the backs and underside being left rough).

7.3.2 Precast Concrete Kerbs

These are manufactured and have:

(*a*) uniformity of size and shape, thus being easier to lay:
(*b*) less weight than the corresponding natural stone kerb;
(*c*) the advantage of cheapness compared with the stone kerb;
(*d*) many different profiles and radiused lengths to suit various road requirements.

But, precast concrete kerbs:

(*e*) are less durable than stone;
(*f*) susceptible to frost damage;
(*g*) spall easily if subjected to a sharp blow.

7.3.3 The Manufacture of Precast Concrete Kerbs

Mix: A typical mix used for the manufacture of precast concrete kerbs would be:

12 mm basalt aggregate	215 kg
3 mm-dust aggregate	144 kg
Sharp sand	255 kg
Building sand	14 kg
Portland cement	99 kg
Total batch weigh	727 kg

Manufacture by Hydraulic Press

The plant is so designed that each batch when mixed is contained in a hopper above the moulding press and precise quantities are weighed out for each kerb mould.

In the forming of the kerb a rather wet mix may be used. The water is squeezed out under the pressure of moulding, but the cement is retained by the filter paper (*see* Figure 7.28). The filter paper is left on the kerb during the initial 24 hours of the curing process.

The kerb is removed from the press and immediately taken from the mould, being stored on pallets for curing.

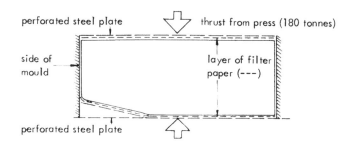

Figure 7.28 *Moulding precast concrete kerb*

Where kerbs have a chamfered face, this remains facing downwards during curing and supported by a tapered wedge of timber.

7.4 Laying Precast Concrete Kerbs

7.4.1 Setting Out the Kerb Line

This has been covered generally in Chapter 2 but further detailed examples are given concerning straight and curved kerb lines.

A straight kerbline (say 200 m in length) may be set out as follows using the line to the front of the kerb with 'wet bed' method.

1 Drive end pins, accurately to line and upright (so that the line is at front edge of kerb).
2 Drive anchor pins, offset from the kerb line.
3 Tie off line to each pin, ensuring that it is at the correct height* at each end pin.
4 Measure accurately the distance between the end pins – to check that it is 200 m.
5 Drive centre pin (upright and to the line), i.e. 100 m from each end.
6 Continue sub-dividing the distance between adjacent pins until they are at 5 m intervals.
7 Bone in centre pin level.
8 Sub-divide each half and continue boning-in the level until all pins have been boned-in.
9 Check the straightness of the line by sighting or eyeing in from the road side of the pins – from each end and make any minor corrections necessary.

*N.B.: If the level is not taken from an existing kerb run then it would normally be an engineer's responsibility to mark off the channel level on the end pins.

Setting Out Curved Kerb Lines

The techniques used for setting out horizontal and vertical curves ('roll over' when the highest point is on the middle pin and 'hollow bone' when the lowest point is on the middle pin) together with setting out a radius have been given. However, this example is useful to demonstrate a practical method to obtain a radiused line when the circle centre is obstructed.

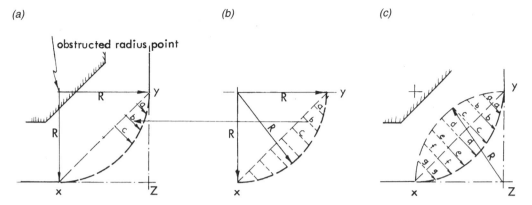

Figure 7.29(a)–(c)   Setting out kerb radii

A kerbline may be set out at a radiused corner where, due to an obstruction, it is not possible to swing a tape from the centre point, as follows:

Set out radius elsewhere as shown in Figure 7.29(*b*).

1 Divide chord 'xy' into equal parts.
2 Measure and transfer co-ordinate lengths a, b, c, etc, to work situation or if possible set out inverse radius from kerb line intercept point Z, proceed as for Figure 7.29(*b*) and transfer co-ordinates.

7.4.2   Forming the Kerb Bed

Following the setting out of the kerb line, the concrete for the kerb bed should then be placed. This should be carried out by the most efficient means available and it could be by shovel, barrow, dumper, lorry or mixer truck. If the last, advantage should be taken of its ability to discharge by chute directly into the trench or 'windrow' position.

Concrete complying with standard mix C.30 could be used. The kerb bed should be accurately placed and will either be formed as a 'windrow' levelled to a line using a depth gauge, or will be poured between formers or shuttering. The kerb bed concrete should be laid on the road sub-base. Subsequently kerbs can either be laid on a 'wet bed', that is, laid on the windrow or concrete foundation before the initial set takes place or, alternatively, the concrete can be left to set and then the kerbs laid on a 15 mm mortar bed ('dry bed' method). If the foundation is to be laid and allowed to harden, then the surface level of the concrete should allow for the thickness of mortar bed. The steel pins should be wrapped with paper to assist in their withdrawal later.

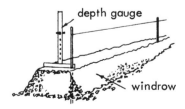

Figure 7.30   'Wet bed' method

The concrete for the kerb should be just wet enough for complete chemical reaction of the cement to occur and it will then be stiff enough to stand without shuttering the sides if the wet bed method of kerb laying is to be used. If the mix is too wet, the concrete will require shuttering to contain it, thus causing extra cost both in time and materials. However, some engineers prefer shuttering or forms to be used on long runs as the saving in concrete over the wet bed is worth the extra trouble.

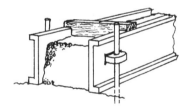

Figure 7.31   'Dry bed' method

7.4.3   Placing the Kerbs

Thought should be given to the positioning of the kerbs prior to laying. They should be placed alongside the works in a line so that they may be fitted into their final position in one movement. Whether stacked in the carriageway (giving some protection to the kerb layers) or on the verge will depend on the position of the line and pins.

The 'Wet Bed' Method

The 'wet bed' or 'green concrete' method of preparing the kerb bed is widely used for short length work as it eliminates the mortar bed operation and allows a quicker pace of working. The bed should be 20–40 mm too high to allow for knocking down of kerbs.

Using a 'wet bed', the various methods of positioning the nylon setting out line on the pins prior to actually laying the kerb are shown. All methods have their attendant advantages and disadvantages (*see* Figure 7.32), which are compared as follows.

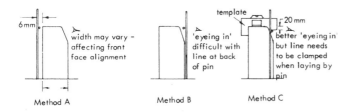

Figure 7.32   Positioning the line

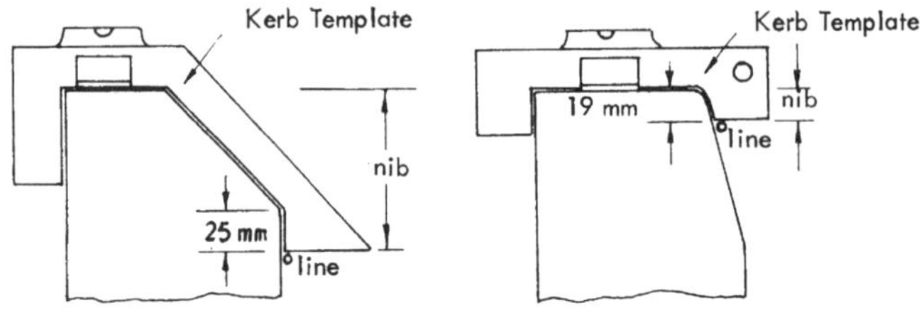

Figure 7.33(a) and (b)   Kerb templates

(a)   *(b)*

Method A (line to rear of kerb)

Advantages

1 All types of standard kerb have regular angled edge to top or rear face.
2 With kerbs stored on carriageway, an easy lift straight into position.
3 Pins protected rather more from accidental damage.

Disadvantages

1 The widths of the kerbs are inconsistent (tolerance of 5 mm). This requires the line to be offset 6 mm.
2 Not the most accurate method.

Method B (line to front face of kerb)

Advantages

1 The front face is the most important face for alignment.
2 On new carriageway construction without any kerb excavation, the kerbs are placed on the verge to avoid restricting the work on the carriageway. The kerbs can then be easily lifted down behind the line.

Disadvantages

1 No square edge exists on the front face of kerbs with splays, batters and arrissed edges. Hence accuracy must suffer.
2 On reconstruction work kerbs are normally placed on the carriageway for ease in offloading and for the protection of the kerb layer. This means lifting kerbs over the line.

Method C (line to front face of kerb, using templates and clamps)

Advantages

1 and 2 As Method B above.
3 Use of templates allows accurate work by semi-skilled kerb layers (see Figure 7.33(a) and (b)).

Disadvantages

1 It is necessary to clamp line to kerb to allow pin to be removed when laying adjacent to pin.
2 As Method B above.

Two laying procedures are described in detail, these being Methods A and C.

Using Method A

(1) Lay the first kerb using a spirit level to check first, that it is to the correct line and level with the line and secondly, that it is level across the top.

(2) Each end of the first kerb must be checked for accurate positioning. If the kerb is too high at one end, it should be tapped at that end with a maul, wooden pick handle or similar instrument, until it is down to the correct level.

If the kerb is too low, the concrete bed should be increased to the correct height. If the concrete has set, a cement/sand filler is required to build up the bed. This is undesirable as it weakens the structure. Hence great care should be taken in placing the concrete bed. Attention to this can save much time and effort in the laying of the kerbs.

(3) Proceed with laying the second kerb; the end adjacent to the first kerb can be checked by eye or by running a straight edge along the two kerbs.
The 'leading' end must be checked with the spirit level for level and line.

(4) When a number of kerbs have been laid, they should be checked by sighting along them and making minor adjustment if necessary.

Using Method C

(1) The first kerb is laid using the correct kerb template. The two checks on line and level as previously described, are provided simultaneously by the template which incorporates a spirit level.

(2) Each end of the first kerb must be checked as before. Proceed with laying the second kerb and continue with 3 and 4 as described in Method A above.

(3) Clamp the line and remove the pin when a pin obstructs the placing of a kerb, or leave out the obstructed kerb and return to place this after the run is completed.

The 'Dry Bed' Method

The method of providing an edge beam, extruded or cast, on which the kerbs are laid after the concrete has hardened involves the expense of form setting. It does have the advantage of controlling the amount of concrete used, thus preventing waste as often occurs in the 'wet bed' method,

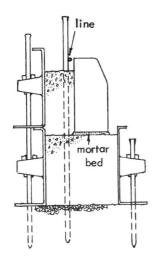

Figure 7.34 Section through dry bed kerb, bed and backing

and also avoids damage to the kerbs by construction traffic. The provision of a mortar bed for the kerbs is necessary and this is of 15 mm thickness of Class 1 mortar. The kerbs are backed by C.30 concrete, or similar.

Epoxy Resin

In recent years an epoxy resin compound has been used to affix kerbs to both rigid and flexible road surfaces.

The 'dry bed' method of constructing a kerb race can first be used and then the kerbs placed with epoxy resin bonding. Because of its strength a concrete backing is not required but it is sometimes provided to give uniformity with adjacent existing sections of kerbing.

The adhesive is supplied in packs of resin, hardener and fine aggregate. The manufacturer's instructions must be carefully followed when mixing, handling and placing this substance, to ensure good results and to avoid health risks.

Precast kerbs or natural kerbs are seldom laid on the final surfacing except in particular cases where it is undesirable to break out the existing carriageway, such as building experimental islands.

7.4.4 Haunching

When the kerb line has been laid and checked, the back of the kerb should be strengthened by the placing of a concrete haunch.

Where a flagged footway is to be laid against the kerb, this haunching should be flat across the top and should finish about 75 mm below the top of the kerb (Figure 7.35).

7.4.5 Tools Used in Laying Kerbs and Channels

Kerb lifters	Rule (1 m)
Wheel barrow	Shovels
Cold chisels	Mauls (see Figure 7.39(*b*))
Cold punches	Square-wooden and joiner type

Pitching tools	Boning rods
Bolster chisels	Lump hammer
Brick and pointing trowel	Straight edge (2–3 m)
Mason's hammer	Road marking crayons
Goggles	Nylon line (plus 4)
Mason's spirit level (1 m)	Steel pins
Boat level (225 mm)	Safety gloves
Tape (30 m)	Templates

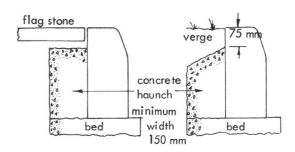

Figure 7.35 Section showing haunching behind kerb

7.5 Terminology

Many of the names and the definitions used in kerbing vary considerably throughout the country. Some standard definitions are given in Figure 7.36.

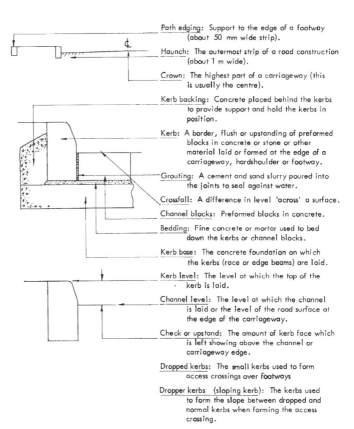

Path edging: Support to the edge of a footway (about 50 mm wide strip).

Haunch: The outermost strip of a road construction (about 1 m wide).

Crown: The highest part of a carriageway (this is usually the centre).

Kerb backing: Concrete placed behind the kerbs to provide support and hold the kerbs in position.

Kerb: A border, flush or upstanding of preformed blocks in concrete or stone or other material laid or formed at the edge of a carriageway, hardshoulder or footway.

Grouting: A cement and sand slurry poured into the joints to seal against water.

Crossfall: A difference in level 'across' a surface.

Channel blocks: Preformed blocks in concrete.

Bedding: Fine concrete or mortar used to bed down the kerbs or channel blocks.

Kerb base: The concrete foundation on which the kerbs (race or edge beams) are laid.

Kerb level: The level at which the top of the kerb is laid.

Channel level: The level at which the channel is laid or the level of the road surface at the edge of the carriageway.

Check or upstand: The amount of kerb face which is left showing above the channel or carriageway edge.

Dropped kerbs: The small kerbs used to form access crossings over footways.

Dropper kerbs (sloping kerb): The kerbs used to form the slope between dropped and normal kerbs when forming the access crossing.

Figure 7.36

7.6 Footways

7.6.1 Footway Surfaces

The surfacing of footways allows the use of many different techniques, the more common being bitumen macadam, modular block paving and paving stones, or flags. Other types of surfacing could be of concrete, precast concrete blocks, bricks or natural stone, the latter types being more usual in urban paved areas.

Well laid precast slabs are usually superior to *in situ* alternatives such as concrete or tarmacadam, both in appearance and in wearing qualities. The cost of flagging is greater initially, but a good flagged footway will last 25–30 years, whereas a bituminous surface will require resurfacing in 5–10 years. The following list gives a broad indication of the initial construction cost of various surfacing finishes.

If the cost of precast slabs is taken as 100 then:

gravel is	35
tarmacadam (60 mm) is	70
in situ concrete is	75
cobbles set in concrete are	240
brick paving is	250/375
York stone is	550
granite setts are	650

7.6.2 Footway Bases

To obtain a good trouble-free surface on a footway, the preparation work must be carefully undertaken.

The sub-grade must be uniform and well compacted by using a vibrating or smooth-wheeled roller, in order to prevent differential settlement. If the soil is cohesive, rolling will be sufficient, but if not, then cement must be added to it by putting it through a concrete mixer with 5% cement and then compacting it by rolling. Particular attention must be given to backfilling any trenches.

Materials used for bases must be stable and immune from the effects of water and frost and well compacted to a uniform density. Loose, coarsely graded clinker, or fine under-burnt ash cannot be compacted properly and should not be used. Granular sub-base material type 1 or type 2 compacted to the appropriate thickness (usually 75 mm) is recommended.

7.6.3 Footway Edging

Precast concrete edgings generally fulfil the same functions as kerbs but are of smaller dimensions to suit the light duties for which they are used. They are particularly suitable for defining the edges of footways and grassed or paved areas. The types of edging are shown in Figure 7.13.

Just as kerbs are used to keep the road substructure dry, so does the edge of the footway require a seal to keep the water out. An edging provides this if placed as low as possible. In certain situations an edging is not essential, such as where paving is laid on concrete or where paving abuts a grass bank or building, but it is usually preferable.

To reduce costs and to cut down on construction time, timber edgings are an alternative to concrete. Timber pegs (50 × 50 × 600 mm) are driven into the ground at a maximum of 1 m spacing and attached to lengths of timber edging. Type 1 footway edging is a minimum of 2 m in length, being 25 × 50 mm in cross-section. Type 2 is more substantial, being 25 × 150 mm in cross-section. All timber used is treated with an approved preservative and is secured to the pegs with 50 mm galvanized clout nails.

7.7 Surfacing Materials and Laying Techniques

7.7.1 Paving Slabs

The range of precast concrete slabs or flags is very wide having not only a series of standard sizes, but also a variety of special shapes.

Standard flags are either 50 mm or 63 mm in thickness and the sizes are:

Flag type	Nominal size (mm)	Work size (mm)
A	600 × 450	598 × 448
B	600 × 600	598 × 598
C	600 × 750	598 × 748
D	600 × 900	598 × 898
E	450 × 450	448 × 448
F	400 × 400	398 × 398
G	300 × 300	298 × 298

As with kerbs, slabs are manufactured as either hydraulically pressed units or open-mould vibrated units.

Hydraulically Pressed Slabs

Three sorts of slabs are manufactured by this process, plain surfaced, exposed aggregate surfaced and moulded designs.

Plain hydraulically pressed slabs are the cheapest form of precast paving and they are available from stock in rectangular, circular, hexagonal and other shapes.

Exposed aggregate types can be expensive due to the labour-intensive nature of the process. The method used is to remove the slab from the production line while it is still 'green' and then apply the appropriate exposure techniques.

Moulded types are possible, but it is essential that the mould chosen is suitable for the manufacturing process and capable of withstanding the pressures involved. *See* Figure 7.28 showing a hydraulically pressed kerb.

Open-mould Slabs

Manufacturing paving slabs by the vibration of open moulds is little more than a 'hand' process and so it is possible to introduce any variation of texture and colour during the process. This type of flag is more expensive than the hydraulically pressed type and does not have the latter's greater strength.

A wide range of treatments can be given in this process with the use of coloured cements and coloured aggregates to tone or contrast with the matrix.

At pedestrian crossing places, tactile slabs are provided, being ramped down to carriageway level, for ease of access to push chairs, wheelchairs, etc.

Tactile slabs are either red or buff in colour (*see* Section 7.8.6 for layouts) and sized TA/E, TA/F or TA/G.

Tactile crossing flag type	Dimension	
	X	Y
TA/E	64	33
TA/F	66.8	33
TA/G	75	37.5

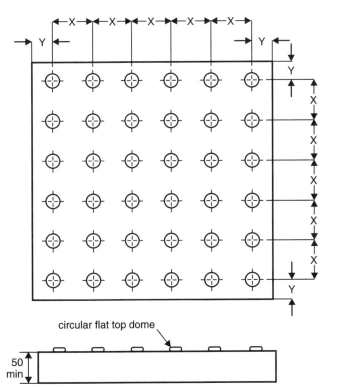

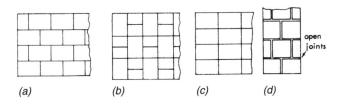

Figure 7.37 Tactile slab

Bonding Patterns with Slabs

Many variations are possible but a few of the more common bonds are shown in Figure 7.38(*a*)–(*d*).

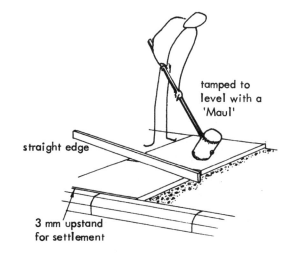

Figure 7.38(a)–(d) Bonding patterns with concrete slabs

Pavings can be laid with an open or closed joint. 'Open' joints are classified as those wider than the 6 mm joint normally used, say 10–15 mm. However, if the joint becomes too wide the mortar will crack away from the pavings.

Methods of Laying

After the preparation of the sub-grade and base (*see* Section 7.6.2) either of three methods of laying precast concrete flags may be used. These are as follows:

Whole bed method

A 25 mm thick bed of lime and sand mortar (proportions 1:8 to 1:10) mixed with water is laid under the whole slab, joints between slabs are slurried in wet with the same mix, or alternatively, a mix of dry cement and sand is brushed into the joints. Another alternative is to sweep neat cement into the joints on still days.

Dry method

Flags are laid on a dry mix of sand and cement or lime (*see* Figures 7.39(*a*) and (*b*)) with tight joints, which are then slurried in with wet 1:3 cement sand mortar. This method is not popular in shopping areas because of the inconvenience caused to pedestrians during and after the slurrying-in process. Hence a dry mix of cement and sand is sometimes swept into the joints.

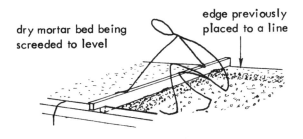

Figure 7.39(a)

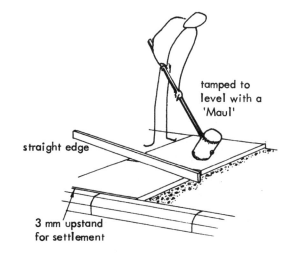

Figure 7.39(b)

Five-spot method

The flags are laid on five spots of mortar, one at each corner and one in the centre of the flag. The spots should be more than 150 mm in diameter and about 20 mm thick. Hydraulic lime and sand mortar is used in mixes varying from 1:3 to 1:8. The flags are tight-jointed and the joint slurried in with the same mix or alternatively a cement sand mortar slurry in mixes varying from 1:2 to 1:12. In some areas the slab is 'buttered' on all edges with cement mortar. They are then fitted tightly together and the joints are pointed. This method of 'buttering' is more appropriate when using natural stone slabs where the sides are not as true as with precast flags.

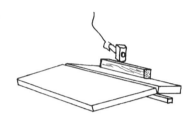

A saw cut is made by a power tool (¹/₃rd of thickness) along a line and the end is trimmed off by a sharp tap

Figure 7.39(c)

General Precautions in Laying

Careless handling and stacking of precast slabs will result in chipping of edges and breakages. If a pinch-bar is used for moving slabs then a pad should be inserted between the bar and slab.

The whole bed method of laying is recommended for the strongest construction and this is essential at footpath crossings and wherever vehicles are likely to run on the paving. At these points the concrete bed should be at least 100 mm in depth and the heavier slabs of 63 mm thickness used.

Finally, concrete edgings should be provided to all edges to prevent the ingress of water to the base which in itself must be of adequate construction as already described.

7.7.2 Flexible Footway

This is often referred to as a 'black top' or bituminous footway, the surfacing will be of, bitumen macadam or asphalt. (N.B.: Fine cold asphalt is often used to resurface existing footways and is referred to in Chapter 14.) A footway constructed from these materials will be cheaper than one which is paved with slabs, but it will be less durable and is more suited for rural use.

It can be laid with a mini-paver but it is more usual to find it laid by hand, which is a labour-intensive job using a technique of spreading and raking. Hand laid material on footways, often has a fluxing agent introduced into the mix, to delay the setting time.

Construction

A typical construction of such a footway is shown in Figure 7.40.

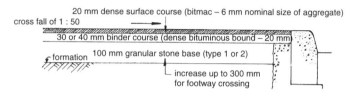

Figure 7.40 Cross-section through footway

Method of Laying

The footway base is formed as described in Section 7.6.2. The binder course material, if over 20 mm stone size, should be spread with a horizontal motion using a shovel or tarmac fork after the material has been tipped in half barrow loads at the front and rear of the footway. Care should be taken to avoid tipping on to the kerbs as these can be badly stained. Hot bitmac tipped on to the existing carriageway may also cause melting of the carriageway surface.

Levels should be established using a straight edge or a line and pieces of brick or wood are often placed in the footway to show this. These are removed as work progresses.

The surface course material should then be shovelled directly to the laying point and raked out to the given level. The rake is operated using a pushing and pulling motion to take out voids, with the material finally being pulled by the rake to the correct level.

It is best to work down hill, unless it is too steep for proper control of rolling. A twin roller is preferable to a single vibratory roller for work on slopes.

If an open graded surface course material is used it may be necessary to seal the surface with bituminous grit. This treatment is applied whilst the surface is still warm and the grit is swept in, or alternatively rubbed in with a squeegee.

Refer to Chapter 5 on flexible pavements for further information on materials and precautions in their use.

Tools Used in Laying

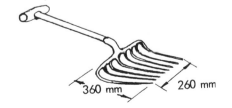

Figure 7.41 Bitmac fork with 8 tines

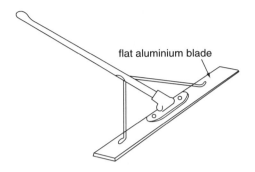

Figure 7.42   Squeegee

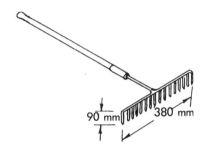

Figure 7.43   Rake with 16 tines

7.7.3   Concrete Footways

These are often constructed in city areas where a good flat surface is necessary for intensive pedestrian traffic together with the capacity to resist damage by vehicles when mounting or parking on the footway. Alternatively, a concrete strip, say 600 mm, is provided alongside the kerbing and adjacent to paving slabs forming the main footway.

The main disadvantage with concrete footways is the difficulty in cutting out to undertake repairs to, or renewals of, services.

Construction

A typical concrete footway is shown in Figure 7.44.

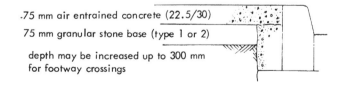

.75 mm air entrained concrete (22.5/30)

75 mm granular stone base (type 1 or 2)

depth may be increased up to 300 mm
for footway crossings

Figure 7.44   Section through concrete footway

Method of Laying

Refer to Chapter 6 for the techniques used and the precautions necessary in laying concrete areas.

7.8   Paved Areas (e.g. Precincts with Special Treatments)

All forms of paving are expensive, so it is usual to restrict the use of precast pavings to limited areas, where their attractive appearance and durability are most desired. An obvious use is the construction of a durable and attractive pathway or carriageway where the most intensive traffic is routed, leading through an area finished with a cheaper and probably less durable finish. This gives an economic and rational landscape with primary paths clearly and visibly differentiated from the rest of the area by means of surface finish.

7.8.1   Concrete Block Paving

This is the modern version of stone setts, retaining the advantages of durability, good appearance, and easy renewal or replacement but being cheaper and having a smoother surface.

They may be used for footway or carriageway and where services are numerous the problem of breaking out and reinstatement of the surface is minimized. Once the footway has been broken into, the blocks can be lifted and reused. A small stock can be held for maintenance purposes.

In spite of the claims of easy renewal, in practice it has been found that the blocks are very difficult to reinstate satisfactorily, i.e. to the quality of the original pavement. This is due to the 'spreading' of the blocks which remain *in situ* when the pavement is opened up: it is often impossible to fit replacements for the ones which have been taken up, in the reduced space remaining.

Manufacture

There is no restriction on the plan shape of a block, but the ratio between length and width is between $1^{1}/_{2}$ and $2^{1}/_{2}$. Blocks should be designed to be picked up with one hand and are usually 60 mm in thickness (80 mm for carriageway use).

Concrete blocks are usually manufactured in vibrating presses and should have a strength of 50 N/mm². The concrete contains about 4.5% of entrained air which makes it resistant to the damaging effects of de-icing salts.

Blocks for use in pedestrian areas should be carefully selected, to ensure they provide a non-slip surface for pedestrians, when dry or wet. In vehicular areas, they should also provide adequate skid resistance.

Bonding Patterns with Rectangular Blocks

The surface of a block footway or carriageway comprises tightly fitted paving blocks, the joints between them being filled with sand particles. These particles prevent a single block from being displaced, so a vertical load applied to a block is transmitted to its neighbours causing the paving to behave flexibly. Horizontal interlock is given to the paving as shown in Figures 7.45 and 7.46.

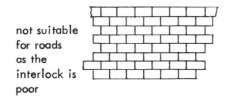

not suitable
for roads
as the
interlock is
poor

Figure 7.45 Stretcher bond

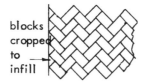

blocks
cropped
to
infill

Figure 7.46 Herringbone

Interlocking Paving

Shaped blocks are designed to interlock to give a more interesting appearance with a stronger bonding. Some common shapes and bonds are shown in Figure 7.47.

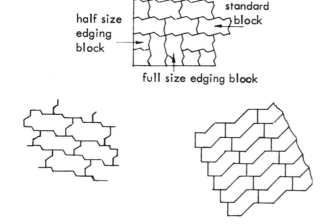

half size
edging
block

standard
block

full size edging block

Figure 7.47 Interlocking paving

Laying Block Paving

This requires an edge restraint which may be provided by precast kerbs or pathway edgings. The edge restraint is laid prior to paving and it is essential that the concrete foundation for the edging does not project in front of the precast edging as this will prevent blocks being properly laid.

The sub-base should be inspected to make sure that it is structurally sound and that its surface level is correct.

The laying course should be about 50 mm in thickness after vibration and consist of washed sharp sand containing not more than 3% of silt and clay by weight and with not more than 15% retained on a 2.36 mm sieve. The sand should be allowed to dry before use and should be sheeted over to minimize moisture changes.

Screeding is done by traditional methods and care in this operation gives a good surface. After the blocks have been placed, the vibration applied will cause them to sink some 12 mm. Thus the sand has to be laid to a surcharge, some of which will be compacted and some will move up into the joints between the blocks (*see* Figure 7.48). A test section will need to be laid to ensure that the surcharge allowed will give the correct final level after vibration. Five or six passes of the vibrator are normally needed to vibrate the blocks to the finished level.

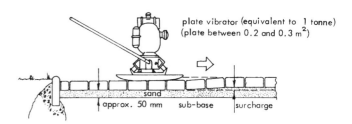

plate vibrator (equivalent to 1 tonne)
(plate between 0.2 and 0.3 m²)

sand

approx. 50 mm   sub-base   surcharge

Figure 7.48 Using a vibrator plate to compact the paving

The joints are then filled by brushing sand over the surface of the blocks and this is vibrated into the joints by a further two or three passes. Surplus sand is then swept away. It is advisable to mix a good weed killer with residual qualities with the sand used for sealing these joints.

7.8.2 Brick Paving

Not all bricks are suitable as paviors and selection should not merely be based on colour and texture. Durability is most important for external paving and Class A or B engineering bricks normally possess this quality, being well burnt and having considerable resistance to abrasion and impact. Red and blue paviors in chequered and slip-resistant varieties are more acceptable for shopping precincts.

A crushing strength over 48.5 N/mm² is usually necessary together with the ability to resist decay due to frost in wet conditions. Paviors used in a corrosive environment should be highly vitrified having a low porosity.

The direction of bedding of paviors; type of mortar; drainage; method of laying and the surface finish are all contributory factors which can influence the durability of brick paving.

Brick Sizes

Sizes vary, but the brick size (or work size of unit) is usually 215 × 102.5 × 65 mm.

Special sizes of 215 × 130 × 33 mm are produced and modular bricks of

$$290 \times 90 \times 90 \text{ mm}$$
$$190 \times 90 \times 90 \text{ mm}$$
$$90 \times 90 \times 90 \text{ mm (specials)}$$

are also manufactured.

Patterns of Laying

There is a wide variety of patterns which can be formed as shown. Alternatively, brick paving can be used in conjunction with paving slabs.

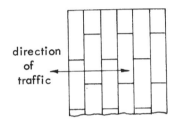

Figure 7.49   Standard brick on edge, stretcher bond

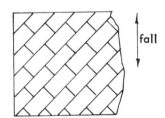

Figure 7.50   Diagonal stretcher bond

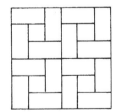

Figure 7.51   Standard brick on flat

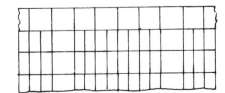

Figure 7.52   Bricks laid with 228 mm square slabs

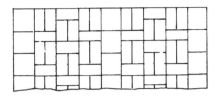

Figure 7.53   Bricks laid with 228 mm square slabs

Laying Bricks

Bricks laid in external situations may, even under normal traffic or weather conditions, tend to shift or spread and hence some edge restraint is desirable. This may consist of a 'soldier' course of bricks set either in the prepared concrete base or in mortar.

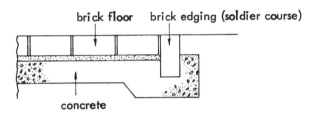

Figure 7.54   Section through brick floor set on concrete base

Where a concrete base is not used then the bricks may be laid on a well-consolidated foundation rolled to the required fall (minimum 1 in 60), the bricks being laid on a bed of lime/sand (1:4) or even ash. The joints should be between 6 and 10 mm and are filled with the same lime/ sand mixture or run in grout. The bedding is normally 50 mm in thickness.

Alternatively, the paving may be laid over a 50 mm layer of sand placed on the sub-base being carefully screeded to the required level. The bricks are 'dropped' on to the sand bed leaving a 10–15 mm joint and grouting is carefully poured into the joints.

All pavings should be allowed to mature undisturbed for at least three days before being used for light pedestrian traffic.

7.8.3   Cobbles

Stone cobbles set in concrete form a surface which is not easy to walk upon and so can be used for areas which could be hazardous for pedestrians to cross or placed to keep people away from grass and plants. Used for a decorative effect (*see* Figure 7.58) they can provide an attractive contrast when laid with slabs.

Cobble patterns often used, include coursed, random and flat parallel designs, as illustrated in Figures 7.55–7.58.

Figure 7.55   Coursed cobbles

Figure 7.56   Random cobbles

Figure 7.57   Flat parallel cobbles

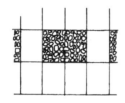

Figure 7.58   Decorative cobbles and slabs

Construction

Typical footway cross-sections are shown where the cobbles are bedded in 1:2:4 mix concrete, the thickness of the bed depending on the cobble size used.

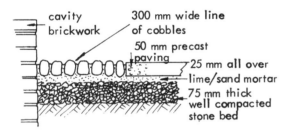

Figure 7.59   Typical footway cross-sections using cobbles

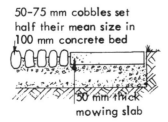

Figure 7.60   Typical footway cross-sections using cobbles

7.8.4 Paved Areas with an Exposed Aggregate Surface

Two methods of producing an exposed aggregate surface on a concrete paved area are described. The choice of method depends on whether the aggregate to be exposed is that used throughout the depth of the slab or is a specially selected and relatively expensive aggregate, limited to the slab surface for reasons of economy.

Method I

In order to produce a textured surface by exposing the aggregate used throughout the slab a concrete mix of the following proportions is suitable:

	Ordinary concrete (kg)	Air-entrained concrete (kg)
Cement	50	50
Sand	78	67
Coarse aggregate (19–9 mm)	200	191

All aggregates between 9 and 5 mm size shall be excluded.

The concrete is spread, compacted and brought to a smooth finish, the surface being brushed with a soft broom to remove surface laitance, care being taken not to disturb any particles of coarse aggregate. The surface is then covered with waterproof paper or plastic sheeting to protect it against drying out too rapidly.

A second brushing with a stiff broom, together with copious spraying with water is given to the surface when the concrete has hardened sufficiently for all surface laitance to be removed without disturbing the coarse aggregate. Brushing and watering must continue until the surface of the coarse aggregate is fully exposed, the mortar being removed to a depth of not more than half of the stone, so that the surface presents a uniform texture.

On completion of the brushing the surface must be protected and cured in the normal way.

Note: The time of the final brushing will depend on the atmospheric conditions, and, to a lesser extent, on the water content of the concrete. It may be up to four hours after compaction of the concrete, and the work should be planned to avoid having to do this final brushing after dark.

If a retarding agent is applied as a spray after the completion of compaction and surface finishing of the concrete, the final watering and brushing can be delayed up to twenty-four hours. Plenty of water must be used to ensure that no trace of the retarding agent is left on the surface.

Great care must be exercised in the use of a retarding agent if a surface of uniform appearance is to be obtained. The retarding agent must be applied at a uniform rate over the surface and brushed and washed off as soon as possible.

Method II

In order to produce a textured surface by spreading specially chosen aggregate over the surface, the concrete area is completed, with the finished surface 6 mm below the tops of the forms. This surface must not be levelled with the scraping straightedge. As soon as a bay has been finished, single-sized aggregates of either 19 mm or 25 mm gauge are spread over the surface, one stone thick, completely covering the concrete beneath. This aggregate is forced into the concrete by means of a vibrating hand tamper until the new surface is level with the tops of the forms and sufficient mortar has been worked up to hold the stones in position. The method of exposure should be as described under Method I brushing being continued until the aggregate is slightly proud of the surrounding concrete.

Note: The single-sized aggregate should be of cubical nature, free from flaky and elongated pieces. The choice between 19 mm and 25 mm single-sized aggregate is a matter of taste. Sizes smaller than 19 mm are more difficult to place and expose, and tend to be less effective in appearance.

These finishes are often used in conjunction with brick-paved and flag-paved areas in pedestrian precincts.

7.8.5 Modular Paving

These slabs are of hydraulically pressed concrete, designed to withstand loads due to vehicles mounting the footway. They are 400 mm square and 65 mm thick, weighing about 24 kg per slab.

Adequate edge restraints, such as precast concrete kerbs, are required for satisfactory results, as the modular slabs are laid on sand bedding, usually in conjunction with concrete block pavings, and are vibrated into position by means of a plate vibrator.

They are easier to lay than the standard flags, being smaller; awkward corners and small areas round street furniture are left to be infilled with concrete block pavings, so that only these small blocks need to be cut.

The modular slabs are supplied with either a smooth or a textured face, the smooth-faced slabs being slightly the stronger because of the cement-rich surface layer that results from the pressing process. This layer is removed in producing the textured surface.

A typical layout of modular paving is shown in Figure 7.61.

7.8.6 Tactile paving surface layouts

The DfT gives guidance on the *Use of Tactile Paving Surfaces* which indicates the various layouts which could help visually impaired pedestrians to negotiate crossings

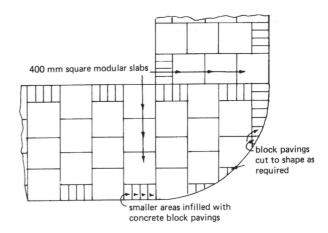

Figure 7.61 Modular paving layout

and a range of special tactile paving surfaces has been developed over the years to assist the mobility of visually impaired people. These surfaces have now been rationalized and each type of surface is intended to provide specific information.

A common form of tactile slab is the sort used for road crossings (*see* Figure 7.37). This type of tactile slab is known as 'blister paving' and is used where the footway meets the carriageway without a kerb or other level change. For controlled crossings only, the paving must be red and at uncontrolled crossings the colour is buff. An uncontrolled crossing is one where there is no active control to halt vehicles (*see* Figure 7.62) Another form is the 'corduroy' surface which is used to warn of hazards such as steps (*see* Figure 7.63). This may be any colour, other than red, which contrasts with the pavement.

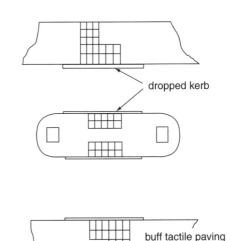

Figure 7.62 Tactile paving at uncontrolled crossings

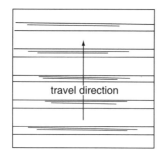

Figure 7.63 Corduroy surface tactile paving for pedestrians – to warn of onward hazard such as steps or level crossings

A third form is the surfacing used at segregated shared cycle track/footway locations. There is a central delineator strip running longitudinally the length of the footway. This is a raised strip as shown in Figure 7.64.

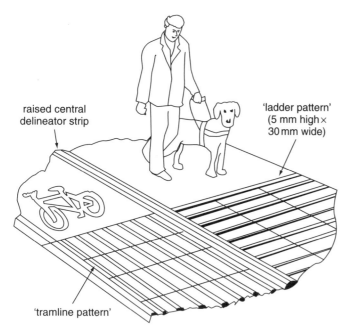

raised central
delineator strip

'ladder pattern'
(5 mm high ×
30 mm wide)

'tramline pattern'

Figure 7.64 Tactile paving at segregated shared cycle track/footway

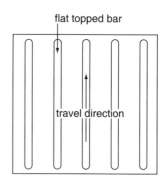

flat topped bar

travel direction

Figure 7.65 Guidance surface tactile paving – guidance paths can guide around obstacles

The profile of the tactile surface comprises a series of flat topped bars which on the pedestrian side make a 'Ladder' pattern and on the cyclist side form 'tramlines'. Guidance paths can help define routes where other cues are not present and the type of tactile paving used is shown in Figure 7.65.

Other special types of tactile paving are used on railway platforms (directional flat topped domed paving), and on light rapid transit and tram platforms (lozenge). For further information, reference should be made to the guidance notes on tactile surfaces.

Revision Questions

1 In a typical batch mix for producing precast concrete kerbs, how many kilograms of Portland Cement would be used?
2 What are the cross-sectional dimensions of type 2 timber footway edging?
3 What proportions are used to mix lime and sand to produce a mortar suitable for laying tactile paving on a 25 mm thick whole bed?
4 What type of crossing would use buff coloured tactile paving?

Note: See page 305 for answers to these revision questions

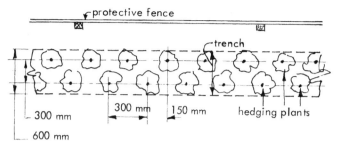

Figure 8.2 Plan

8.1.2 Methods of Planting

Dig adequate holes for plants (Figure 8.3). Insert plant, spreading roots carefully. Shake the plant as soil is shovelled back into the hole (to ensure that soil is mixed with root system) (Figure 8.4). When the hole has been slightly overfilled, firm the plant by treading with foot (keeping leg straight). Make sure that the heel does not break roots from root ball or scrape bark from the stem of the plant (Figure 8.5).

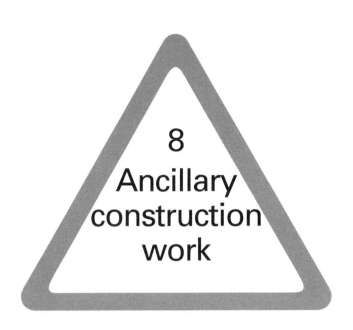

8.1 Hedging

The main requirements for a hedge are:

that it must be long-lived;
that it is healthy, and not susceptible to pest disease;
that it is hardy (i.e. frost resistant, etc.);
that it must be capable of withstanding cutting;
that it should be close-branching and twiggy;
that it is cheap to produce. The cost is directly related to the ease of propagation of the plants.

8.1.1 Layout of Boundary Hedges

Hedges, which are planted to form the highway boundary, should be protected by a fence.

The details for planting, which are applicable to quickthorn, beech, hornbeam, and privet, are shown in Figures 8.1 and 8.2.

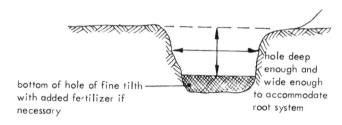

Figure 8.3 Method of planting

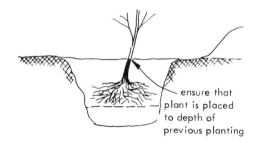

Figure 8.4 Method of planting

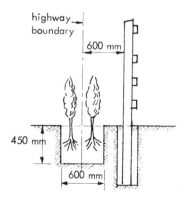

Figure 8.1 Sectional elevation

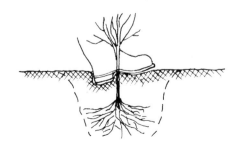

Figure 8.5 Method of planting

8.1.3 Hedging Plants

Quickthorn (*Crataegus monogyna* or *Crataegus oxycantha*). In roadwork jobs, it is almost always quickthorn which is used.

Some alternatives are:

Hornbeam (*Carpinus betulus*)
Beech (*Fagus sylvatea*)
Holly (*Ilex aquifolium*)
Privet (*Ligustrum ovifolium*)

Lawson's cypress and Leylandii are quick growing evergreens suitable for hedging.

Leylandii will grow 600 mm in a year in poor soil. It is also easily and readily propagated by cutting.

N.B.: Yew (*Taxus baccata*) is poisonous and must not be used for boundary hedging.

8.1.4 Reasons for Lack of Success in the Establishment of a Hedge

Many hedging plants are delivered in bundles of 50 or 100. If a length of, say, 1 km of hedge is to be planted, the bundles are broken open and the plants strung out over the planting length.

At the end where planting is started, the plants are still damp on planting. Further along, the root systems, having been exposed for longer are drying out and the plants are unable to recover from this. The symptom of this is a hedge where almost 100% successful (growing) planting is apparent at one end but the success rate becomes steadily worse as the length extends. Hence, it is most important to keep roots damp until planting. If planting is delayed, the plants should be heeled in.

Lack of 'firming' or inefficient firming can result in the death of the hedging plants. Voles, rabbits and hares can kill plants (from both above and below ground). The hedging plants may be choked by weeds. In particular, they must be kept clear of weeds for the first two years after planting. In protecting from weeds, great care must be taken when applying herbicides that the spray does not accidentally blow on the hedging plants.

8.1.5 Cutting

Hawthorn (quick) hedging, in particular, develops too much top growth and hence requires frequent cutting, otherwise it tends to develop tree-like trunks and loses its lower growth and, hence, its value as a hedge. This remedial work is done by the process known as hedge laying, in this (*see* Figures 8.6 and 8.7):

all rubbish is first cut out (e.g. weeds, brambles, etc.);
all major growth that is not required is cut out;
the main growths to be retained are cut nearly through at an angle and near to the ground (say about 150 mm above);

Figure 8.6 Hawthorn hedging before laying

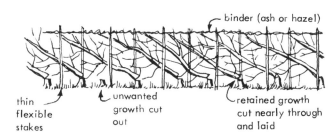

Figure 8.7 Hawthorn hedging after laying

these cut growths are laid longitudinally at a slight angle to the horizontal and supported by stakes at frequent intervals;

the tops of the stakes are linked by a pliable binder (of ash or hazel).

8.2 Planting Small Trees of About 2–3 m in Height

1 Check the location of the proposed planting (e.g. if in a verge ensure that it is not to be planted under overhead cables, etc.).

2 Check the root system of the tree to be planted, to ensure that it has not been too severely pruned or damaged.

3 Preferably plant when the tree is dormant (normally, November to March).

4 Weather conditions must be suitable (i.e. do not plant during frosty weather).

5 Dig an adequate hole, i.e. deep enough and wide enough to accommodate the root system when spread out (Figure 8.8).

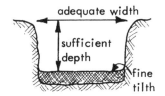

Figure 8.8   Method of planting a small tree

6 The bottom of the hole should be forked to a fine tilth, and fertilizer added if necessary (usually about 0.25 kg of bone meal per tree).

7 Drive in wooden stake to support tree.

8 Insert tree in prepared hole, spreading roots carefully.

9 Shake tree as soil is shovelled back into the hole, to ensure that the soil is well mixed with the roots. (The soil excavated should be liberally mixed with peat before backfilling: this will assist in the rapid growth of the fine roots.)

10 The tree should be so placed that its depth of planting is the same as it was before being moved (i.e. plant to mark on trunk). *See* Figure 8.9.

11 Slightly overfill the hole, then 'firm' by treading with the foot, keeping the leg straight, and taking care not to damage either the roots or bark. (*See* Figure 8.10.)

12 Strap the trunk of the tree to the stake, using two plastic (adjustable) tree ties.

13 Place a mulch of 50 mm of peat over the backfill, with a ridge of peat in a ring around the hole. (*See* Figure 8.11.)

14 Water the tree at regular intervals until it has become established. That is particularly important when a tree is moved whilst growing/in leaf.

N.B.: If the tree is to be moved whilst in leaf, the roots should be well watered before it is lifted and the root ball kept well together by wrapping tightly with hessian on lifting. The wrapping is removed only when the tree is placed in the hole in its new location.

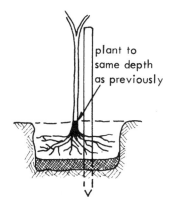

Figure 8.9   Method of planting a small tree

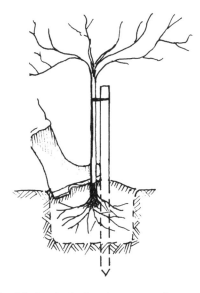

Figure 8.10   Method of planting a small tree

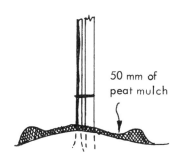

Figure 8.11   Mulch for newly planted tree

8.3   Planting Semi-Mature Trees 4–5 m or More in Height

1 The tree will have been lifted ready for replanting with its ball well secured in hessian, or similar wrapping.

2 A hole is dug about 450 mm wider and 200 mm deeper than the root ball.

3 The bottom of the hole is forked to a fine tilth and bone meal fertilizer mixed in (0.25–0.5 kg of bone meal).

4 The tree is carefully positioned in the hole, the root ball wrapping removed, and the hole backfilling begun.

The backfill should be such as to leave the surface of the root ball about 50 mm above the surrounding ground.

5 The ground should be thoroughly firmed around the roots.

6 The tree is then supported by three wire ties, fixed at about 60 degrees to the horizontal and anchored to angle irons driven well into the ground. These ties are

tensioned over branches (protected by plastic sleeves) and the ties are wound tight (at least six complete turns).

7 The ground under the tree is then covered with a mulch of 50 mm of peat, ridged around the perimeter to assist in retaining water.

8 The trunk of the tree should be wrapped to prevent the bark from drying out. The usual wrapping is hessian strip 150 mm wide.

9 Regular checks should be made to ensure that the ties remain tight and that the trunk wrapping is secure.

10 Ties and wrappings may need to be retained for two seasons of growth.

Alternative methods of support are possible. One of these uses underground securing, as shown in Figure 8.13. This method is useful for conifers which are not usually as easy to support with wire ties as are the deciduous trees.

N.B.: Precautions for planting need to be as for small trees.

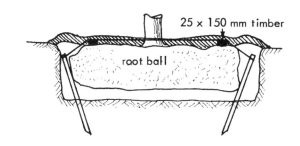

Figure 8.13 Alternative method of supporting a semi-mature tree

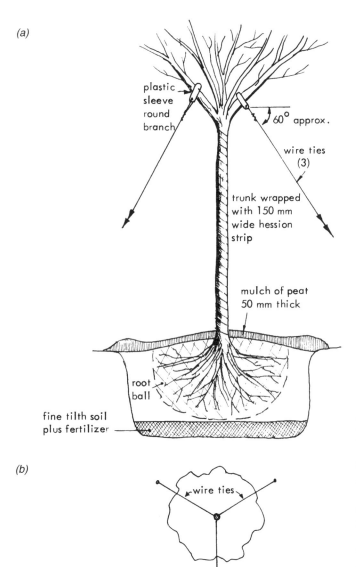

Figure 8.12(a) and (b) Planting and supporting a semi-mature tree

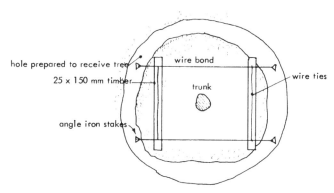

Figure 8.14 Alternative method of supporting a semi-mature tree

8.4 Seeding of Verges, etc.

Where an area of ground is to be seeded, provided that the soil is in good condition with an adequate amount of top soil, it is necessary only to level the area, dig over the top soil (either by hand or mechanically, depending on the size of the area), and rake it down to a fine tilth.

Add about 30 g/m² of bone meal (long-term) fertilizer and rake well into the soil. Roll it lightly, grading the surface to eliminate noticeable lumps and hollows and harrow the surface.

Sow grass seed. The rate of application of the seed should be about 70 g/m², Tread the seed into the surface and then cover with a thin layer of top soil.

Seeding is best done in spring or early autumn and then only when the ground is damp. This is the traditional method. An alternative method is given below.

For protection of the seed as well as providing fertilizer the prepared area is sprayed with a mixture of grass seed, chopped straw and bitumen. The bitumen provides a sealant to keep the moisture in the soil to assist germination and both the chopped straw and bitumen act as fertilizers for the seed.

This method has the added value of protecting the seed from birds, so that the rate of application of the seed can be less than when using the traditional method.

Where the area to be grassed is of impervious soil, then the preparation of the ground is more involved. In this case, the top 200 mm of the impervious material should be removed (skimmer equipment, or front loader mechanical excavators/loaders are suitable for this purpose, or the material may be bulldozed aside if it can be disposed of on the site). Then, spread a 75–100 mm thick layer of ash or sand over the area and level it, topping it with 100 mm of good quality top soil. This soil should contain fertilizer, such as bone meal at the rate of 30–60 g/m².

Rake the top soil to a fine tilth and level it. Roll it with garden roller or light pedestrian roller until it is firm enough to walk on. Lightly harrow the surface to receive seed. If a suitable outlet can be found, the area should be drained by a shallow herringbone land drain system.

8.5 Fencing

There are many types of fencing, but the majority of the types used in ancillary works connected with roadworks, mainly highway boundary fencing, are of the wooden post variety.

The more common types are as follows.

8.5.1 Post and rail fencing

Either three or four rail fencing structures may be erected using hardwood posts and softwood rails. Rails are nailed on the field side of the posts to prevent livestock pushing rails off the posts.

The more simple types are illustrated in Figures 8.15 and 8.16 with the alternative design shown in Figures 8.17 and 8.18. All the timber must be treated with preservative before erection of the fencing.

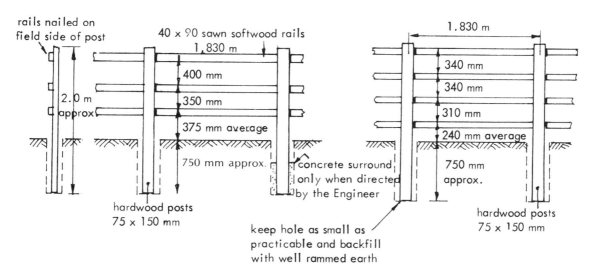

Figure 8.15 Three rail fence

Figure 8.16 Four rail fence

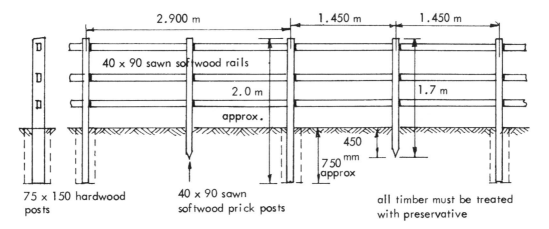

Figure 8.17 Alternative type of three rail fence

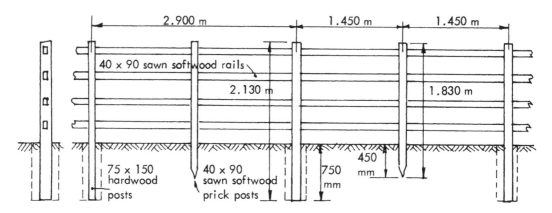

Figure 8.18   Alternative type of four rail fence

8.5.2   Post and wire fencing

(a)   Woven wire fence

The wire is supplied in large rolls and strained (i.e. stretched) between strutted straining posts. It is then additionally supported by driven intermediate posts (with pointed ends) to which each horizontal wire should be stapled (*see* Figure 8.19).

(b)   Strained wire fence

In this case each horizontal strand is separately fixed, the plain wires being stretched between straining posts and stapled to the intermediate posts. The barbed wire strands are carried straight by the straining post to which they are stapled as well as to the intermediate posts (*see* Figure 8.20).

With the post and wire fencing, the posts are usually round in section.

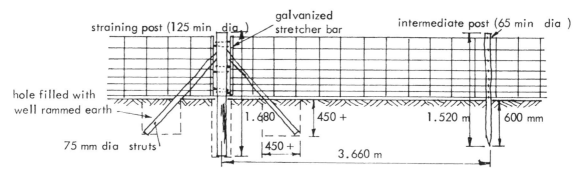

Figure 8.19   Woven wire fence with round wooden posts

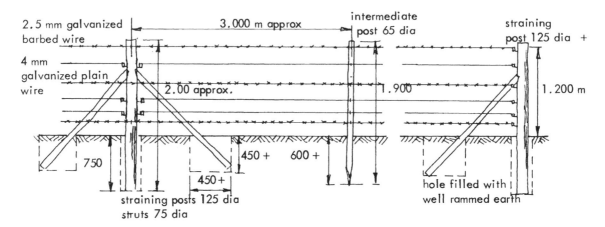

Figure 8.20   Strained wire fence (6 strand) with round wooden posts

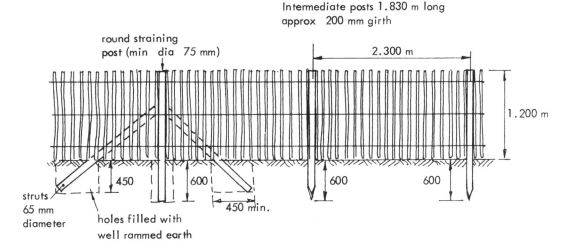

Figure 8.21 *Cleft chestnut paling fence with round wooden posts*

8.5.3 *Chestnut Paling Fencing*

Cleft chestnut paling fencing is supplied in rolls with the palings wired together near the top, middle, and bottom. Straining posts, strutted and intermediate driven posts are required as shown in Figure 8.21.

8.6 Fencing Posts

Other materials are used for fencing posts, particularly precast concrete and steel. Using these other materials, slightly different types of fencing can be erected, but the main principles of erection are similar.

8.7 Erection of Fencing

It is most important that the line of a fence be carefully set out prior to erection and post holes dug after being positioned. Post holes are usually dug with a special type of spade or with an earth auger.

Posts should be carefully aligned to the fence line and well-rammed soil packed around the base of the post, checking that the post is truly vertical. As the erection of the line of posts is continued each new post should be carefully 'eyed in'.

If the post holes are checked for depth using a gauge stick, there should be little difficulty in lining through the tops of posts on reasonably level or evenly sloping ground.

Where there is a change of gradient along the fence line, care must be taken to align the tops of the posts. The fencing will not present a pleasing appearance unless there is an accurate line or regular curve to the line of the tops of the posts.

When fixing fence rails, the post should first be measured and the level of the top of each rail marked on it.

Rails are fixed to posts by nailing and there is always a need to support the post during nailing; usually this is done by holding a sledge hammer against the back of the post opposite to where the nail is being driven. This provides a guard against the levering action of the hammering of the nail, which tends to knock the post out of the vertical and to loosen the backfill in the post hole.

8.8 Snow Fencing

Snow fencing is a device for the control of drifting snow and is encountered in Britain particularly in the mountainous parts of Scotland and Wales, together with the Pennine chain and other similar high ground, such as the Lake District and Cheviot Hills. Its use may also be necessary in flat areas of open country.

It is not possible to treat the subject in detail, but additional information can be obtained from the Road Research Laboratory Booklet No. LR 362 'Snow Fences'.

The basic thinking behind the location of snow fencing to protect a length of road is the acceptance of the idea that freshly fallen snow will be loose and will drift; therefore, the fence is located away from and upwind of the carriageway so that the drift will build up, mainly downwind of the fence and that the eventual snow drift will not reach as far as the carriageway.

From practical experience, it has been found that these are the main points:

1 By erecting a snow fence at a suitable distance to the windward of a road to be protected from drifting snow, the depth of the snow on the road can be reduced to a depth almost identical to that of the undrifted snow.

2 In the British Isles, fences 1.5 m high are normally adequate.

3 The optimum density ratio for a snow fence is 0.5 (i.e. a fence where the gaps in it are equal in area to the solid part).

4 A snow fence should be constructed with a gap of 150–200 mm between the bottom of the fence and the ground (this allows the snow to creep over the surface of the ground and under the fence and it forms a larger drift, than would be the case if there were no space under the fence).

5 The distance (L) at which a snow fence should be erected to the windward of the area to be protected is related to the density ratio (D) and the height of the fence (M) within a practical range of D and H, by the expression:

$$L = H(32 - 18D) \text{ (dimensions in metres)}$$

Hence, if the fence is 1.5 m high with a density ratio of 0.5, the distance it should be from the road is:
$L = 1.5(32 - 18 \times 0.5) = 1.5(32 - 9) = 1.5 \times 23 = 34.5$ m.

6 Cleft chestnut paling is the cheapest satisfactory form of snow fencing for use in the British Isles.

7 A portable fence made from strips of polyethylene (density ratio 0.5) has been found to be satisfactory.

8 Blower fences have not proved worthwhile (these consist of an open-sided sloping roof, which is intended to deflect the wind and so increase the wind velocity as it reaches the road, so blowing the snow off the carriageway).

9 Snow fences are most effective when erected upright and perpendicular to the direction of the snow-bearing wind.

10 Roads in cuttings should have a cutting side slope of 1 in 6 to 1 in 8 to provide adequate storage space (to windward of the road) for the drifted snow.

11 Cuttings deeper than 6 m are able to keep free of drifts due to the wind eddies formed in the cutting, which deposit snow on the windward side. Where there is considerable likelihood of snowfalls over prolonged periods, consideration should be given to:

(*a*) providing multiple rows of snow fencing; theoretically two rows of fencing can double the 'storage' capacity of the drifts when compared with a similar single fence;

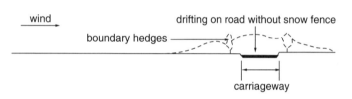

Figure 8.22(a)   Snow drift formations: without snow fence

N.B. Distance of snow fence
from carriageway must
be greater than maximum
length of leeward drift

Figure 8.22(b)   Snow drift formations: with snow fence

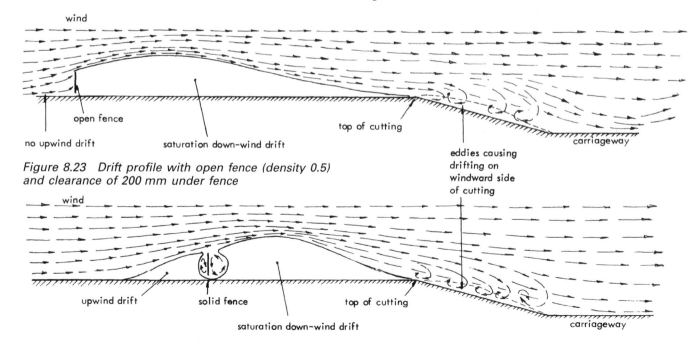

Figure 8.23   Drift profile with open fence (density 0.5) and clearance of 200 mm under fence

Figure 8.24   Drift profile with solid fence (density 1.0) and clearance of 200 mm under fence

(*b*) providing shelter belts of trees and bushes. These should be of 10 to 12 rows at 2 and 3 m apart and the wooded area becomes a 'reservoir' for snow.

8.9 Arrangement of Snow Fences

Figures 8.25–8.28 show possible arrangements of snow fences in relation to the wind direction. It must be noted that these are only an indication of the solution to the problems of snow drifting: further information is available from Transport Research Laboratory.

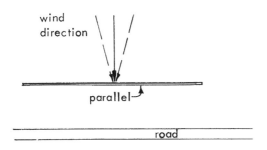

Figure 8.25 Wind at right-angles to road

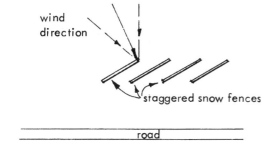

Figure 8.26 Wind at angle of about 70° to road

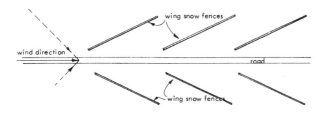

Figure 8.27 Wind blowing along road

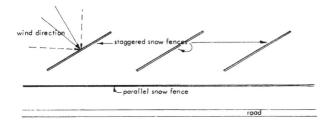

Figure 8.28 Combination of snow fences

8.10 Brick Walling

The roadworker is likely to be concerned with brick walling as a highway boundary mainly in urban areas. The construction of brick walls is usually associated with road improvement schemes where the realignment of the carriageway causes existing walls to be demolished and new brick walls to be built along the altered frontages. This is normally termed 'accommodation works'.

This walling is generally of half-brick (112 mm) thickness with piers, or of single brick (225 mm) thickness. Piers should be at not more than 3 m intervals.

Occasionally a cavity wall may be encountered. This is 275 mm thick with two half-brick walls separated by a 50 mm cavity, but linked by wall ties across this gap to provide stability (*see* Figure 8.29).

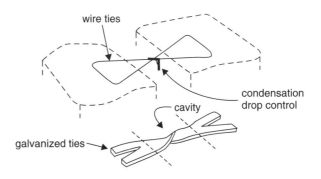

Figure 8.29 Ties for cavity walls

Where walling is either high (say 2 m or more), or is structural and load bearing, reference should be made to specialist information.

Brick walls normally require:

1 an adequate foundation: usually a concrete footing;
2 a damp course: to prevent rising damp from the ground;
3 a coping to protect the top of the wall from weather (this should also be set on a 'damp proof' course).

Brickwork requires careful bonding in order to produce a structure capable of holding together under load, by giving good weight distribution through the wall (*see* Figure 8.30).

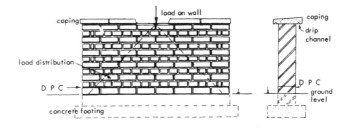

Figure 8.30 Typical brick wall

The two most important bonds used in walling are English and Flemish bonds.

8.10.1 English Bond

This consists of one course of headers and one course of stretchers placed alternately (*see* Figure 8.31(*a*)).

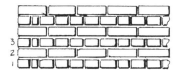

Figure 8.31(a)   English bond

The first course in a 225 mm wall is usually a header course, but this starts with a header and closer, which together form three-quarters of the length of a stretcher, all the remaining bricks being headers. When the stretcher course is then started (second course: Figure 8.31(*a*)) the first stretcher will reach to the centre of the second header in the first course, thus giving the quarter bond required in each course of headers and stretchers.

An alternative method is shown in Figure 8.31(*b*) using a $^3/_4$ brick stretcher to start the second course having laid a plain course of headers as the first course. All cross joints (vertical) should be kept in correct perpendicular position as shown.

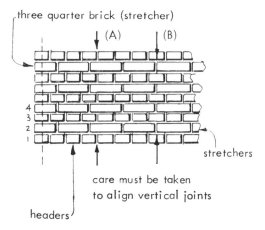

Figure 8.31(b)   English bond (alternative)

A variation of the English bond is the English garden wall bond shown in Figure 8.32. This has a bond of three stretchers and one header course.

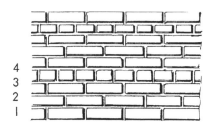

Figure 8.32   English garden wall bond

8.10.2 Flemish Bond

This consists of alternate headers and stretchers in each course, bonded as shown in Figure 8.33. This bond requires a header and closer at the start of the first course followed by alternating stretchers and headers. The second course starts with a stretcher and then alternates headers and stretchers throughout. The closer used is a half brick wide, cut longitudinally.

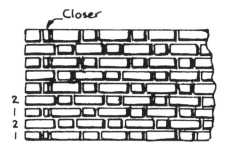

Figure 8.33   Flemish bond

A variation of the Flemish Bond is the Flemish Garden Wall Bond. This has three stretchers and one header repeating in each course, as shown in Figure 8.34.

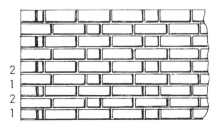

Figure 8.34   Flemish garden wall bond

N.B.: The jointing of these bonds is achieved by the placing of mortar to provide joints between bricks at the time of construction. Surplus mortar is struck off or the joint is raked out during construction and pointing is carried out afterwards when the mortar joints have set.

Some of the forms of jointing used are shown in Figure 8.35.

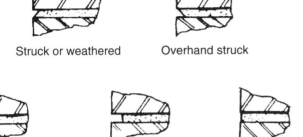

Struck or weathered   Overhand struck

Curved or keyed   Raked or recessed   Flush

Figure 8.35 Types of jointing

In existing brickwork, pointing consists of raking out mortar from the joints to a depth of at least 13 mm and then filling with specially prepared mortar, finishing as for jointing.

8.10.3 Mortar

The mortar on which the bricks are bedded is as important as the bricks themselves.

The functions of a mortar are to:

1 provide an even bed so that the load is evenly distributed;
2 make the wall weather and water proof;
3 bond the brickwork into one solid mass and so help to resist lateral forces.

The mortar should be readily workable and sufficiently resilient to allow minor structural movements to be accommodated as well as expansion and contraction due to temperature changes. The mortar should be durable, but not stronger than the bricks which it is to bond together.

The main mortar designations and contents are tabulated below:

Mortar designation (strength)	Cement/ lime/ sand	Masonry cement/ sand	Cement/ sand + plasticizer
I (strongest)	1 : 0.25 : 3		
II	1 : 0.5 : 4–4.5	1 : 2.5–3.5	1 : 3–4
III	1 : 1 : 5–6	1 : 4–5	1 : 5–6
IV	1 : 2 : 8–9	1 : 5.5–6.5	1 : 7–8
V (weakest)	1 : 3 : 10–12	1 : 6.5–7	1 : 8

8.10.4 Brick Chambers (for Catchpits and Manholes)

The layout of the bricks for the bonding in such a chamber may be as shown in Figure 8.36 although other bonds are possible.

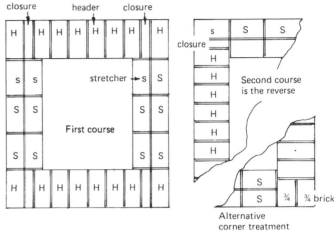

Figure 8.36 Brick manhole construction

8.11 Retaining Walls

There are many cases where the full width required for an embankment or cutting is not available. In such cases a retaining wall can be constructed parallel to the carriageway to retain the earth bank and so save overall width.

A retaining wall may be of reinforced concrete, mass concrete, masonry or brick, but in every case its function is to resist the pressure exerted by the soil which it supports and to transfer this force on to the ground on which it is built.

Unreinforced walls are commonly called 'gravity walls' as they depend for their stability on their own weight. The calculation of the forces involved and the detailed design is beyond the scope of this book.

In very simple terms, the forces acting on the retaining wall are shown in Figure 8.37 and these consist of:

1 the dead weight of the wall itself;
2 the weight of the earth behind the wall (tending to stabilize it);
3 the lateral force on the wall due to the load behind it, which causes a tendency for the wall to overturn about the toe of the wall.

The resultant force (i.e. the single force which would have the same effect as these three forces) is shown as 'R' and the line of action of this force, for real stability and safety, should pass through the 'middle third' of the base of the wall. This ensures that none of the concrete in the wall is subjected to tensile stressing.

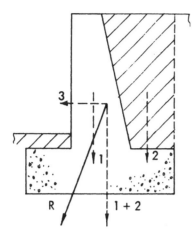

Figure 8.37 Retaining wall

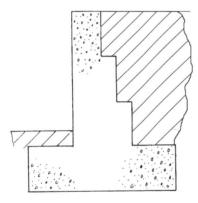

Figure 8.38 Alternative stepped shape

These mass concrete walls impose very heavy loads on the sub-soil due to their own weight and, as a rough guide, the width of the base is usually between 0.25 and 0.5 of the height.

Careful checks must be made at the design stage to avoid:

(*a*) the wall sliding along the sub-soil;
(*b*) the load on the sub-soil being too great for it to withstand.

Reinforced concrete retaining walls use less concrete than the equivalent mass concrete walls, the steel reinforcement providing the structural strength of the walls. They may be of:

(*a*) *the cantilever type* in which the wall is cantilevered from the base as shown in Figure 8.39.

Where there is severely restricted space in front of the line of the wall, it may be constructed as a simple 'L' shape, but this does present additional difficulties over stability. The provision of a toe beam is a valuable anti-sliding precaution (*see* Figure 8.40).

Provision must also be made for the drainage of the soil retained behind the wall, in order to prevent the build-up

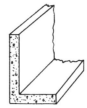

Figure 8.39 Cantilever retaining wall

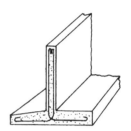

Figure 8.40 Retaining wall with toe

of fluid pressure, which would result in fluid force or thrust on the wall and this would be much greater than the force due to the soil. To allow this drainage to take place weep holes are provided through the wall.

(*b*) *the counterfort type* as shown in Figure 8.41 in which the wall spans between the triangular shaped counterforts.

Brick retaining walls may be provided where the wall is not more than about 1.5–2.0 m high (Figure 8.42). This type of wall will require a concrete base and good quality bricks and workmanship. A central mat of light steel reinforcing fabric is useful in helping to tie the brickwork into the base. Again weep holes must be provided for drainage.

Masonry (stone block) walls can also be built as retaining walls in a similar manner to the brick walls. It is, however, more usual to find the brick and stone used only as a facing to a concrete wall (Figure 8.43). In this case, the concrete provides the strength, the brick or stone being a decorative feature on the front face of the wall (*see* Figure 8.43).

Where retaining walls are of considerable length, expansion joints are needed. These should be at not more than 10 m intervals and a feature can be made of them, such as a vertical groove in the surface which can match other

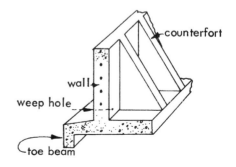

Figure 8.41 Counterfort retaining wall

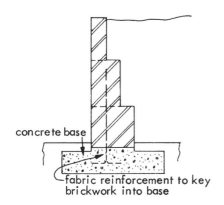

Figure 8.42 Stepped brick retaining wall

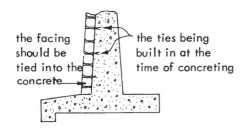

the facing
should be
tied into the
concrete

the ties being
built in at the
time of concreting

Figure 8.43 Stone faced retaining wall

vertical grooves which accommodate the weep holes, so that water staining is contained within the groove and does not spoil the overall appearance of the retaining wall.

8.12 Shuttering for Walls

Great care must be taken in preparation and fixing of shuttering for concrete retaining walls.

On small-scale jobs the shuttering is usually of timber, but larger works often make use of steel panel shuttering. These steel panels are generally square and flanged on all four sides for ease of bolting together to form the complete wall face area. The panels must be adequately supported by a framework of beams and braced to prevent distortion when the wet concrete is poured (*see* Figure 8.44).

The inside face of the panels should be painted with a release agent before erection of the shuttering to prevent the concrete from sticking to the shuttering and so causing surface damage to the wall.

Information on release agents is given in Section 6.18.

8.13 Reinforcement for Walls

The reinforcement shown in Figure 8.44, whether bars or mesh, must be fixed with appropriate 'spacers' placed to ensure the correct positioning of steelwork when the concrete is vibrated. A typical 'spacer' is shown in Figure 8.45.

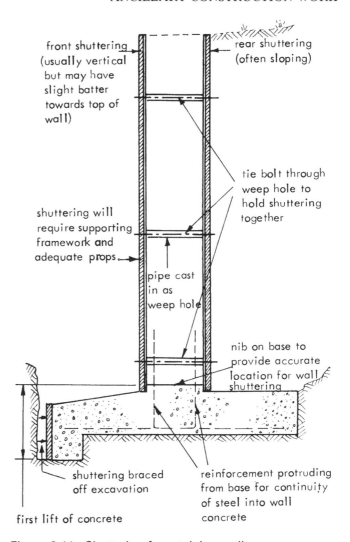

front shuttering (usually vertical but may have slight batter towards top of wall)

rear shuttering (often sloping)

tie bolt through weep hole to hold shuttering together

shuttering will require supporting framework and adequate props

pipe cast in as weep hole

nib on base to provide accurate location for wall shuttering

shuttering braced off excavation

reinforcement protruding from base for continuity of steel into wall concrete

first lift of concrete

Figure 8.44 Shuttering for retaining wall

Figure 8.45 Reinforcement spacer

Revision Questions

1 When planting large trees how might you ensure they survive the first few storms?
2 With a typical wooden post/rail fence, which side should the rails be nailed and why?
3 List five forms of jointing used in brickwork bonds?
4 A retaining wall is designed to resist the pressure upon it and exert it where?

Note: See page 305 for answers to these revision questions

9.1 Safety and the Law

Since 1875, many Acts of Parliament have been passed to ensure that minimum standards are set for safe working conditions throughout industry. Some Acts have come about as an attempt to create and improve standards for the safety, health and welfare of employees and other Acts were necessary as, and when, the advance of technology brought new hazards into industry.

One of the earliest pieces of legislation that was wholly concerned with safety was the Explosives Act in 1875. This controlled the use and storage of explosives as some spectacular accidents had caused considerable public alarm. The Boiler Explosions Act came a little later and was followed by a series of Acts, some of which were, the Employment of Women, Young Persons and Children Act 1920, the Radioactive Substances Act 1948, the Mines and Quarries Act 1954, the Agriculture Safety, Health and Welfare Provisions Act 1956, the Factories Act 1961, the Construction Regulations 1961, the Public Health Act 1961, the Offices, Shops and Railway Premises Act 1963, and so on, culminating in the Health and Safety at Work Act 1974.

This last Act came about because, in recent years, it had become apparent that the piecemeal growth of legislation concerned with safety did not cover all employees and had concerned itself rather with places of work. It was time for a review and the Robens Committee on Safety and Health at Work 1972 examined this situation and its recommendations were subsequently incorporated into the Health and Safety at Work Act 1974. The Committee had found that there was a common law, many Acts of Parliament,

many regulations and that all this was considered by many to be counter-productive. It said that a new philosophy was needed and so the Act is aimed at people and their activities rather than purely premises and processes. The new philosophy shows that all persons at work have responsibilities for safety whether they be employers, employees or self-employed. Consider some of these responsibilities.

So far as is reasonably practicable an employer must provide:

(a) safe plant and systems of work;
(b) safe handling, storage and transport of goods;
(c) information and training;
(d) safe places and access to work;
(e) a safe and healthy environment.

An employee must take reasonable care of his own safety and that of others and to co-operate with others in order to ensure that there is compliance with statutory duties relating to health and safety at work.

The earlier Acts of Parliament have parts which are still effective, but this Act supplements all these and imposes much more severe penalties than were previously possible. In certain instances, unlimited fines and the imprisonment of a person up to two years, could be the penalty for the wilful contravention of this Act of Parliament.

This 'enabling Act' has led to the creation of several sets of regulations which concern the activities present in roadwork. Some of the most important are the Control of Substances Hazardous to Health (COSHH) Regulations, the Noise at Work Regulations 1989, the Construction (Head Protection) Regulations 1989 and the Electricity at Work Regulations 1989. The effects of these regulations are outlined later in this chapter.

More recently, with the UK's participation in the Treaty of Rome, Health and Safety Directives from the EU have been enacted as Regulations within the UK. These not only underpin the general requirements made by the Health and Safety Act 1974, but specify a technique of 'risk assessment' as the method of ensuring safe places of work and safe procedures. Some of the main regulations which concern roadworks are as follows:

1 Management of Health and Safety at Work – this gives the general principles of risk assessment the purpose and practice.
2 Workplace (Health, Safety and Welfare) – this covers the Health and Safety requirements at the workplace, i.e. depots, offices, workshops, etc.
3 The Provision and Use of Work Equipment – these cover the provision of safe work equipment and its safe use.
4 The Personal Protective Equipment at Work – this requires employers to provide appropriate PPE and training in its usage, if the risk cannot be controlled by other means.
5 The Manual Handling Operations – this requires employers to make an assessment of the risks to the health of their employees when lifting or handling loads.

6 Construction (Design and Management) – this is directed at all construction work and requires that all health and safety considerations are examined prior to any work commencing. A Safety Supervisor has to be appointed and safe methods of work planned for all large contracts.

7 Construction (Health, Safety and Welfare) – this replaces previous construction legislation, particularly the 1961 Construction Regulations.

8 Lifting Operations and Lifting Equipment – concerns all arrangements for lifting, placing and moving loads on site.

Large organizations need to appoint Safety Officers and these advise both employer and employees on all aspects of safety in the workplace.

9.2 A Safe Place of Work

As safety can only be achieved by co-operation between management and all individuals on site, it follows that the site supervisor will be the key person to ensure the initial provision of a safe site and then its maintenance as a safe place of work. The site supervisor will need to:

(a) ensure that all practices on site conform with the appropriate legislation (checking that risk assessments have been carried out on all hazardous activities);

(b) attempt to improve on the minimum legal requirements by, say, encouraging a clean orderly site, with well stacked materials and clear access routes;

(c) identify potential hazards that could exist on the site and take the necessary precautions;

(d) be vigilant in ensuring that young persons under 18 years and older persons given tasks new to them, are adequately trained and supervised in their work.

N.B.: A place of work does not have to be a fenced off area of land, but could equally be a plant operative's cab or a coned off portion of the highway.

9.3 Safety on Highway Sites

Some five aspects of safety should be amongst those considered when taking precautions or making risk assessments on highway sites. These are the:

(a) safety of the travelling public (*see* Section 9.4);

(b) protection of highway personnel from the travelling public;

(c) safety of personnel from potential hazards on a particular site;

(d) safety of an operative from another's actions;

(e) protection of an individual from the consequences of their own actions.

Items (a) and (b) will require a system of temporary roadwork signing to be established at and on the approaches to the site (*see* Temporary Signing of Roadworks, Section 9.4). An examination of the other items (c), (d) and (e), follows.

9.3.1 The Safety of Personnel from Potential Hazards on Site

Much information on this is detailed in the Construction (Health, Safety and Welfare) Regulations 1996 and covered again in the Construction (Design and Management) Regulations 1994.

(a) Falls and Injuries

To prevent persons from falling

1 Keep the site tidy and get rid of rubbish. Store materials methodically and establish clear routes through the site.

2 Provide hand rails for stairs and gangways over trenches.

3 Fence or cover all openings in floors, holes in ground and similar hazards.

4 Provide good lighting on stairs, passages and other access routes.

5 Use only well-constructed ladders properly secured, and where necessary, suitable scaffolding for all work that cannot be done safely from the ground or from a building.

6 Check that any safeguards, such as guardrails, toeboards, hand rails and covers over openings in floors, which have been provided are kept in place.

To prevent materials from falling

1 Ensure that materials are properly stacked and that they are not likely to be blown or knocked over.

2 If persons are working regularly in places where they are liable to be struck by falling materials, provide a strong protective cover. Similar protection is advisable over passageways in common use. Provide safety helmets.

3 Materials should not be thrown down haphazardly from a height. Lower them or provide a chute.

4 Pipes should be placed at an angle of 45° to the run of the trench and sufficiently far back to prevent them being accidentally knocked into the excavation.

To prevent head injuries

On many sites there is an obvious risk of head injury from falling or swinging objects or from striking the head against something. For this reason the Construction (Head Protection) Regulations 1989 were introduced to make the wearing of hard hats compulsory on construction sites where there was foreseeable risk of head injury. Employers must ensure that employees wear hard hats when engaged in operations involving work in excavations, partly completed structures, near machinery/plant with swinging booms and work on a scaffolding. Hence only certain personnel at risk on site, e.g. scaffolders, might be required to wear hard hats or an entire area could be designated a 'hard hat' site, with the appropriate signing, where all present would need to wear hard hats.

(b) Electricity on Site

Electricity is always a potential hazard on site and the Electricity at Work Regulations 1989 are intended to ensure that almost all workplaces are subject to the same legal requirements relating to electrical safety. An employer must take precautions in order to minimize the risks to personnel working with electrical equipment or near live conductors. This means electrical equipment/systems must be checked and adequately maintained and safe working practices adhered to when working near underground and overhead conductors on sites.

1 Supply for sites

When temporary lighting is installed or portable tools are used the risk of serious injury can be lessened by using a transformer to reduce normal supply voltage of 230 V to 110 V or less. The transformer unit is connected to the incoming supply and any portable tools or equipment are then plugged into the transformer. Care is always necessary to ensure that cables are properly terminated and that live wires are not exposed to casual contact. Another device in use is the residual current operated circuit breaker (RCB) which is embodied in a 230 V outlet socket or, alternatively, in the plug on the actual cable connected to the portable tool. The RCB provides better overcurrent protection and protection against small current leakage. Use of battery operated equipment removes the risk of electrocution and must be considered an option.

Having correctly set up a supply to a hand tool it is vital to prevent cables lying in water or being left in such a position as to become damaged or to trip passers by. Nowadays, portable generators for use on sites supply 110 V or less, but early models may supply higher voltages and care must be exercised if, for example, a set is used to power temporary traffic lights.

2 Services in vicinity of site

Prior to site work beginning and throughout the whole of the construction period, precautions must be taken to prevent accidental contact with live underground or overhead cables. Where possible the safest way to prevent this happening is for arrangements to be made to disconnect or divert the Electricity Supply Company's cables. During the design stage of highway construction work, information as to the whereabouts of electricity supplies is sought and together with surveys made on site will usually reveal the position of any plant.

3 Underground cables

These are normally buried in the ground, but spurs to users' homes, etc., have no defined depth and can be just below the surface, so special care needs to be taken. The presence of a cable or other service can normally be recognized by a layer of sand 150 mm thick laid over the cable.

Clay tiles or plastic boards 1.2 m long and 100 mm wide may also be used to cover the cable, with information about the voltage written on them. Plastic tape giving this same information may also be found. The tape is usually in addition to sand cover and should be at least 150 mm above the cable.

Some cables are armoured, being wrapped in steel tape. Others have a light covering of hessian over the lead or aluminium sheath, but many are merely plastic coated, this being easily damaged by mechanical excavators or pierced by a jack hammer or pick axe.

Cable markers indicating the position of a buried cable are found in exceptional cases but normally there is no external indication that a live cable is beneath the ground surface. Pilot holes can establish the direction of a cable run but it must be remembered that the depth can fluctuate. Equipment is available for locating buried cables and pipes and ranges from metal detectors to devices which respond to:

(a) signals from an energized power cable;
(b) radio frequency signals re-radiated from a metal pipe or conductor below ground;
(c) signals induced into a buried metal pipe or cable from a portable generator.

See Chapter 11 for location methods.

Clearly, if there is any doubt as to the whereabouts of an existing cable then assistance must be obtained from the Electricity Company.

When demolishing a building, first make sure that the electricity supply has been disconnected as the removal of electricity meters does not necessarily mean that the supply cable has been cut off from the mains.

Do not handle or attempt to alter the position of any cable and if a cable is damaged or uncovered, notify the Electricity Company immediately. A site foreman should keep readily available a list of emergency telephone numbers for all services.

4 Overhead electricity lines

These are normally un-insulated and if a metallic object is brought into contact or even into close proximity with an overhead conductor then an electric current will discharge through the object to earth. It follows that before work commences on a site the position of all overhead lines should be noted. If a line passes overhead, then precautions should be taken to prevent crane jibs, scaffold poles, or vehicle tipper bodies getting too close. One method is to erect 'goal posts' which will channel plant along a particular route and through the posts.

Additional warning signs should be erected where a plant operative's view of the posts could be obstructed.

As it is difficult to estimate the height of overhead lines by observation it is preferable that vehicles are not driven below lines if an alternative route is available.

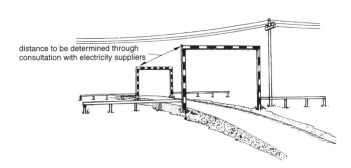

distance to be determined through consultation with electricity suppliers

Figure 9.1 Fences and 'goalposts' guarding approach to overhead lines

Tipping soil or material beneath overhead lines should be avoided as this reduces the ground clearance and increases the risk of contact. Also, it is preferable not to load or off load equipment or stack materials below, or in the vicinity of, overhead lines.

Leaving long metallic objects such as lighting columns parallel to the ground and insulated by trestles prior to erection beneath high voltage overhead lines, can result in a high voltage being induced in the object. A person touching this object could then earth an electric current and instances are known of serious shocks being received. The greatest risk when working under power lines is from tipping vehicles, lorry mounted loaders, mast lighting, mobile elevating work platforms (mewps) and single telescopic boom extending work platforms (cherry pickers).

(c) Excavations

Many dangers arise from excavation work and there are numerous regulations in existence designed to prevent workers from working in a hazardous situation.

The principal danger is that of the collapse of the sides of an excavation; 25% of all reported collapse incidents in recent years were fatalities. No one can be sure of the strength of the walls of an excavation and almost all ground will collapse under certain conditions. An excavation should be properly supported or the sides should be sloped

back to a safe angle as soon as the excavation reaches a depth where workers could be buried or trapped should a collapse take place. The Construction (Health, Safety and Welfare) Regulations require that adequate and suitable material is to be used to prevent danger from falls or dislodgement of the sides of an excavation or materials adjacent to it where a person is liable to be struck or buried by a fall of earth.

It may well be necessary to timber even shallow excavations.

Workers particularly at risk are those who take a chance and do work in advance of timbering, such as holding a boning rod from within the trench or 'bottoming up' the base of the trench. The first person in to erect the timbering or other forms of support requires protection by some temporary device. The methods and equipment used for trench support are shown in detail in Chapter 4 on Drainage.

Prior to commencing an excavation, the area must be checked with detection equipment (*see* Chapter 11, Section 11.26) and the route taken by the various underground services indicated with paint marks on the surface. After the surface has been broken by compressor tools or an excavator bucket, careful probing with hand tools must be undertaken in the vicinity of services before recommencing to excavate with power tools. An excavator bucket must not be used within 0.5 m of the suspected location of gas pipes or power cables. Detection equipment should be repeatedly used as the excavation progresses.

When excavating it is possible to dislodge or pierce a gas pipe and in some instances gas has been smelled percolating through the ground from old corroded pipes in the vicinity of an excavation. In every case the Transco emergency service must immediately be contacted and precautions taken to prevent the use of equipment that could 'spark off' an explosion.

(d) Noise

Personnel can easily be exposed to the risk of damage to their hearing when operating machinery on a construction site. The Noise at Work Regulations 1989 put legal obligations on employers to prevent damage to the hearing of workers from excessive noise at work. The employer is required to carry out a noise assessment where excessive exposure is likely, a rough guide being where people have to shout or have difficulty in being heard clearly by someone about 2 metres away. Action must then be taken to reduce the risks present.

Noise is usually measured in decibels (dB), shown on a meter filtered to reduce very low and very high frequencies as a dB(A) reading. Note that, as the decibel scale is logarithmic, an increase of 3 dB is actually a doubling of the noise intensity. A decibel scale of typical noise levels gives the following values:

0 The threshold of hearing
10 A quiet whisper

30 The tick of a watch
50 A normal conversation
70 A smooth-running machine
80 A noisy street – hearing damage starts
80 The first action level – an employer must provide hearing protection and attempt to reduce the intensity
85 The second action level – an employer must ensure that hearing protection is worn and ear protection zones are signed
100 A pneumatic drill
120 A rivet gun
130 The pain threshold

Clearly, the dangers of increasing deafness require all personnel to use methods of reducing the risks of excessive noise exposure. In addition to using hearing protectors, measures for quietening or muffling noisy equipment/tools or even of replacing them with quieter types need to be considered for most sites.

(e)   Use of Hazardous Substances

Many hazardous substances are used in or generated from work activities on construction sites. In roadwork, workers may be exposed to considerable health risks when using tar binders, bridge deck sealants, epoxy resins, etc., or from fume or dust generated when burning off yellow mastic road markings. The Control of Substances Hazardous to Health (COSHH) Regulations came into force in October 1989 and apply to any work activity involving exposure to harmful substances (other than lead, asbestos and ionizing radiations, which have specific regulations). Under these regulations, an assessment must be made of the risks present when any hazardous substance is used or produced. A sealant with a low risk level when used in the open air may present a serious hazard in a confined space, so the method of use must be considered when assessing the level of risk. Protective equipment may need to be used when applying such substances and its use must then be monitored to ensure that safe working practices are adhered to.

The results of assessments must be passed on to the workforce through briefings/training and management must periodically review the efficiency of the measures taken.

9.3.2   The Safety of a Roadworker from Another's Actions

Many dangers arise from the thoughtless actions of workers unconcerned with the safety of their workmates. Some of the more common and those resulting in a high incidence of accidents are listed below:

(a) Treading on nails projecting from loose timber is a very commonplace site accident. When timber framing or shuttering is dismantled, nails should be removed or knocked flat or the timber should be stacked where it cannot be trodden on.

(b) Failure to replace guarding equipment after carrying out work which necessitates its removal.

(c) Careless operation of machinery without ensuring proper safeguards such as help in reversing, or use of a banksman with cranes or earthmoving machinery.

(d) Failure to report defects in machinery and equipment, resulting in unsafe plant being used by others.

(e) Creating makeshift and unsafe situations such as sub-standard electrical connections and trailing cables, etc.

9.3.3   The Protection of Individuals from the Consequences of their own Actions

When workers injure themselves through their own actions the cause is usually either ignorance of the work process involved or just carelessness. It follows that a worker must be trained or made familiar with any techniques before being allowed to begin work. Many accidents could be prevented with the use of protective equipment, such as:

(a) High visibility and reflective jackets for operatives working on the highway.

(b) Protective spectacles, goggles or visors when working with hammers and chisels or cutting machinery on concrete or stone.

(c) Face masks when cutting concrete or stone with disc cutters.

(d) Helmets when working below other personnel or in situations where objects are liable to fall from above.

(e) Ear muffs or plugs when using compressed air tools such as breakers and drills. Also when working with piling equipment.

(f) Gloves in manual activities such as kerbing, etc. Also for operating vibrating equipment, together with the provision of information on health hazards.

(g) Protective boots with steel caps and soles to prevent injuries through crushing and from treading on nails.

If plant and equipment is not used sensibly then a considerable risk is involved. For example, ladders should be set at the correct angle (one out for every four up) and always secured, preferably at the top but failing this by a person standing at the bottom, to prevent slipping.

Regulations require employers to assess manual lifting activities for the need for special safeguards to minimize the risk of back injury, etc. This will obviously necessitate, in many instances, the provision of lifting equipment for certain tasks at present carried out manually.

Finally, many highway workers are injured by unthinkingly reaching or stepping out from a protected work area into the path of an oncoming vehicle.

The techniques of protecting a road site are made clear in the next section.

9.4 Temporary Signing of Roadworks

Highway authorities, contractors and statutory undertakers have both a civil law, liability and indeed a requirement under the Health and Safety at Work Act 1974, to warn road users of obstructions on the highway caused by road works. They also have an obligation to remove the signs when the work is complete, as these in themselves constitute an obstruction and failure to comply with this requirement is an offence liable to a fine under the Highway Acts.

Furthermore, Highway Authorities also have powers under the New Roads and Streetworks Act 1991 and Highways Act 1980 to enforce safety measures on road works carried out by statutory undertakers. However, statutory undertakers should, and normally do, consult the Police and the Highway Authority in good time about the safety measures required in any specific situation.

Roadworks invariably affect traffic using a road by causing temporary restrictions, which may result in either partial or complete obstruction of the highway, whether on the verge, the hard shoulder, the footway, cycleway or carriageway. These works may affect the safety and free movement of vehicles, cyclists and pedestrians (particularly persons with mobility impairments). All reasonable steps should be taken to ensure that the effects of the roadworks on the public are reduced to a minimum, especially by giving clear and early warning of any obstructions.

The way in which a roadwork site is signed depends largely upon the following factors:

9.4.1 Correct Signs

The revised Traffic Signs Regulations and General Directions Act 2003 stipulates the design (shape, etc.) and sizes of both temporary and permanent signs. Any signs other than those described in the Act are illegal and a person erecting such other signs is liable to prosecution.

9.4.2 Positioning the Signs

The revised Chapter 8 (1991) 'Traffic Safety Measures and Signs for Roadworks and Temporary Situations' within the *Traffic Signs Manual* gives information as to the best positioning and use of temporary signs at the approaches to the works site, on site and at the ends of the roadworks. The chapter can be considered as a Code of Guiding Principles or good practice, and its recommendations should be carried out when the signing of any roadworks is being undertaken.

Chapter 8 should be read in conjunction with the NRSWA Code of Practice for the signing of utility and local authority roadworks. Also, the Highways Agency guidance for works on high speed roads which was released in 2002 should be consulted.

Risk assessments in accordance with the Management of Health and Safety at Work Regulations, are required in the activities of setting out, maintenance and removal for temporary traffic management.

9.4.3 Types of Roadworks

There are three types of work for which different standards of signing and marking are applicable. They are termed type A, type B and type C and are briefly described as follows (*see* Figure 9.2):

Type A works are those which remain in all traffic flow and visibility conditions. Static signs placed on the highway are used.

Type B works are those which are allowed to remain in operation when there is little traffic and visibility is good. Static signs placed on the highway are used. However, if the works cannot be completed during daylight, then normal traffic conditions must be reinstated overnight, or additional signing installed to bring it up to type A standard.

Type C works are similar to type B, but the traffic signs used can be vehicle-mounted (usually requiring some static signs in addition). These works are carried out in good visibility and generally only in daylight. They include continuous mobile operations, as well as those which involve movement and periodic stops, such as gully emptying or line marking. They also include minor works, carried out either from a single vehicle or from a small number of vehicles.

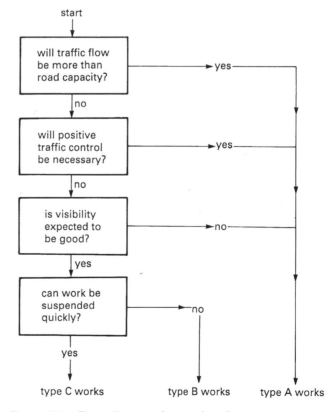

Figure 9.2 Flow diagram for roadwork types

Usually work will fall into type A category, other than in the case of minor work on lightly trafficked roads or certain mobile operations. The following principles relate to type A works unless otherwise specified.

9.5 Principles of Good Signing at Roadworks

The need to provide early and adequate warning of any obstruction cannot be emphasized too strongly.

Signs are intended to:

(*a*) *warn* the driver of a hazard ahead;
(*b*) *inform* the driver as to the nature of the hazard;
(*c*) *direct* the driver around the hazard;
(*d*) *protect* the personnel working, from the vehicles passing by the site (*see* Section 9.5.5, Safety Clearances and Zones).

Contractors should display on site, for the benefit of the passing public, an information board giving the name of the contractor/undertaker together with an emergency telephone number.

9.5.1 Common Signs

The first sign a driver should see is the *warning* sign 'roadworks ahead' (Figure 9.3(*a*)). This is normally followed by an *information* sign showing the road narrows and is partially obstructed either on the left or the right-hand side (Figure 9.4). Near the obstruction, mandatory keep right (or left) blue and white arrow signs (Figure 9.5) direct the driver past the coned off obstruction. Finally, the driver passes the first sign repeated together with an 'end' plate indicating the end of the roadworks (Figure 9.3(*b*)).

This procedure is varied when direct traffic control is necessary. A sign indicating traffic lights or the sign indicating traffic control (for Stop/Go boards) (*see* Figure 9.6) is positioned after the first warning sign described above together with the addition of a 'single file traffic' plate under the road narrows sign (Figure 9.7).

Figure 9.6

If traffic lights are used a white paint line or a sign (Figure 9.8) is placed to stop traffic in advance of the lights. A sign is preferable to a line when it is a temporary site. If a complete traffic lane is closed then the red and white board (Figure 9.9) should be used in conjunction with the blue and white arrow. Note: Stop/Go boards (Figure 9.10) must be 900 mm in diameter.

Figure 9.7   Figure 9.8

Figure 9.9 Barrier

Figure 9.3(a)
Roadworks ahead

Figure 9.3(b)
End of roadworks

Figure 9.4 Road narrows on left

Figure 9.5 Keep right

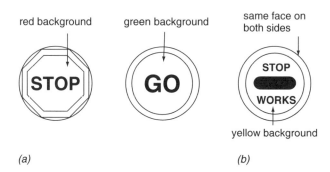

(a)   (b)

Figure 9.10(a) and (b)

9.5.2 Traffic Control

Some form of traffic control is required when single file traffic becomes necessary. Give and take, Stop/Go boards, portable traffic lights and priority working are described, together with the conditions controlling their use.

Single file traffic will be necessary if the available width of unobstructed carriageway is below 5.5 m (6.75 m if there is a bus route past the works site). In these circumstances the width is further reduced to 3.7 m by bringing out the cones and implementing some form of traffic control. The desirable minimum width for single file traffic is 3.25 m. However, single file traffic can 'shuttle' (self-control by traffic) past the roadworks in certain situations on very lightly trafficked roads where less than 20 vehicles may be counted over a 3 minute period (i.e. 400 vehicles per hour). Otherwise, the decision as to whether Stop/Go boards will be used must depend mainly on whether the work will be kept open throughout the hours of darkness or, if only a daylight site, on the density of traffic flow and length of site.

Stop/Go boards may be used when the relationship of the two-way traffic flow to the length of the site is no more than the limits shown in Table 9.1. If the site is not more than 20 metres in length then only one Stop/Go board is usually necessary. If the work is of extremely short duration, then the stop/works board may be used to halt traffic in both directions (see Figure 9.10(b)).

Site length (m)	Maximum two-way flow	
	Vehicles/3 minutes	Vehicles/hour
100	70	1400
200	63	1250
300	53	1050
400	47	950
500	42	850

Table 9.1   Site length related to traffic density of Stop/Go boards

Portable traffic signals may be used for most sites which are 300 metres or less in length. Using portable traffic signals requires authorization from the Highway Authority and they must be correctly set up (see details in Section 11.24). Traffic lights and Stop/Go boards are considered to be direct forms of traffic control. An indirect form of traffic control, sometimes used on lightly trafficked roads where both ends of the site are clearly visible, is known as 'priority working'. This may be used where the two-way traffic flow is less than 42 vehicles/3 minutes and the works extend for no more than 80 m (inclusive of taper). The signs shown in Figures 9.11 and 9.12 indicate priority to traffic from one direction.

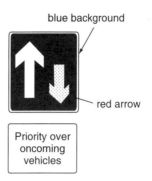

Figure 9.11   Priority over vehicles from the opposite direction

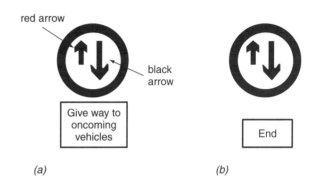

Figure 9.12   (a) Priority to vehicles from the opposite direction and (b) indication of end of priority

9.5.3   To Determine Sizes and Distances when Positioning Signs and Cones

Figures 9.13 to 9.16 give the general layouts for various situations. The detailed information needed to find the sizes and placement distances of signs together with spacing of cones is given in Tables 9.2 and 9.3.

An additional 'Lane Closed' barrier is required at the end of the 45° taper in Figures 9.14, 9.15 and 9.16, due to the distance L between the end of the taper and the barriers at the roadworks.

9.5.4   Notes on Signs and Cones etc.

(*a*) Signs and cones must be covered with reflective material and hence have to be kept clean otherwise their clarity, and reflective qualities are lost.

(*b*) Cones are normally spaced at 9 m centres, but 3 m centres are usual on low speeds roads. No less than 2 cones should be used in any length between tapers.

(*c*) Traffic cylinders can be fitted into the base of existing reflecting road studs or into purpose-made bases. These are only used to separate opposing flows of traffic.

(*d*) Flat traffic delineators (blades which fit into heavy bases) have the same elevation as cones and may be used as such on dual carriageway roads.

(*e*) 'Lead in tapers' used with positive traffic control and all exit tapers, should be 45° to the kerb line having 4 cones evenly spaced.

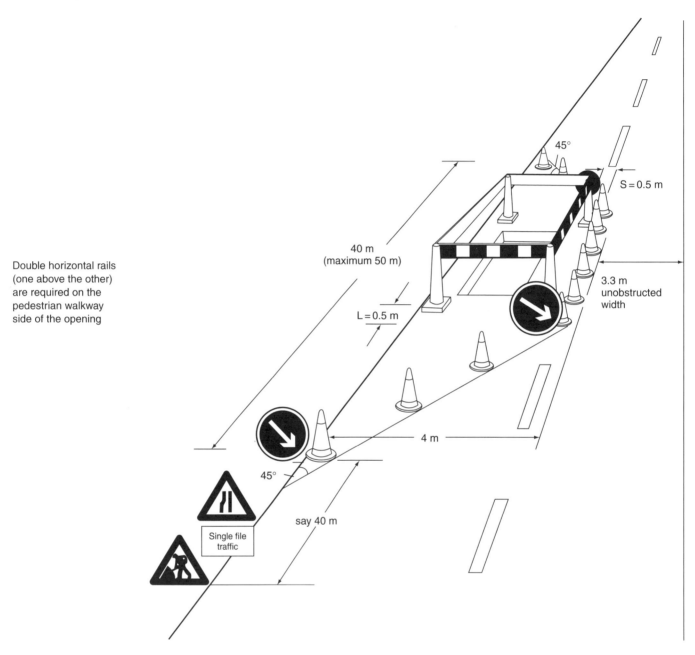

Double horizontal rails (one above the other) are required on the pedestrian walkway side of the opening

Figure 9.13   Give and take system
Case 1 Daylight working – example

Data:
1   Two-way traffic flow of 15 vehicles/3 min (i.e. less than 20)
2   Approximately 12 heavy goods vehicles/hour (i.e. less than 20)
3   Speed limit 30 mph
4   Length of works 40 m, width 4 m
5   Both ends of site clearly seen by drivers

Note: Signs only shown on one side of road.

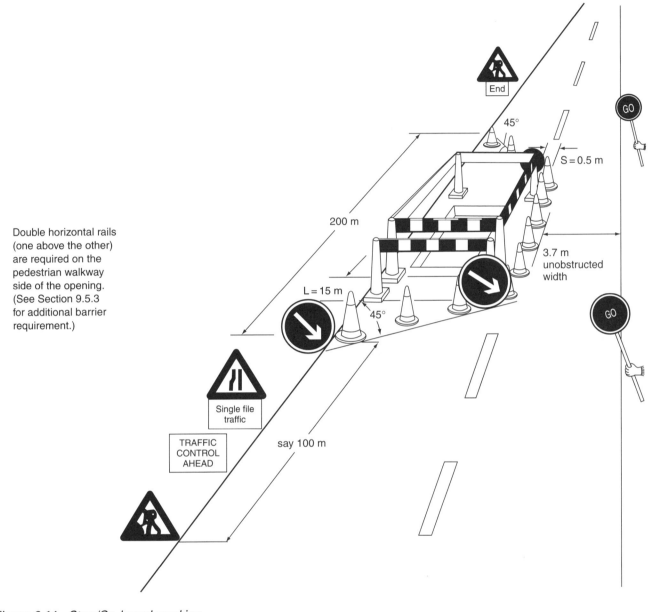

Double horizontal rails (one above the other) are required on the pedestrian walkway side of the opening. (See Section 9.5.3 for additional barrier requirement.)

Figure 9.14   Stop/Go board working
Case 2 Daylight working – example

Data:
1   *Two-way traffic flow of 50 vehicles/3 min*
2   *Speed limit of 40 mph*
3   *Length of works 200 m*

Notes: *Two Stop/Go boards are needed as site is over 20 m long.*
    Roadworks end board is placed if length of works exceeds 50 m.
    Signs only shown on one side of road
    Stop and Go boards are particularly useful for mobile works, but are not generally considered for long term works, because of the cost comparison between two operatives and the hire of temporary traffic lights.

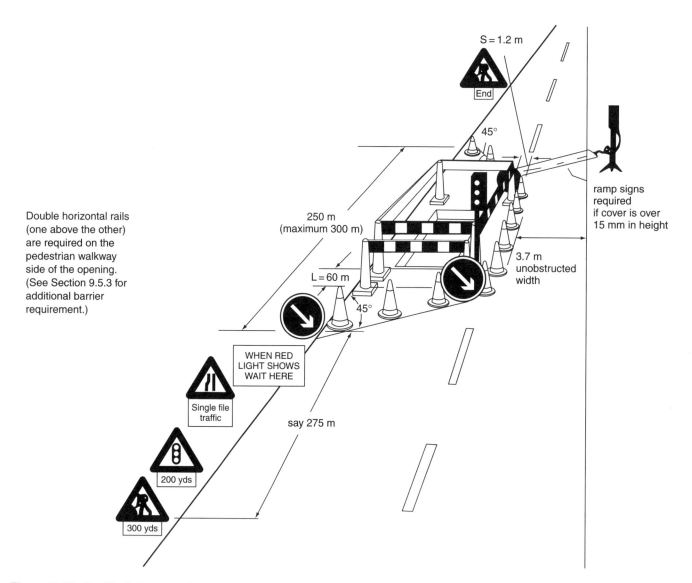

Double horizontal rails (one above the other) are required on the pedestrian walkway side of the opening. (See Section 9.5.3 for additional barrier requirement.)

S = 1.2 m

45°

ramp signs required if cover is over 15 mm in height

250 m (maximum 300 m)

3.7 m unobstructed width

L = 60 m

45°

WHEN RED LIGHT SHOWS WAIT HERE

say 275 m

Single file traffic

200 yds

300 yds

Figure 9.15   Traffic light control
Case 3 Day and night working – example

Data:
1   Two-way traffic flow of 80 vehicles/3 min
2   Speed limit 50 mph
3   Length of works 250 m

Note: Signs only shown on one side of road.

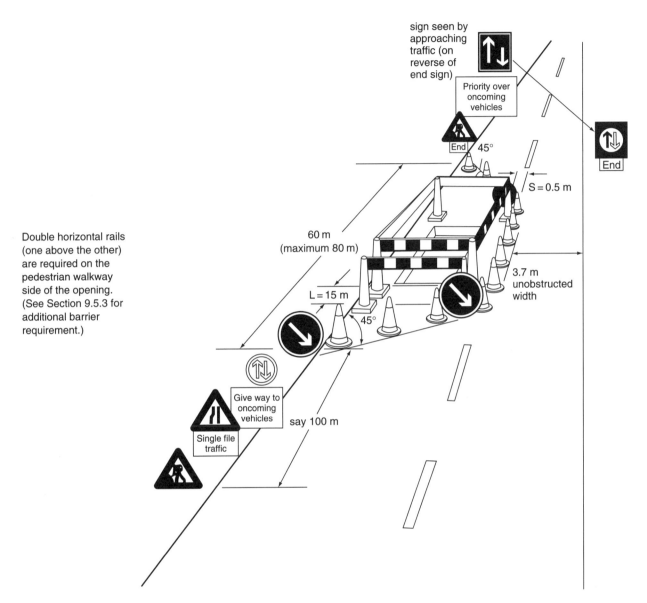

sign seen by approaching traffic (on reverse of end sign)

Priority over oncoming vehicles

End 45°

S = 0.5 m

Double horizontal rails (one above the other) are required on the pedestrian walkway side of the opening. (See Section 9.5.3 for additional barrier requirement.)

60 m (maximum 80 m)

3.7 m unobstructed width

L = 15 m

45°

End

Give way to oncoming vehicles

Single file traffic

say 100 m

Figure 9.16   Priority working
Case 4 Daylight working – example

Data:
1   Two-way traffic flow of 30 vehicles/3 min
2   Speed limit of 40 mph
3   Length of works 60 m
4   Both ends of site clearly seen by drivers from 60 m before cones, extending to 60 m beyond site

Note: Signs only shown on one side of road.

Type of road		Width of hazard (metres)						
		1	2	3	4	5	6	7
All purpose single-carriageway road, urban, restricted to 30 mph or less	Length of taper (T) in metres	13	26	39	52	65	78	91
	Minimum no. of cones	4	4	6	7	9	10	12
	Minimum no. of lamps at night	3	3	5	6	8	9	11
All purpose single-carriageway road, restricted to 40 mph or less	Length of taper (T) in metres	20	40	60	80	100	120	140
	Minimum no. of cones	4	6	8	10	13	15	17
	Minimum no. of lamps at night	3	5	7	9	12	14	16
All purpose dual-carriageway road, restricted to 40 mph or less	Length of taper (T) in metres	25	50	75	100	125	150	175
	Minimum no. of cones	4	7	10	13	15	18	21
	Minimum no. of lamps at night	3	6	9	12	14	17	20
All purpose single-carriageway road, with speed limit of 50 mph or more	Length of taper (T) in metres	25	50	75	100	125	150	175
	Minimum no. of cones	4	7	10	13	15	18	21
	Minimum no. of lamps at night	3	6	9	12	14	17	20
All purpose dual-carriageway road, with speed limit of 50 mph or more	Length of taper (T) in metres	32	64	96	128	160	192	224
	Minimum no. of cones	5	9	12	16	19	23	26
	Minimum no. of lamps at night	4	8	11	15	18	22	25

Table 9.2 Details of lead-in tapers for roadworks not having direct traffic control

Type of road	Minimum siting distance (D) of first sign in advance of works (metres)	Minimum clear visibility to first sign (metres)	Minimum size of signs (mm)	Minimum height of cones (mm)
All purpose single-carriageway road, urban, restricted to 30 mph or less	20–45	60	600	450
All purpose single-carriageway road, restricted to 40 mph or less	45–110	60	750	450
All purpose dual-carriageway road, restricted to 40 mph or less	110–275	60	750	450
All purpose single-carriageway road, with speed limit of 50 mph or more	275–450	75	750	450
All purpose dual-carriageway road, with speed limit of 50 mph or more	725–1600	105	1200	750

Notes: 1 On roads with speed limits of 50 mph or more all advance signs should have plates giving the distance to the works in yards or miles.
2 The range of siting distance (D) is given to allow the sign to be placed in the most convenient position bearing in mind available space and visibility for drivers.

Table 9.3 Siting distances and sizes of signs for roadworks

Permanent signs made temporarily redundant by the roadworks should be covered, but the temporary road signs must not accidentally obscure other permanent signs.

9.5.5 Safety Clearances and Zones

(*a*) The 'works area' is the actual excavation, chamber opening, etc., where work is being undertaken.

(*b*) The 'working space' is the space around the works area needed to store tools, excavated material, equipment, plant and enough room for movement of personnel.

(*c*) The minimum sideways clearance between a working space and that part of the carriageway which is being used normally by traffic should be 1.2 m for high speed roads and 0.5 m for other roads.

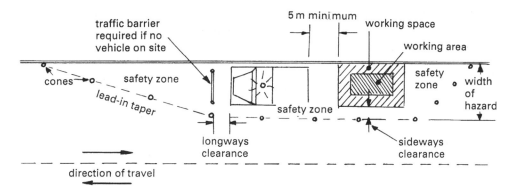

Figure 9.17 Safety clearances and zones (with works vehicles)

(*d*) The minimum longways clearance between the end of the lead-in taper of cones and the working space varies with the speed limit.

Speed restriction (mph)	Minimum longways clearance, L (metres)
30	0.5
40	15
50	30
60	60
70	100

(*e*) These clearances, together with the coned tapers, constitute a safety zone into which men and plant should not normally enter and in which materials should not be deposited.

(*f*) Excavations must be fenced off along the sides where pedestrians might have access. Use 'pedestrian barriers' and link them where necessary to form a continuous barrier.

(*g*) Signing and coning on the carriageway is also necessary when working on the verge, footway, island or central reserve within 0.5 m of the carriageway (1.2 m for high speed roads).

9.5.6 Illuminating Roadworks

(*a*) After darkness, lamps having a yellow body and showing a steady amber light, should be placed mid-way between cones to mark out the works. Common practice is now to utilize a cone as a lamp stand, which effectively halves the cone spacing.

(*b*) Where there is good street lighting and a speed limit of 40 mph or less, lamps which flash at 55–150 times per minute may be used if within 50 m of a street lamp. In all other cases, static lamps should be used.

(*c*) High intensity flashing warning beacons which give an intermittent amber light must only be used to draw attention to the particular hazard existing on a site. They are not to be used to mark out the works.

(*d*) If a sign is placed within 50 m of a street lamp and the speed limit is 40 mile/h or more – the sign must be individually illuminated.

(*e*) Stop/Go boards must be directly illuminated if used at night.

9.5.7 Miscellaneous Notes

(*a*) Railway Level crossings – if it is intended to work in close proximity to, or on a level crossing, then reference should be made to Chapter 8 of the Traffic Signs Manual. Also, it is essential that contact is made with the appropriate Rail Authority, prior to work commencing on site.

(*b*) Supplementary plates are attached to roadwork signs to give:

1 *Distances to* roadworks on high speed roads only, i.e. traffic speeds over 50 mph (*see* Figure 9.18(*a*)).

2 *Distance over* which hazard extends, i.e. Figure 9.12(*a*) (used at site of roadworks) and also Figure 9.18(*b*).

3 The nature of the work being undertaken, i.e. 'grass cutting', 'gully emptying', 'line painting', etc. See Figure 9.18(*d*).

4 Other information, i.e. Figure 9.7, Figure 9.12(*b*) or an arrow to indicate which direction the roadworks are when they are near a road junction. See Figure 9.18(*c*).

(*c*) Road closure and diversion signs should only be placed under the direction of an engineer.

The 'Road Closed' sign is used to indicate that the road beyond the point at which it is placed is closed to traffic. The 'Road Ahead Closed' sign is used at the entrance to a road which, because of roadworks cannot be used by

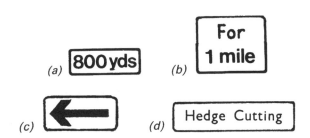

Figure 9.18(a)–(d) Additional information plates

through traffic. *See* Figures 9.19 and 9.20. Diversion signs may be required as they have to be used if there is only one alternative route available to traffic. Diversion signs (black letters on yellow background) are shown in Figures 9.21(*a*), (*b*) and (*c*); 9.22 and 9.23.

(*d*) A warning sign 'Ramp Ahead' should be placed not less than 30 m before a ramp and a 'ramp' sign must be at the ramp. A sign marked 'Temporary Road Surface' should be used to warn drivers of an unfinished road surface. (*See* Figure 9.24.)

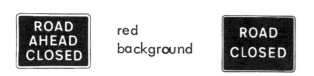

red background

Figure 9.19   *Figure 9.20*

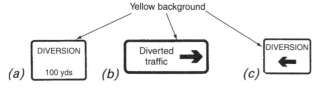

Figure 9.21 (a) Sign indicating distance to start of traffic; (b) sign telling diverted traffic to turn right; (c) repeat sign for diverted traffic

Figure 9.22 Direction sign for diverted traffic   *Figure 9.23 Diversion ends sign*

Figure 9.24 Yellow background

9.5.8 Safety of Pedestrians

(*a*) Pedestrians must be provided with an alternative safe route if the footway is closed. The temporary pedestrian way should never be less than 1 m in width; wherever possible, 1.5 m should be provided.

(*b*) Make sure that pedestrians are not diverted on to an unguarded carriageway.

(*c*) Special care is necessary when pedestrians are forced on to the carriageway and properly constructed

rigid barriers (clearly marked in red and white chequers) should be used between the pedestrians and traffic flow and around the working area. (*See* Figure 9.25.) The appropriate sideways clearance must be allowed between the barriers and the cones to give a safety zone on the traffic side.

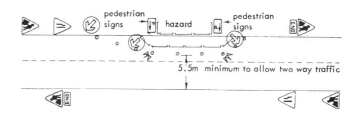

Figure 9.25 Layout of temporary footway in carriageway diversion up to 400 m maximum

(*d*) Pedestrians should be provided with the signs shown in Figures 9.26 and 9.27. Additional signs are 'Pedestrians look left/right'. Two signs should be used at each location. They should be double faced and should be provided at each edge of the carriageway. Pedestrians should not be directed with road signs, i.e. blue and white arrows, etc.

(*e*) When pedestrian crossings are temporarily taken out of use the sign shown in Figure 9.27 is used.

(*f*) Visually impaired persons: all excavations on or near footways must be guarded with secure barriers to give warning and protection to these persons.

(*g*) Provide ramps for wheelchairs, etc., if necessary on the temporary footway.

Figure 9.26 This is used where the new route is not obvious (arrow reversible)

Figure 9.27 This sign is attached to the beacon posts and faces across the carriageway

9.5.9 Check List for Signing Roadworks

Ensure that a risk assessment has been undertaken prior to commencement of works.

1 Find out if roadworks will require traffic control and whether the work will remain open during the night.

2 If traffic control is not required, find the length of the coned taper and determine the number of cones required.

3 If traffic control is required, determine the type of control and if necessary gain authorization, then find the type and number of signs and cones required.

4 Decide how pedestrian traffic will be maintained and if necessary make provision for this with the appropriate signs and barriers.

5 Collect equipment and have some idea of the appropriate spacing distances before arriving on site.

6 Wear high-visibility garments and use your vehicle as protection (flashing beacon to be switched on).

7 Set out 'Roadworks Ahead' sign first and work in the direction of the traffic flow.

8 Use the footway or verge if it is necessary to carry signs to their location. Place signs on the verge if possible or if not place on the footway or carriageway with a cone at foot (lamp at night). Weight signs with sandbags.

9 If site is to remain open at night, decide how many lamps are needed.

10 Ensure that any misleading permanent signs have been covered.

11 Now drive through the job to check the signs from the road user's point of view.

12 If circumstances change, alter the signs, cones and lamp positions to suit.

13 Make sure signs, cones and lamps are regularly cleaned, maintained and replaced.

14 Ask for advice if situations occur which are more complicated and you are uncertain as to the layout.

15 Finally, remove all signs, cones and lamps when the work is completed.

9.5.10 Minor Works on Lightly Trafficked Roads (often type B works)

The advice already given covers most work situations, but for minor works on lightly trafficked roads when using a conspicuous vehicle, a lower standard of signing may be appropriate.

1 When there is a conspicuous vehicle on the approach to the works, a traffic barrier need not be placed provided that drivers can see the vehicle clearly from at least 25 m.

2 If the vehicle has a roof-mounted beacon in continuous use and

(a) the total two-way traffic flow is less than 20 vehicles/min

(b) less than 20 heavy goods vehicles pass the site per hour

(c) the speed limit is 30 miles/h or less and

(d) the beacon can be clearly seen from at least 50 m in either direction

then it is unnecessary to use the 'Roadworks Ahead' and 'Road Narrows Ahead' signs in advance of the works. However, the taper of cones together with the 'keep right/left' signs must be used.

3 If the available width for traffic is reduced below 5.5 m, 'shuttle working' (i.e. 'give and take' without control) may be used for single-file traffic using the 3.15 to 3.7 m carriageway. However, the length of the works from the start of the lead-in taper to the end of the coning must not exceed 50 m and the drivers approaching from either direction must be able to see both ends of the site.

9.5.11 Mobile Works (type C works)

These include continuous mobile operations, as well as those which involve movement and periodic stops, and short-duration static works, often to public utilities, carried out from a single vehicle or a small number of vehicles.

(a) Using a Single Vehicle

Carry out work when there is good visibility and during low traffic periods.

The vehicles should be conspicuously coloured, have one or more roof-mounted beacons operating and display a 'keep right/left' sign to drivers approaching on the same side of the carriageway.

Additional static signs must be used:

1 when the works vehicle can't be clearly seen (i.e. bends or hills present);

2 in traffic build-up;

3 when the vehicle is slow moving or needs to make periodic stops;

4 where space for traffic moving past the vehicle is restricted.

These signs in both directions are 'Roadworks Ahead' with appropriate plate and 'Road Narrows' sign with 'Single File Traffic' plate. Work should not be carried out further than 1 mile from these signs.

(b) Using more than One Vehicle

Mobile lane closure techniques may be used when the work can't be carried out at normal road speeds in the right-hand lane of a two-lane dual carriageway or the middle and/or right-hand lane(s) of a three-lane dual carriageway.

The technique uses a convoy of vehicles carrying road signs. Further details are given in Chapter 8 of the *Traffic Signs Manual*.

9.5.12 Motorways and High-speed Dual Carriageways

Signing standards are more rigorous for these roads, but as far as the number and size of signs are concerned, *Traffic Signs Manual* Chapter 8 standards should be adopted, although special arrangements usually exist for coning on motorways.

Because of the complexities of the specialized signing layouts required on these roads, only a brief outline of some of the more important aspects is given.

Reference should also be made to the supplementary guidance released by the Highways Agency in 2002, for roadworks on high speed roads.

Signs, Signals and Road Markings

Signs should be provided at all sites to inform the public about the nature of the work being undertaken and appropriate advisory speed limit signs should be erected at all lane closures. Mandatory speed limits are only likely to be effective if there is an obvious police presence. Advance warning signs should be provided at 3 miles (if queuing is expected) and in all circumstances at 2 miles with motorways and 1 mile for major dual carriageways. It is important to remove any lengths of road markings that conflict with temporary traffic lane diversions. Old markings should be deleted or covered over and new ones provided.

Motorway signals should only be used when setting up/dismantling a roadworks site, otherwise they should be kept in reserve to deal with possible incidents.

Some additional signs used to give advance warning of closure of traffic lanes are shown in Figure 9.28(*a*), (*b*) and (*c*), together with a sign indicating that a diversion of traffic to the other carriageway is ahead in Figure 9.28(*d*).

Typical layouts in Figure 9.29 show the positioning of signing.

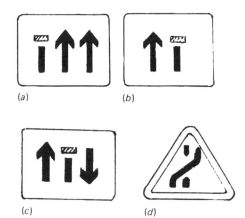

(a)   (b)

(c)   (d)

Figure 9.28

Rotating reflector delineators are now in use. These are simply rotating amber reflectors mounted on top of traffic cones. They are not road danger lamps but are often used between lamps in the line of cones on the outer edge of a safety zone. Their use needs authorization by the DfT.

Setting Up and Dismantling Signing

Procedures follow the principle that all setting up of signing is carried out with the traffic flow, commencing with the first advance sign and finishing with the 'end of roadworks' sign.

Figure 9.30(*a*) shows a closure on the lane adjacent to the central reserve. In method A the signing vehicle is driven along the hard shoulder, allowing personnel to place signs and cones progressively both sides of the carriageway. Alternatively, method B may be used, in which the vehicle crosses the carriageway after the taper is placed and hence more easily off-loads cones as it is driven along the outer lane.

Removal of signs and cones is carried out in two steps, with the coned area being removed against the traffic flow (*see* Figure 9.30(*b*)), followed by the collection of signs with the traffic flow (*see* Figure 9.30(*c*)). Police assistance is often necessary when both setting up and dismantling signing, particularly when vehicles collecting signs are carrying out the U-turn manoeuvre.

The illustrations only show setting up and dismantling signing in straightforward lane closure configurations, whereas more complex situations involving contraflow or junctions (access and egress points) need to be considered individually and planned by police and motorway engineer.

Traffic Management Considerations

As roadwork schemes are always a compromise between getting the work done as quickly as possible and keeping the traffic flowing, it is essential to plan the activities and ensure liaison between police and adjacent Highway Authorities.

Maintenance works should be undertaken in the shortest time, taking up the minimum of road space and when a length of road is closed the opportunity should be taken to carry out all other maintenance on that carriageway.

Work in peak hours and on peak days should be avoided if possible and dual three lane carriageways should not be reduced to one lane except as a last resort and then only in off-peak periods.

Careful attention should be given to arrangements for safe access and egress by works traffic.

Traffic flows at all sites should be continually monitored so that if problems develop, the police can be asked to take appropriate action, which could include activating the emergency matrix signals on motorways.

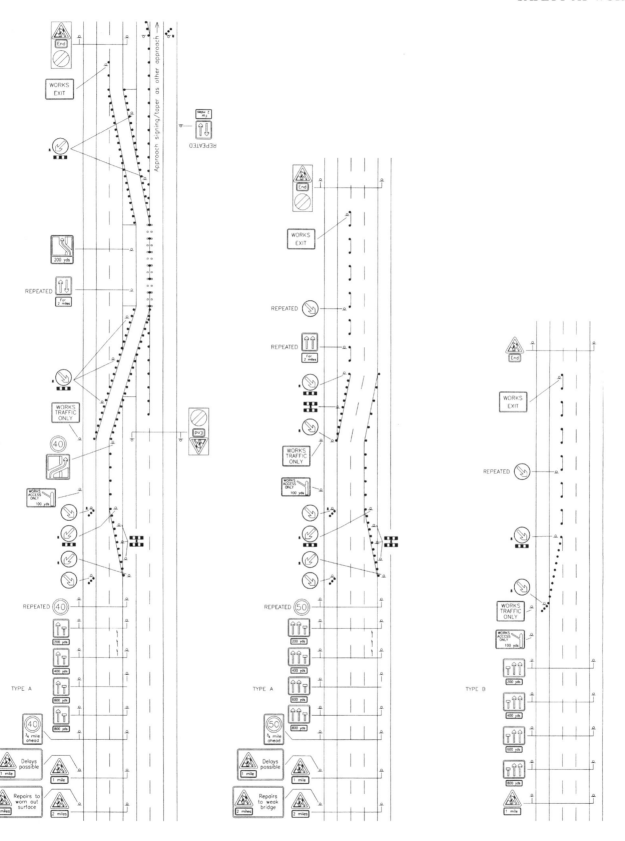

Figure 9.29   (left) High-speed dual carriageway – one carriageway closed (middle and right) Motorway – one lane closed

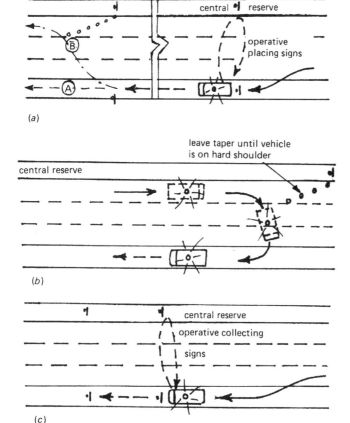

(a)

(b)

(c)

Figure 9.30 (a) Placing signs and cones (b) Removing cones, etc. from outer lane (c) Removing advance signing

9.6 Spillage of Chemical Substances on the Highway

Roadworkers are often first on the scene after a chemical spillage has occurred as the result of a road accident. The dangers of approaching the spillage of an unknown substance cannot be overstressed.

The following guide may be helpful.

1 Approach no closer than 200 m unless it is obvious that the tanker is carrying a material which is not harmful to personnel, such as a milk tanker. Otherwise it should be assumed that the spilt material is a dangerous chemical. Any person who has to approach the vehicle in order to identify its contents should do so from upwind and uphill.

2 Endeavour to keep other people and vehicles the same distance away from the spillage.

3 Arrange for the police to be called.

4 Do not attempt to clear the substance until the police or surveyor has taken advice as to how the substance should be treated and what safety and other site precautions have to be taken.

The police control any such incident and co-ordinate the activities of the other emergency services. The Fire Service has the basic responsibility of containing the spillage and where practicable rendering it harmless. Highway personnel may be called in to assist in containing the spillage and preventing it from seeping into the drainage systems and waterways until the chemical can be retrieved with specialist equipment. In these circumstances detailed instructions will be provided. Temporary road signing may be needed to set up diversionary routes.

To make the maximum information available to emergency services, vehicles carrying chemical substances, i.e. tankers, display placards giving information as to whether the contents are explosive, toxic, corrosive, flammable, radioactive or a combination of all five.

The European Agreement concerning the International Carriage of Dangerous Goods by Road (ADR) require vehicles carrying dangerous goods to be placarded bearing the following information (*see* Figure 9.31):

1 The Emergency Action Code (for UK registered transport) or a Hazard Identification Number (HIN) for transport operating within the EEC.

2 The United Nations applied number to a given chemical or substance (found in ADR documents).

3 An emergency phone number where specialist advice can be obtained (UK only).

4 An internationally recognised warning diamond displaying the nature of the hazard carried.

5 A space for the manufacturer's logo (symbol) or name (UK only).

The Emergency Action Code is interpreted through the use of a reference card (*see* Figure 9.32) produced by the Fire Service Inspectorate and held by members of the emergency services and certain highway personnel.

In addition to a placard shown on the vehicle the driver must carry a Transport Emergency Card (Road) (TREM Card), inside the vehicle outlining: load definition, nature of danger, personal protection, general and any additional or special actions to be taken by driver, what to do in the event of a fire, first aid and any other additional information.

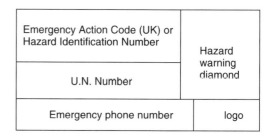

Figure 9.31 Vehicle placard

EMERGENCY ACTION CODE			
Code	1 COARSE SPRAY 2 FINE SPRAY 3 FOAM 4 DRY AGENT	NB: Treatment necessary to prevent or deal with fire	
P	V	LTS	DILUTE
R			DILUTE
S	V	BA & FIRE KIT	DILUTE
T			DILUTE
W	V	LTS	CONTAIN
X			CONTAIN
Y	V	BA & FIRE KIT	CONTAIN
Z			CONTAIN
E		PUBLIC SAFETY HAZARD	

Figure 9.32 Emergency Action Code Card

Figure 9.33 Example of placard

Figure 9.34 Types of Hazchem diamond warning plate

NOTES FOR GUIDANCE

DRY AGENT – Water must not be allowed to come into contact with the substance at risk.

V – Substance can be violently or even explosively reactive, including combustion.

LTS – Liquid-Tight chemical protective suit with breathing apparatus (BA).

DILUTE – May be washed to drain with large quantities of water.

CONTAIN – Prevent, by any means available, spillage from entering drains or water course.

E – People should be warned to stay indoors with all doors and windows closed but evacuation may need to be considered. Consult Control, Police and product expert.

An Example in Identification

From reference card, 2WE is interpreted as:

2 – requiring fine spray to be used;
W – showing liquid-tight chemical protective suit with BA to be used;
E – people warned to stay indoors or evacuate.

The warning diamond in Figure 9.33 shows the nature of the hazard and some common examples are illustrated in Figure 9.34.

9.7 Safety in Sewers

The Confined Space Regulations 1997 consolidate the general requirements of Section 30 of the Factories Act 1961 and the general duty of care under Sections 2 and 3 of the Health and Safety at Work Act 1974.

The new regulations follow the current trend in health and safety legislation in seeking to eliminate the need for hazardous activities by finding alternatives to workers being sent into a confined space, unless it is not reasonably practicable to do otherwise.

Where the work does require entry, this should be made clear in a risk assessment and the control measures must be clearly stated.

The main considerations for work in a confined space are:

(*a*) breathable air
(*b*) adequate supervision
(*c*) adequate accessibility for entering and leaving safely
(*d*) making sure that further hazards are not introduced whilst work is in progress
(*e*) means are provided for removing an unconscious worker in an emergency
(*f*) adequate information, instruction and training has been provided for all those involved in the work.

A confined space is defined as 'any place, including any chamber, tank, vat, silo, pit, trench, pipe, sewer, flue, well or other space where there is a reasonably foreseeable specified risk'.

These specified risks are:

(*a*) injury due to fire or explosion
(*b*) loss of consciousness due to an increased body temperature (heat stress)
(*c*) asphyxiation.

When entering a sewer or similar situation, either to begin work or just to carry out an inspection, it is necessary that a strict procedure should be followed in order to minimize the risks. In some instances it is necessary to introduce a 'permit to work' system.

9.7.1 Safety Procedure Before Entering a Sewer or Confined Space

Ensure that a responsible person at the depot knows exactly where and when the gang will be working. All personal and gang equipment should be checked before leaving the depot. Typical equipment is listed below:

1 Personal Equipment

Bump hat/Safety helmet
Sweat band (for neck)
Lamp (intrinsically safe, cap or hand)
Safety belt or harness
Cotton based PVC gauntlets
Protective clothing
Thigh boots (rubber) with studded soles
Breathing apparatus: 'saver' set (Figure 9.35).

2 Gang Equipment

Detector lamps or detection equipment (to give warning of dangerous gases and of lack of oxygen).
Lead acetate papers and wetting agent (glycerol) to detect hydrogen sulphide (if using detector lamps).
Breathing apparatus: working or rescue set (Figure 9.36).
Safety apparatus:
    two 16 m × 16 mm life lines
    one 16 m × 8 mm handline
Lifting harness
Drag sheet
First aid kit
Hand lamp
Traffic signs and barriers.

3 Detection Lamps

(a) Spiralarm

This lamp must be lit 45–60 minutes before use. It is a flame type which burns with a higher flame in flammable atmospheres, causing a red warning lamp to light. When there is insufficient oxygen the lamp goes out. (*See* Figure 9.37.)

(b) Ringrose Mk 1

This is an electrical type which gives a warning light in the presence of flammable gases. This lamp will not show when there is insufficient oxygen unless this is due to displacement by a detectable gas. (*See* Figure 9.38.)

4 Electronic Detection Equipment

Equipment of this type has largely replaced detection lamps and electrical monitors which use detect cards.

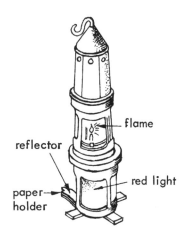

Figure 9.37 Spiralarm warning lamp

Figure 9.35 'Saver' set

Figure 9.36 Working or rescue set

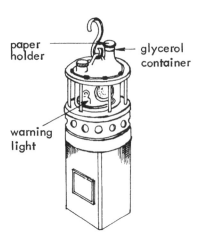

Figure 9.38 Ringrose warning lamp

Electronic 'gas' detectors are now available as compact and accurate instruments, but are subject to developing faults in a very demanding work environment.

A typical multi-gas monitor will test for explosive gases, high and low levels of oxygen and toxic gases.

1 A sensor tests for explosive gases, i.e. hydrocarbon gases and vapours including methane, petrol vapour, propane. This is done on a continuous two-minute cycle, but a manual override button can give a readout in 10 seconds.

2 Another sensor detects low and high levels of oxygen and tests continuously, giving a visible and audible alarm at less than 19% or more than 23% oxygen in air.

3 Toxic gases, such as hydrogen sulphide, etc., are monitored for both absolute level and time-weighted average. An alarm is given if the short-term exposure limit is exceeded or if an average level over the time for which the instrument has been switched on is exceeded.

The equipment, as shown on the left in Figure 9.39, is ready for use when the supply button is switched on and the automatic check sequence is completed. The autocheck illuminates each indicator lamp in sequence and completes the procedure by sounding the audible alarm for 2 seconds. If a fault exists the fault indicator together with the indicator for the faulty test circuit will operate and a continuous alarm sounds.

An additional meter is often used which will test specifically for carbon monoxide (Figure 9.39, right). This is a relatively simple device which when switched on gives a readout of carbon monoxide as parts per million (ppm) in air. The meter will alarm at 50 ppm giving both an audible and a visual alarm. Other single-function 'pocket-sized' meters are available for monitoring oxygen levels or detecting hydrogen sulphide.

Compared with lamps, no warm-up period is necessary and the meters are ready for use when switched on. The meters are lowered progressively down a manhole, pausing to measure at intermediate levels, before finally testing at a point 300 mm from the bottom of the manhole or surface of water. Both visible and audible alarms would operate if dangerous concentrations of gases were present.

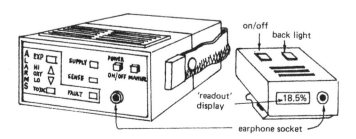

Figure 9.39 Multi-gas monitor and (right) carbon monoxide monitor

The meters should be attached to the lifting harness of the lead person, to remain with them whilst inside the confined space.

9.7.2 Site Procedure

(1) On site, safety barriers must be set up around manholes and warning signs placed where necessary.

(2) The ganger should locate the nearest telephone and check that it is in working order. (If it is not a public 'phone, permission should be obtained to use it.)

(3) The manhole which is to be entered should then be opened, together with the next manhole upstream and downstream (to create a draught and flush out foul air).

(4) Detector equipment should be used as previously described or alternatively detector lamps (already lit at the depot) should be checked and used as follows:

A lead–acetate paper should be moistened with glycerol/distilled water solution. It should be lowered with the lit detector lamp into the manhole to a point as near to the bottom as possible. It should be left in this position for at least 2 minutes, and carefully watched to see whether the red light shows. After 2 minutes plus, the lamp and paper should be pulled up and inspected. If all is well, the lamp will be alight and the paper should not be stained (brown stain indicates the presence of hydrogen sulphide).

(5) To try to detect unusual conditions, the air issuing from the open manhole should be sniffed. IF IN DOUBT, STAY OUT.

(6) Weather conditions must be considered (e.g. workers must not enter a sewer immediately after a severe storm).

(7) If the ganger considers the sewer to be safe, a top person (or more) should be posted and it must be ensured that this person clearly understands the following duties. The 'top person' must:

(*a*) *keep contact* with the workers underground. Every 5 minutes the working team should advise their 'bottom person' that they are all right. The 'bottom person' should pass this message to the 'top person'. All messages must be acknowledged. If the 'top person' fails to hear from the working team, he or she must call the rescue service;

(*b*) *never leave* the sewer unattended;

(*c*) *not go into* the sewer at all, without leaving a responsible person on top as a replacement;

(*d*) *report* signs of possible rain (upstream). In this event the ganger will call the team out of the sewer.

When entering the sewer the workers must always go as a team. The following person should wait until the person in front calls, before starting to go down.

9.8 General Precautions Whilst Working in Sewers

The first person down must take the detecting equipment. Should the flame be extinguished, the lamp must not be relit underground. All matches and lighters should be left on the surface.

Care should be taken in descending by ladder or step irons. A lifeline should be used when descending in difficult or dangerous conditions and also when moving in sewers having steep gradients or of large diameter. Remember when stepping off the ladder or step irons, that a large, fast flowing sewer can cause a person to lose their footing and be carried away. Also, if a sewer is almost level or blocked, there will be stagnant sludge in the bottom which can hold trapped gases. Workers should move slowly and carefully.

In the event of a worker having a feeling of sickness, dizziness, eye irritation, faintness, weakness, lack of control of limbs, breathlessness or noticing any unusual smell, the gang should immediately leave the sewer.

If conditions are normal, existing guard chains or bars should be checked and also it must be ensured that guard bars or chains are fixed downstream when working in large diameter sewers.

As a gang moves further away from the point of entry, additional members may be required to act as links between the gang and the 'top person'. Telephones or communication sets should be used if available.

When working in a narrow manhole with benching, operatives should not stand in the pipe invert, but straddle the sewer with feet on the benching and, if possible, wedged in position, In narrow sewers, stand with the feet apart, knees bent and braced against the walls, with one shoulder against the wall and the weight of the body forward. Then it is possible to move forward using a shovel or stave to assist balance.

Conditions in a sewer can change very quickly and the signs of possible risks are as follows:

(a) Flooding

Advance warning from the 'top person' or noise of approaching water; possibly an increasing flow of air in front of the flood water. (Waiting to see if the water rises may leave a person with insufficient time to get out.)

(b) Deficiency of Oxygen

A feeling of faintness or dizziness and a shortage of breath. A reduced lamp flame.

(c) Explosive Conditions

Danger can arise from operating a petrol or a diesel engine underground. A naked light must not be used in a sewer and it should be ensured that only spark-proof electrical equipment is below ground.

(d) Gas Build-up

Presence of strange smells and a feeling of sickness, dizziness, eye irritation, etc. Exhaust fumes can be sucked underground from vehicles or engines operating near an open manhole.

(e) Possible Infection

When persons are working on ironwork or a structure which has been fouled by sewage, the structure should be scrubbed or cleaned down with a high pressure hose and a strong disinfectant used. Any cut or scratch should be washed, treated with antiseptic and covered with a waterproof plaster.

On finishing work all equipment should be collected and tools lashed ready for lifting. The 'top person' should call the team out, one at a time, and check that everyone is out of the sewer. The tools should be hauled out and the rim of the manhole greased and the lid closed. Boots and gloves should be washed and then hands, neck and face in a thorough manner.

Finally, the depot should be informed that all the team are out of the sewer and then conditions in the sewer should be reported. This should include broken rungs, step irons, chains, bars, masonry, etc., and any abnormal circumstances such as presence of gas or low oxygen levels. The presence of rats and any damage to equipment or of personal injury (including minor cuts and scratches) should be noted.

9.9 Emergencies

Detailed rescue procedures exist for workers employed in sewers, but the following points are vital and should be understood by all engaged in highway drainage work.

9.9.1 Collapse of a Person in a Sewer

If a person collapses for no apparent reason, it must be assumed that person has been gassed. You should then put on your 'saver' set (if available) and place the collapsed person's mask on, but only if still breathing and provided that the mask can be replaced before it runs out, say a maximum of fifteen minutes.

Then prop up the casualty to prevent any possibility of drowning and report to the 'top person'. Get out carefully, as quick movements might stir up more gas and you will certainly use more oxygen.

Do not go back for the casualty unless you have breathing equipment such as the working or rescue set. From the time of collapse, there may be no more than ten minutes to get this person out and if you stay or go back without breathing equipment, you too will collapse. Then there will be two people to get out, not just the one.

Most authorities rely heavily on the fire and ambulance services but anything you can do to prepare for the rescue before the ambulance and fire brigade arrive will help.

Obviously, if you can make the rescue yourself *without undue risk* you may save a life and reduce the possibility of further injury to others.

9.9.2 Injury to a Person in a Sewer

If a person is injured when down a sewer but the sewer conditions are normal, you may remain or return for the casualty, but all the usual precautions must be taken.

9.9.3 What You Must Tell the Rescue Services

(*a*) On the telephone:

Say where you are.
Say who you are.
Say, as far as you know, what has happened.
Say where your base manhole is and if you know them, the nearest manholes to the casualty.

(*b*) When the rescue teams arrive:

Tell them what happened.
Tell them how long ago it happened.

Tell them what conditions are in the sewer, and which way it flows, and be prepared to give any useful advice.

9.9.4 Getting the Casualty to the Manhole

If you go down for a casualty who has, or may have been gassed, you must *wear breathing equipment which you have been trained to use.*

Do not attempt to use any breathing equipment of a type with which you have not been trained.

Unless the casualty is much nearer to the upstream manhole, it is usually best to drag that person downstream, headfirst, if possible. In this way, the effluent flow helps in the moving and sewage is less likely to splash over the individual's head.

Revision Questions

1 What must an employer do when noise levels reach 90 decibels?
2 What is the definition of 'Type A' works in respect of traffic management?
3 On an all-purpose single carriageway road with a speed restriction of 40 mph or less, and a hazard width of 4 m, what would be the length of the lead-in-taper?
4 List the process for signing roadworks? (Some 15 aspects should be considered.)

Note: See page 305 for answers to these revision questions

Materials: Flexible Pavements

Currently all British Standards are being revised to meet European Standards. Readers will need to be aware of new BSEN Standards.

10.1   Bituminous Roadbuilding Materials

All bituminous road surfacings are composed mainly of aggregate, which may be crushed natural rock, slag or gravel with sand or stone fines. The aggregate plays an important part in the behaviour of the surfacing. Many aggregates have sufficient mechanical strength to be used for roadmaking, even in the wearing course which has to carry the direct traffic load, but the majority of these, however, tend to polish under heavy traffic, which causes the surface to become slippery and unsuitable for traffic use.

The shape of the particles may also limit the usefulness of an aggregate for roadmaking. If the material is produced by crushing, a considerable degree of control of the shape is possible. Aggregate from even a single source can vary considerably and frequent checks of quality must be made to maintain a consistent supply of road material.

Although the proportion of bituminous binder in surfacing is much less than that of the aggregate, the quantity and quality of the binder both have a marked effect on the behaviour of the surfacing.

Bituminous binders are of two main types.
These are:

1 tar (which is a product of the carbonization process);
2 bitumen (which is either natural asphalt or a product of petroleum oil processing).

Both types of binder are produced in a wide range of viscosity to suit the many differing processes used in bituminous road building.

One of the main functions of a bituminous binder is to act as an adhesive, both between aggregate particles and also between the aggregate and the underlying road structure. For good adhesion, water and dust need to be excluded.

A road surfacing must be capable of resisting the deforming stresses of traffic at high road temperatures and of resisting brittle fracture at low road temperatures. Hence, the binder must be able to withstand these extremes of conditions and also be sufficiently fluid to be pumped or sprayed and to coat mineral aggregate. The viscosity of bituminous binders and its variation with temperature is of great practical importance.

The mechanical properties of the bituminous road mixture (binder plus aggregate) depend not only on the strength, shape and size of the aggregate and on the properties of the binder, but also on the compacted structure of the aggregate particles. The mixture of materials has considerable internal friction properties as well as viscous flow properties which again are dependent on the temperature.

All bituminous binders change on exposure to weather, becoming progressively harder and more brittle at low temperatures. This creates the greatest problem with open-textured materials, such as coated macadams. The binders in the interior of really dense mixtures (e.g. rolled asphalts) are almost completely protected from the weather with a resultant greater life expectation for this type of material.

Weathering can improve the surface in dense bituminous surface courses, by increasing its skid resistance.

10.2   Roadmaking Aggregates

Almost all roadmaking aggregates are obtained from either natural rocks or from slags derived from metallurgical processes. Natural rocks occur as massive outcrops or gravels (which are derived from rock outcrops) *see* Figure 10.1. They are classified into three main groups.

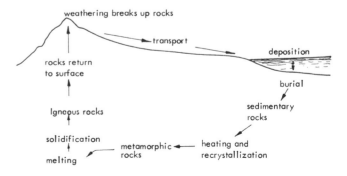

Figure 10.1   The rock cycle

(a) Igneous

Mainly crystalline and formed by the cooling of molten rock (acid, intermediate and basic).

(b) Sedimentary

Formed either from the deposition of insoluble granular material from disintegrating existing rock or from inorganic remains of marine animals or by crystallization of soluble minerals from solution (this is rare).

(c) Metamorphic

Igneous or sedimentary rocks subjected to great heat, or to great pressure and heat, become transformed into minerals and textures different from those of the original rock.

Gravel is formed from any natural rock and frequently contains a wide variety of types of rock. There are examples of gravel of a single type of origin rock, such as the quartzite gravels of the Midlands.

The common feature of all gravels is that they are a mixture of pebbles or stones of various sizes, of rounded or irregular shape and smooth surfaces (due to rubbing together under the action of sea or river water). For roadmaking purposes, the larger pebbles are often passed through a crusher to provide a proportion of angular material with freshly crushed surfaces.

Sand is almost entirely the final residue of grains of the most resistant minerals resulting from the weathering of rocks. The main mineral in sand is normally quartz. Grain size of sands may vary from about 2 mm to dust.

Blast furnace slag is the most widely used slag for road building. It is composed of minerals resembling some of those found in igneous rocks, but its texture varies from glassy to honeycombed.

10.3 Rock for Roadmaking Purposes

The British Standard Group Classification of roadmaking rocks is used. This is a classification of eleven groups:

10.3.1 Artificial Group

Slags, clinker, burnt shale, etc. are by-products of industrial processes.

10.3.2 Basalt Group

Basalt, dolerite, basic porphyrite, andesite, basic and intermediate igneous rocks of medium or fine grain size, including some less strongly metamorphosed equivalents form this group.

Most easily recognized are feldspars and ferromagnesian minerals (dark coloured and heavy).

10.3.3 Flint Group

Flint and chert are very fine grained sedimentary rocks (less than 0.2 mm size) varying from white to black and light in weight. Flint is widely distributed as gravel providing the main local source for eastern and south-eastern England.

10.3.4 Gabbro Group

Gabbro, basic diorite, basic gneiss, etc., are basic and intermediate igneous rocks of coarse grain size (over 2 mm) and including their metamorphic equivalents (similar in colour and density to the basalts).

10.3.5 Granite Group

Granite, quartz–diorite, granite–gneiss, etc., are acid intermediate igneous rocks of coarse grain size. Usually light in colour; specific gravity below 2.80. Widely distributed and extensively used as roadstone.

10.3.6 Gritstone Group

Sandstone, ruff, etc. are mainly siliceous sedimentary rocks, of medium or coarse grain size, cemented with fine grain material on occasion, may contain minerals from igneous rocks. Medium or light in colour, specific gravity below 2.80.

10.3.7 Hornfels Group

This group consists of thermally metamorphosed rocks (with the exception of marble and quartzite). Fine to medium grain size, medium to dark in colour. Dense, hard rocks, not widely distributed.

10.3.8 Limestone Group

Limestone, dolomite, marble are sedimentary rocks composed mainly of calcium or magnesium carbonate. Medium to fine grain size, medium to light colour, moderate weight. Widely distributed, extensively used for roadstone.

10.3.9 Porphyry Group

Porphyry, granophyre, microgranite, felsite, etc. are acid intermediate igneous rocks; medium or fine grain size, medium to light colour, medium weight. Widely distributed and extensively used as roadstone.

10.3.10 Quartzite Group

Quartzite, quartzitic sandstone and ganister are siliceous sedimentary or metamorphic rocks composed almost entirely of quartz. Colour light, medium to fine grain size, medium weight. Occur both as massive rock and as gravel, and are extensively used.

10.3.11 Schist Group

Laminated rocks, such as schist, phyllite and slate. Rarely used as roadstone.

10.3.12 Properties

The properties of most importance in a roadmaking aggregate are its resistance to crushing, impact, abrasion and polishing, its specific gravity and water absorption and its grading and particle shape.

Not all of these properties are required for every roadmaking application. For instance, high resistance to abrasion and polishing is not required for a binder course aggregate, but is essential for surface course material.

The Transport Research Laboratory issues a list of the major sources of roadmaking aggregate in Great Britain, this information includes the type of stone worked, its grouped classification and colour.

Roadstone must be tested in accordance with a series of British Standard Tests, which include:

1 aggregate abrasion test (Micro-Deval test)
2 accelerated polishing test (Polished Stone Value)
3 specific gravity and water absorption test
4 sieve analysis
5 flakiness test
6 resistance to fragmentation test (Los Angeles test)
7 cleanness
8 resistance to freezing and thawing
9 volume stability of blast furnace slag.

In general, aggregate tests are used to check whether the stone is within the specific limits, or to compare different roadstones in order to determine which is the most suitable for a particular use.

When a range of aggregates is available, the aggregates having the better physical properties are naturally preferred for use in the surface course of all the more heavily trafficked roads. 'Good' stone is, however, so limited in the quantity available, that it should not be used in situations where a poorer quality stone would suffice.

10.4 Bitumen

10.4.1 Derivation

Bitumen is understood to be 'a viscous liquid or solid material, black or brown in colour, having adhesive qualities, consisting essentially of hydrocarbons, derived from petroleum or occurring in natural asphalt and soluble in carbon disulphide'.

The major proportion of road bitumen used in this country is derived from petroleum by a refinery distillation process. This is sometimes referred to as residual bitumen or as straight-run bitumen.

10.4.2 Rock Asphalt

This usually consists of a rather porous limestone, or more rarely sandstone, naturally impregnated with bitumen with approximate proportions of 90% stone, 10% bitumen.

It is now normally used only in mastic asphalt which is an expensive surfacing and, therefore, used only to a limited extent.

See BS 1446 for more information about mastic asphalt.

Blending of different rocks may be necessary to obtain the desired binder material.

10.4.3 Lake Asphalt

This is a naturally occurring asphalt in which the mineral matter is finely divided and dispersed through the bitumen which is the major constituent. The whole mass is capable of flowing. Trinidad Lake asphalt is the only type used in roadmaking in Britain, being used, when suitably fluxed, mainly in rolled asphalt surface courses.

10.5 Sources of Refinery Bitumen

Bitumen, produced by industrial refining of crude oil, is known as 'residual' bitumen, 'straight-run' bitumen or 'steam-refined' bitumen. The term 'refinery bitumen' is, however, the most descriptive and appropriate.

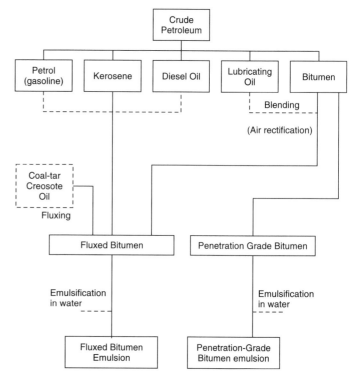

Figure 10.2 Various types of bitumen and their preparation from petroleum

Bitumen is produced from selected crude oils by the distillation of petroleum. The distillate is obtained by first heating to about 350°C under atmospheric pressure to drive off light fractions, e.g. gasoline, kerosene and gas oil.

10.6 Penetration-grade Bitumen

Straight-run and air-rectified bitumens produced as described above can, by varying distillation and rectification conditions, be made into a wide range of viscosity grades. In the UK, the recognized grades, in terms of penetration at 25°C (for 100 g load, maintained for 5 seconds) are

20/30 30/45 35/50 40/60 50/70 70/100 100/150
160/220 250/330
(these penetration values are in 1/10ths of mm).

Some of these penetration-grade bitumens are made by fluxing a harder grade with a softer grade or by fluxing with an oil of high boiling range.

These types of bitumen are used as binders with roadstone and these mixed materials must be laid while still warm, within a few hours of mixing.

Details of the penetration test are given in Section 10.20.

10.7 Fluxed Bitumens

For use in surface dressing, and similar applications, it is necessary to have a binder that can be handled at relatively low temperatures and which can be mixed with aggregate that is either cold or only warmed sufficiently to make it surface-dry.

Fluxed bitumens are prepared by blending penetration grade bitumen with a volatile oil, e.g. creosote during manufacture in the refinery. This slows down the rate at which the binder will set until the volatile oil evaporates.

Viscosity values are obtained by testing in the standard tar viscometer (*see* Section 10.21) and are expressed as seconds of flow in the standard tar viscometer at 40°C.

Deferred set binders. These are produced by blending a less volatile oil, e.g. gas oil, with a bitumen during manufacture in the mixing plant. The deferred set binder allows the bituminous material to be used even when it is cold, e.g. depot stock.

10.8 Bitumen Emulsions

Bitumen emulsions consist of globules of bitumen dispersed in water. This 'mixture' is achieved by the use of emulsifying agents (soap type substances). The bitumen is carried by the water; this obviously means that the resulting mixture is 'thinner' than bitumen, needs no heat (except for some high binder content grades) and therefore is easier to apply.

When reference is made to a percentage, such as 'tack coat 30%', this means that the bitumen content is 30%. Therefore a 60% emulsion is 'thicker' than a 30% emulsion.

When reference is made to an emulsion 'breaking', this is the change from a liquid to a coherent film (the suspended particles congeal together). The first sign of breaking is a change in colour from brown (emulsion) to black (bitumen). The rate of break is dependent upon the following factors:

(*a*) The composition of the emulsion.
(*b*) The rate of evaporation of the water, which in turn is dependent upon wind conditions, relative humidity, atmospheric temperature, rate and method of application.
(*c*) The porosity of the surface to which the emulsion is being applied, and the consequent removal of water by capillary attraction.
(*d*) The chemical and physical influence of the aggregate with which the emulsion comes into contact.
(*e*) The mechanical disturbance of the emulsion/aggregate system during laying and rolling or by the action of traffic.

There are two types of emulsion – anionic and cationic.

10.8.1 Anionic Emulsions

These have alkaline emulsifiers and the break relies principally on loss of water by *evaporation*, therefore difficulty can be experienced in obtaining the 'break' during periods of wet weather.

Anionic emulsions are classified as follows:

(*a*) Class A1: Labile. An emulsion with rapid break on application and normally unsuitable for mixing with aggregate; used cold.
(*b*) Class A2: Semi-stable. An emulsion with sufficient stability to permit mixing with certain grades of aggregate before break occurs; used cold.
(*c*) Class A3: Stable. An emulsion with stability for mixing with aggregates; used cold.
(*d*) Class A4: Slurry seal. With the addition of water, sufficiently stable to form a free-flowing slurry with the aggregate specified and capable of sustaining this condition throughout the laying procedure adopted. Setting time of the mix may be within a few minutes or extended as desired. Special proprietary emulsions may include additives used at the time of laying.

10.8.2 Cationic Emulsions

These have acid emulsifiers and the break is controlled by *chemical coagulation* and not loss of water by evaporation. Therefore these emulsions are particularly suitable for use in wet weather.

Cationic emulsions are classified as follows:

(*a*) Class K1: Rapid acting. An emulsion characterized by rapid deposition of binder on contact with road surfaces and aggregates followed by early resistance to rain. Unsuitable for mixing with aggregates; high binder content grades used hot, other grades cold.

(*b*) Class K2: Medium acting. An emulsion in which the rate of deposition of binder is sufficiently delayed to permit mixing with certain clean coarse aggregates, before breaking to form a continuous adhesive film without stripping; used cold.

(*c*) Class K3: Slow acting. An emulsion in which the rate of deposition of binder is sufficiently delayed to permit mixing with certain fine aggregates before breaking to form a continuous adhesive film without stripping; used cold.

10.8.3 Summary

Due to their rapid break and adhesive properties cationic emulsions are the most suitable for all the year round application.

As all emulsions contain water they should be protected from extremes of temperature. If stored in drums, turn the drum over occasionally.

Under no circumstances mix the cationic with the anionic; this will result in coagulation. If using spraying equipment clean the pipes, pump, etc. thoroughly with diesel oil daily and always before changing to another type.

For many years, there has been a need for a bituminous macadam, able to be cold laid and capable of withstanding the actions of both weather and traffic.

It is the development of bitumen emulsions which has lead to the use of cold laid material for repairs and reinstatements.

10.8.4 Foamed Bitumen

Foamed bitumen is produced by the controlled introduction of water into hot bitumen. The foam produced has a very high surface area and extremely low viscosity, making it ideal for coating aggregate. The foam-mix process was originally developed as a means of stabilizing marginal aggregates but has since developed into a widely accepted maintenance and construction technique. The foam-mix process is commonly used in the *in situ* recycling of roads and also the coating of recycled/secondary aggregates. Mix materials can be stockpiled for a period before being used; this results in minimum wastage.

10.9 Mastic Asphalt

This is rock asphalt, ground to powder, to pass a 2.36 mm sieve and a minimum of 5% must pass a 0.063 mm sieve.

The powdered rock asphalt is mixed with 13–16% of asphaltic cement, heated to between 175° and 215°C. If not used immediately it is cast into blocks weighing 25 kg (approx.) with 20–30% filler added.

Mastic asphalt is generally only used on bridge decks (for water proofing) and occasionally in city streets which carry very heavy traffic densities.

10.10 Aggregates in Bituminous Materials

Aggregates in bituminous materials are defined by size as:

(*a*) Coarse aggregate, which should be crushed rock, slag or gravel substantially retained on a 3.35 min BS sieve. This material provides the main interlocking structure.

(*b*) Fine aggregate which should be crushed rock, slag or sand substantially passing a 3.35 mm BS sieve. This fills (or partially fills) the voids in the coarse aggregate and provides the surface texture.

(*c*) Filler to be crushed rock, slag or Portland Cement at least 75% passing a 75 gm BS sieve. This assists in filling small voids, increases the viscosity of the binder and thus reduces run off of binder from the aggregate.

10.11 Tar

Due to the decline of the coal industry, tar is now rarely used.

Tar used for road purposes is a product of coal by the carbonization process and it must be refined before it can be used (*see* BS 76).

There are three stages in the preparation of this material from coal:

(*a*) carbonization of coal to produce crude tar;
(*b*) refining, or distillation, of the crude tar;
(*c*) blending of the distillation residue with distillate oil fractions to produce the desired road tar.

10.12 Hardness of Tar

The hardness of the residual base tar (or pitch) can be altered by varying the extent of the distillation.

The ring and ball test is used to determine the 'hardness'. Pitch is described as 'hard' if the softening temperature is above 85°C; 'medium-soft' if the softening temperature is between 73°C and 82°C; 'soft' if the softening point is below 65°C.

Two grades are commonly used for road tars:

1 Softening point (ring and ball) 30°C upwards.
2 Softening point (ring and ball) about 80°C.

See Section 10.22 (ring and ball test).

Types of tar
Type S: this is used for surface dressing and binder courses.
Type C: used for tarmacadam surface courses and carpets.

10.13 Modified Binders

Straight run bitumens are not always suitable for all applications, in particular, they lack the properties necessary to withstand the stresses caused by present day heavy traffic loads. This has led to the use of binders which have been chemically modified to improve both the quality and length of life of pavements.

The following are examples of modified binders:

Epoxy resin: this is an early example which, when blended with bitumen produces a surface dressing binder of great strength (Shell grip).

When Sulphur is added, it increases the workability, also enhancing the stiffness of the binder by allowing higher penetration values to be used. However, experience has shown that the mixtures have a tendency to become brittle.

Ethylene Vinyl Acetate (EVA) is a common thermoplastic co-polymer used to increase the viscosity of the binder, the vinyl acetate structures being more 'rubbery'.

Styrenic block co-polymers are commonly used thermoplastic rubbers producing strong elastic mixes of which SBS (styrene–butadiene–styrene) is an example.

10.14 European Terminology

In endeavouring to standardize terms throughout Europe, the word 'Asphalt' is to be used, in a generic sense, to cover all asphalts and coated macadams.

Asphalts with a descriptions such as 'hot rolled' or 'stone mastic', will still be used.

10.15 General Description of Bituminous Materials

The following are general descriptions of bituminous materials which are in current British Standards.

10.15.1 Introduction

Coated macadams are defined as 'road material consisting of graded aggregate that has been coated with a tar or bitumen, or a mixture of the two and in which the intimate interlocking of the aggregate particles is a major factor in the strength of the compacted roadbase or surfacing'.

Hence, tarmacadam is coated macadam in which the binder is wholly or substantially road tar and bitumen macadam is coated macadam in which the binder is wholly or substantially bitumen.

10.15.2 Classification

Coated macadam is classified by:

(a) Nominal size. This is the size of the largest pieces of aggregate in the material, subject to a small permitted amount of oversize.

(b) Grading. The surface texture is controlled by the nominal size and fines content (fines are those particles which pass the 3.35 mm or 2.36 mm BS sieve).

The types of grading are:

1 *Open graded*: where the fines content is low, generally less than 15%
2 *Medium graded*: this applies only to 6 mm size surface course
3 *Close graded*: applies only to 10 mm and 14 mm size surface courses, formerly called 'dense'
4 *Fine graded*: applies only to 3 mm size fine graded surface course (formerly called fine cold asphalt)
5 *Dense*: this applies to any material which is relatively impervious
6 *Pervious*: surface course mixtures open grading permits drainage from the surface.

10.15.3 Identification

Coated macadams for roads and other paved areas are listed in BS 4987 as follows:

Group one:	roadbase mixtures
	40 mm size dense roadbase
	28 mm size dense roadbase
Group two:	binder course mixtures
	20 mm size open graded binder course
	40 mm size single course
	40 mm size dense binder course
	28 mm size dense binder course
	20 mm size dense binder course
Group three:	surface course mixtures
	14 mm size open graded surface course
	10 mm size open graded surface course
	14 mm size close graded surface course
	10 mm size close graded surface course
	6 mm size dense surface course
	6 mm size medium graded surface course
	3 mm size fine graded surface course
	Coated chippings for fine graded surface course
Group four:	pervious surface course mixtures
	20 mm size pervious surface course
	10 mm size pervious surface course.

10.15.4 Dense Tar Surfacing: BS 5273

14 mm nominal size
10 mm nominal size.

10.15.5 Hot Rolled Asphalt: BS 594

Roadbase with 60% and 65% coarse aggregate content using crushed rock, gravel or slag.

Binder courses with 55% and 65% coarse aggregate content using crushed rock, gravel or slag.

Surface courses with 0%, 15%, 30%, 40% coarse aggregate content using crushed rock, gravel or slag with an additional 55% coarse aggregate content for crushed rock only.

Coated chippings (for application on surface courses only) using 20 or 40 mm nominal size chippings.

10.15.6 Types and Uses of Traditional Pre-mixed Bituminous Materials

Figure 10.3 gives an approximate guide to the types and uses of the various flexible road materials.

10.15.7 Developments in Bituminous Materials

It is vital, to deal with the demands of traffic, to examine methods of extending the life of a pavement, thus reducing the frequency of repairs being needed, or the time taken in carrying out those repairs.

Traditional materials in BS 4987 and BS 594 cater for most situations, but their life has shown that they are not suited for heavily trafficked routes. The introduction of modified binders has played a large part in strengthening existing materials.

Where hot rolled asphalt has been used, severe problems have occurred with tracking and deformation and it is being replaced by mastic asphalt.

Traditionally, mastic asphalt was a fluid, hand floated material, unsuitable for machine laying. A machine laid material has been used in Germany for several years and which is known in Britain as 'Stone Mastic Asphalt'.

Hot rolled asphalt is being replaced by Stone mastic asphalt, which has a high stone and binder content. The possible 'run off' of the binder is overcome by the addition of cellulose or rock wool fibres. It has the advantage of high stability and, due to the stone content, does not require coated chippings to provide skid resistance.

Porous asphalt, previously known as pervious macadam, has been formulated to contain 15–20% air voids after compaction as with other pervious materials, surface water is carried away on the underlying dense binder course. It also has the advantage of reducing traffic noise.

Most new developments tend to be introduced by private companies, hence trade names are used to promote them. Thin surfacings are typical examples and include

Thick slurry dressing	Ralumac
Multiple surface dressing	Surphalt
Paver laid surface dressing	Safepave
Thin stone mastic asphalt	Masterpave, Megapave
Thin polymer-modified asphaltic concrete	Axoflex, UL-M, Hitex

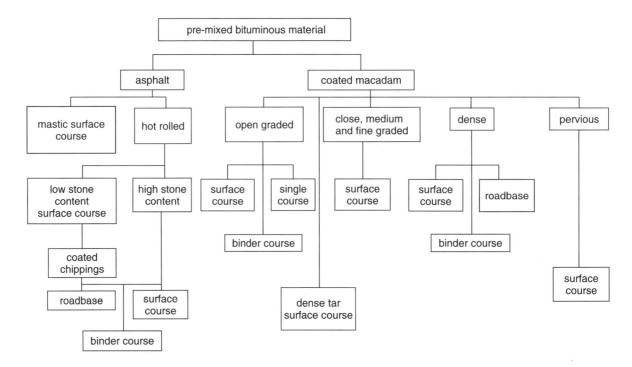

Figure 10.3 Flexible road materials and their uses

Thin surfacings tend to be fast to lay, produce a durable finish and often have low noise and spray properties. The thicker applications are used to provide a regulating layer.

10.16 Required Working Temperatures for Bituminous Materials

The best guide to the temperature at which a bituminous material should be laid is the viscosity of the binder in the mix. 'Viscosity' is the tackiness of the bitumen (e.g. water) has a very low viscosity, being very runny; while treacle has a much higher viscosity). The higher the viscosity or the more solid the binder or bitumen in the road material, the higher the temperature at which it has to be mixed (to make the bitumen liquid), and the higher the temperature at which it has to be laid to make it workable. The higher the viscosity the harder the material will set. Hence, to find the required temperature, ask for the viscosity of the binder.

Laid material will cool to the setting temperature in a very few minutes, thus limiting the time available for compaction.

Testing of Bituminous Materials

10.17 Sampling Techniques

The quality of a bituminous road material is mainly determined at the mixing plant, there is close control of aggregates (for size grading, shape and strength) and of the binder – one of the more important points being the mixing temperature. To preserve the quality, the freshly mixed material is transported in insulated and sheeted lorries, so that it arrives on site ready for placing and compaction in exactly the right condition.

Once the binder has started to set, if the material has not by then been placed and compacted, there will be a noticeable loss of strength. This loss is due to the binder becoming less 'sticky' and the coated material being too 'solid' to compact fully.

Tests are carried out to check the materials, mainly by sampling. These samples may be taken:

1 at the mixing plant;
2 at the site before laying;
3 after laying.

The easiest sampling point is at the mixing plant. It must be remembered that it is difficult to get a good sample out of a large mass of material in a lorry and, due to the transportation, there will probably be some segregation of the mix.

Sampling after laying requires cores being carefully cut out of the completed road structure (*see* Figure 10.4). This is an expensive process and it is difficult to obtain a good core sample. In addition, the road structure then has to be made good where the core was removed.

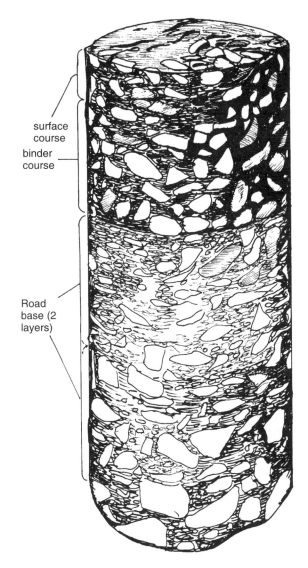

surface course
binder course

Road base (2 layers)

Figure 10.4 Typical core sample from a flexible carriageway

10.18 Selecting Samples

To select a suitable sample from a stock pile of stone, or a sample of bituminous material which has had the bitumen removed, two methods are available.

Quartering

This can also be called the 'Hot Cross Bun' method. For this, a large sample of material is reduced in quantity by arranging it into a circular heap and dividing it, by diameters at right angles, into four approximately equal parts. Two diagonally opposite parts are removed, the remaining two quarters being thoroughly mixed, rearranged in a circular heap, and the process repeated until a sample of the required size remains.

Riffling

The material in the heap, from which the sample is to be taken, is passed through a riffle box of a suitable size. This divides the sample into halves. The process is repeated until the required size of sample remains.

In order then to test whether the stone is correctly graded, a set of standard sieves is needed. The sample is riddled through the set of sieves, the stone retained on each successive size of mesh being weighed and, thus, the proportions and grading of the stone can be determined and checked against the specification.

10.19   Standard Test Sieves

A series of sieves of different mesh sizes is required in order to sample an aggregate, by dividing the sample into sections according to size. The sample is placed in the top sieve, covered and shaken. After shaking, the quantity of material retained in each sieve is weighed to check against the specified size grading.

Details of standard sieves are given in BS 410. This requires sieves to be mounted in frames, circular in shape, which can be locked together and fitted with a receiver and cover (*see* Figure 10.5(*a*)). The design must be such that the seal between frames prevents the material being sieved from lodging – in other words the sieves, cover and receiver must nest together and assemble snugly with one another. The receiver and cover are to prevent loss of material when the nest of sieves is being shaken.

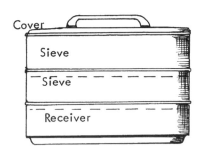

Figure 10.5(a)   Standard test sieves

The number and sieve size is dependent on the grading of a particular granular material specification. The UK has adopted the following sieve sizes for coarse and fine aggregate, which is defined as the 'basic set' plus two additional sieve sizes, as shown below (*see* Figure 10.5(*b*)).

Grading of granular material is now presented in different classifications. Consider an example e.g. **Gc 90/10**

where:

G = Grading category
c = coarse
90 = Minimum % passing 'D' (oversize control)
10 = Maximum % passing 'd' (undersize control)

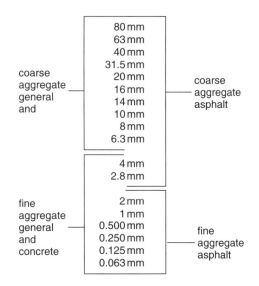

Figure 10.5(b)   New sieve sizes (BSI National Guidance Published Documents)

This means that a sample of 1000 gm, if passed through a sieve of 63 mm (oversize control) would result in 90% passing (900 gm) and after passing through other sieves i.e. 31.5 mm and 20 mm, would have 10 gm passing the 10 mm sieve (undersize control) into the receiver.

10.20   Penetration of Bitumen

A very accurately made needle is used to penetrate samples of bitumen under a standard load for a limited time to see how far it will go into the sample in a limited time. The load used is 100 g which is applied vertically to the bitumen sample for 5 seconds at a temperature of 25°C.

The apparatus used is called a penetrometer and it includes a device for measuring the distance by which the needle penetrates the bitumen sample. Most penetrometers have an automatic timing device (*see* Figure 10.6).

The procedure for carrying out the test is as follows:

1 Heat a sample of 300 g of bitumen in an air bath to a temperature of 110°C, stirring it constantly for 15 minutes to remove any water which may be present.

2 Raise the temperature of the bitumen sample to 80°–90°C above the softening point of the material, stirring until the sample is completely fluid and free from air bubbles. This heating should be limited to a maximum of 30 minutes.

3 If the material contains visible foreign solid particles, it should be sieved (0.250 mm mesh).

4 Pour the bitumen sample into a flat bottomed metal sample container, filling to within 2 or 3 mm of the rim (these containers are 55 mm diameter and either 35 mm or 57 mm deep). Prepare two samples.

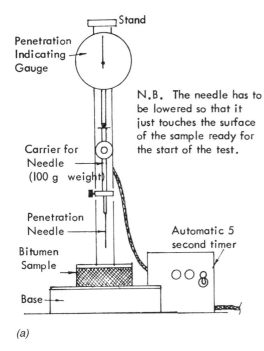

(a)

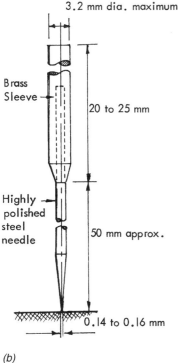

(b)

Figure 10.6   Automatic penetrometer

7 Clean the penetration needle and fix it into the needle holder and guide.

8 Load the needle holder with 100 g (±0.25 g).

9 Place one of the samples on the table of the penetrometer (using a transfer carrier).

10 Lower the needle slowly until its tip just makes contact with its image on the surface of the sample.

11 Zero the dial reading on the penetrometer.

12 Release the needle holder for 5 seconds (±0.2 second).

13 Record the depth of penetration to 1/10 mm.

14 Repeat the test for a total of three readings on the one sample.

For comparison of different bitumens, the greater the penetration distance, the softer is the bitumen and the lower its viscosity.

10.21   Viscosity of Bitumens and Tars

10.21.1   *Viscosity of Bitumens*

Bitumens which are too soft to be tested on a penetrometer have their viscosity measured on a standard tar viscometer.

Briefly, the test consists of:

1 A sample is placed in the container in the water bath (Figure 10.7) at a temperature of 40°C.

2 The stopper is released and the time taken for 100 ml to flow into the measuring jar is timed in seconds.

3 This result is referred to as the viscosity; 50 seconds, 100 seconds and 200 seconds are common grades.

From this it can be seen that the longer the time in seconds, the more viscous (thicker) the bitumen.

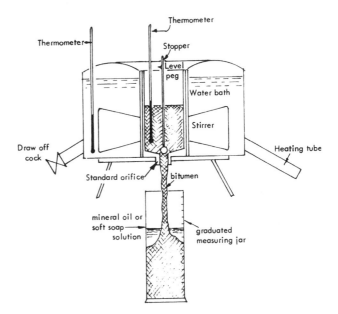

Figure 10.7   Standard tar viscometer

5 Protect the samples and containers from dust and allow them to cool for 1 hour in a room at a temperature of 18°C to 25°C.

6 Place the sample in a constant-temperature water bath for one hour (35 mm tin)/one and a half hours (577 mm tin) at a temperature of 25°C ± 0.1°C.

10.21.2   EVT of Tar

This test measures the viscosity of a tar on a scale known as EVT (Equi Viscous Temperature). In simple terms, this means the temperature at which a tar must be, if it is to have the same time of flow through the viscometer as other tars, i.e. they all have equal viscosity.

The test equipment is the standard tar viscometer, with the time set at 50 seconds. Hence, the temperature at which the sample flows into the measuring jar in 50 seconds gives the comparison with other tars.

The lower the viscosity, the softer the tar.

Table 10.1 shows tar EVTs, but, as already stated, these are rarely used nowadays.

Material	Equi-viscous temperature (°C)	
	Summer	Winter
Binder course (surface course open textured)	30–40	27–35
Surface course (medium textured)	37–42	34–38

Table 10.1   Examples of uses of tar

10.22   Softening Point of Tar or Bitumen: Ring and Ball Test

This test determines the temperature at which a given tar or bitumen reaches a certain degree of softness. The brief procedure is as follows:

1 The tar or bitumen is melted and poured into a standard brass ring placed on a flat plate.
2 When the sample has cooled, a standard sized steel ball is placed on the sample and it is suspended in a water bath.
3 The water temperature is raised at a rate of 5°C per minute.
4 The temperature at which the bitumen/tar softens sufficiently to allow the ball to pass through the ring and drop on to a retaining plate is noted.

Hence, the softening point is the temperature at which the ball fails through the sample on to the plate which is 25.4 mm below the ring.

10.23   Polished Stone Value

This is a laboratory obtained value for which the 'accelerated polishing machine' is used.

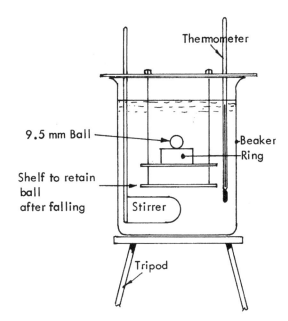

Figure 10.8   Ring and ball apparatus

10.23.1   Accelerated Polishing Machine

This consists of a lever-loaded rubber-tyred wheel which is brought into contact with a loaded road wheel to which a number, often 14, of small curved metal mould boxes can be fitted to complete the circumference.

Specimens of road stone are cast into these sections and, when set, they are mounted on the road wheel. This is then rotated at 320 rev/min (constant speed) with the loaded rubber-tyred wheel running on it. In order to speed up the natural rate of wear, water and/or emery powder is applied to the surface of the road wheel during rotation.

At the end of the test, the specimens are removed and their skid resistance measured by means of the portable tester (*see* Section 10.23.2). To make full use of the test, readings should be obtained before the test run, as well as after. Different stones can be tested simultaneously for direct comparison between them, under the same test conditions.

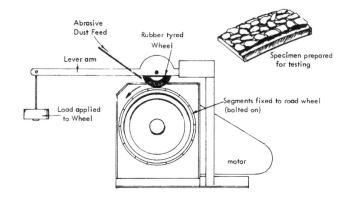

Figure 10.9   Accelerated polishing test apparatus

The test is essentially an artificial one, as no account is taken of time and conditions, such as rain, snow, spillage of oil, rubber deposits, etc.

10.23.2 Skid Resistance Tester

This is used on the actual road to obtain a skid resistance value. The principle of operation is of a pendulum arm, which carries a friction shoe on its outer end, being released from the horizontal position to rub over a specified length of road surface. The energy absorbed by this reduces the 'follow-through' swing. The amount of this swing is measured against a calibrated scale on the basis that the shorter the swing, the rougher the surface and, hence, the higher is the skid resistance of that surface.

The procedure for using the apparatus is as follows:

1 Set up the tester so that the centre column is vertical (this is done by adjustment of the three levelling screws on the base).

2 Adjust the height of the pivot so that the arm swings freely through its arc without touching the road surface (check that it swings right through to the zero on the scale).

3 Lower the pivot height so that the friction foot is in contact with the road surface over the precise distance as shown by the gauge which is placed alongside.

4 Raise the pendulum arm to the starting position and engage the retaining catch.

5 Water the road surface where the test is to be taken.

6 Check that the maximum swing indicator is set to the vertical position.

7 Release the pendulum arm.

8 After it has swung through, catch it to prevent a back swing which could disturb the apparatus and reading.

9 Take and note the reading on the scale.

10 Repeat the test to obtain a minimum of three readings. If these are reasonably consistent, they may be

accepted. If not, repeat until three consistent readings have been obtained.

10.23.3 Texture Depth Test

Other tests include the measurement of texture depth (surface roughness) by means of the sand patch method or by use of the Mini Texture Meter. These are described in Chapter 5, Section 5.22.4.

Materials: Rigid Pavements

10.24 Materials used in Rigid Pavements

Concrete is used to construct the road slab for a rigid pavement. It is laid on a sub-base, generally of stone or similar material, with a separation membrane, between the slab and sub-base, of polythene sheeting.

The concrete in the upper part of the road slab has to be air entrained, as well as being finished in such a manner as to provide a non-skid surface.

As the concrete forms the equivalent of the surface course, binder course and roadbase, it is the main material considered in this section.

10.25 Concrete

Concrete is a mixture of aggregate, cement and water. The aggregate is normally in two parts:

(a) coarse aggregate, e.g. gravel, granite chippings.
(b) fine aggregate, e.g. sand, granite dust.

The cement forms the binder for the mix. Water is added to cause the chemical reaction to change the dry cement to an adhesive. Too little water causes incomplete reaction and loss of strength as a result. Too much water tends to reduce the effect of the cement and, again, weaken the concrete.

The mixing of the materials is most important. Ideally every particle of aggregate should be coated with cement before addition of the water.

The strength of a correctly designed mix of concrete comes mainly from the coarse aggregate. The fine aggregate proportion should be just sufficient to fill all the voids. Hence the strength of the concrete is dependent upon the:

1 proportions of cement, fine aggregate and coarse aggregate in the mix;
2 efficiency of mixing;
3 crushing strength of the coarse aggregate;
4 cleanliness of the aggregates;
5 type of cement used;
6 proportion of water used (this is normally quoted as a water/cement ratio);

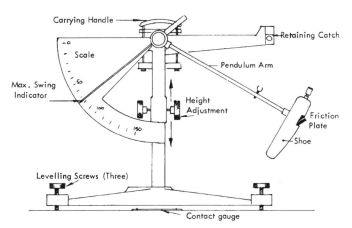

Figure 10.10 Portable skid resistance tester

7 degree of compaction obtained;

8 efficiency of curing the concrete.

Concrete mixes are based on the average strengths of concrete test cubes and, once the required strength has been decided, the mix must be designed to meet this requirement. Before accepting a mix design, a trial mix is required and, based on the results of this, modification of the mix design may be made.

Concrete mixes are generally specified by the weights of the various materials used. The different mixes are identified by their 'grade', the grade figure being the minimum strength which the concrete may be expected to have (this is known as the 'characteristic strength'), e.g. a grade 30 concrete mix has a crushing strength at 28 days of 30 N/mm^2.

A comparison of various grades of concrete in pavements is given in Table 10.2.

In roadworks, the concrete is often delivered ready-mixed in mixer trucks and it is the responsibility of the concrete supplier to ensure that the mix supplied is of the correct grade. The selection of the grade is the responsibility of the engineer.

Details of some mixes are given in Table 10.3, which is based on the quantities of materials required to produce one cubic metre of fully compacted concrete in each case.

Pavement layer	Designed mix grade
1 Surface slabs:	C40
Unreinforced concrete (URC)	
Jointed reinforced concrete (JRC)	
Continuously reinforced concrete	
pavement (CRCP)	
2 Continuously reinforced concrete	C40
roadbase (CRCR)	
3 CRCP ground beam anchorages	C30
4 Wet lean concrete 4) for roadbases	C20
5 Wet lean concrete 3) for sub-bases	C15
6 Wet lean concrete 2) as required in	C10
7 Wet lean concrete 1) Appendix 7/1	C7.5
8 Cement Bound Material Category 1 (CBM1)	
9 Cement Bound Material Category 2 (CBM2)	For road bases
10 Cement Bound Material Category 3 (CBM3)	and sub-bases
11 Cement Bound Material Category 4 (CBM4)	

Table 10.2 Pavement layers – grades of concrete

10.26 Aggregates

These are the most variable part of the concrete mix.

Cleanliness of the aggregates is very important. Coarse aggregates can be checked for this by visual inspection.

Concrete grade	Nominal max. size of aggregate (mm)	40		20		14	
	Workability	Med	High	Med	High	Med	High
	Limits to slump that may be expected (mm)	50–100	100–150	27–75	75–125	10–50	50–100
10	Cement (kg)	210	230	240	260	—	—
	Total aggregate (kg)	1900	1850	1850	1800	—	—
	Fine aggregate (%)	30–45	30–45	35–50	35–50	—	—
15	Cement (kg)	250	270	280	310	—	—
	Total aggregate (kg)	1850	1800	1800	1750	—	—
	Fine aggregate (%)	30–45	30–45	35–50	35–50	—	—
20	Cement (kg)	300	320	320	350	340	380
	Total aggregate (kg)	1850	1750	1800	1750	1750	1700
	Sand*						
	Zone 1 (%)	35	40	40	45	45	50
	Zone 2 (%)	30	35	35	40	40	45
	Zone 3 (%)	30	30	30	35	35	40
30	Cement (kg)	370	390	400	430	430	470
	Total aggregate (kg)	1750	1700	1700	1650	1700	1600
	Sand*						
	Zone 1 (%)	35	40	40	45	45	50
	Zone 2 (%)	30	35	35	40	40	45
	Zone 3 (%)	30	30	30	35	35	40

*Sand is fine aggregate resulting from the natural disintegration of rock.

Table 10.3 Concrete mix proportions

Lumps of clay, etc. even dust or silt will weaken the concrete. Vegetable matter (grass, leaves, etc.), cigarette ends and similar foreign matter can slow the setting of concrete.

Fine aggregates can be checked for cleanliness by rubbing between the hands. If the palms of the hands stay clean, the sand is likely to be clean enough for concreting. If the hands are stained then the silt test should be carried out.

10.27 Admixtures to Concrete

The selection of an admixture should always be with a particular purpose in mind. The justification for using an admixture is either:

(*a*) to improve the properties of the concrete; and/or
(*b*) to effect overall economies in the cost of the concrete.

Admixtures are used primarily for the following purposes:

1 water-reducing;
2 retarding (i.e. extending the setting time);
3 accelerating (i.e. reducing the setting time);
4 air-entraining (improving durability where concrete is exposed to alternate wetting and drying conditions);
5 waterproofing.

Water reduction is achieved by using an admixture which enables the workability of the mix to be held constant with a significantly lower water content. This reduction in the amount of water in the mix also increases the strength of the concrete produced. A secondary effect can be to reduce the amount of cement in the mix – the additive maintaining the workability and strength of the plain mix, but using less cement. By reducing the amount of water used, there is also a reduced tendency for 'laitence' or 'bleeding' – the tendency for water to rise to the surface of the mix on compaction; this improves the surface hardness and durability.

It should be remembered that the function of water in the mix is to:

(*a*) hydrate the cement; and
(*b*) to provide the required amount of plasticity of the mix to enable it to be placed and fully compacted (i.e. workability).

The *amount of water* required for the hydration of ordinary Portland cement is about 12 kg per 50 kg of cement.

The 'total' water (i.e. added water plus water in aggregates) in a mix varies between 20 and 30 kg per 50 kg cement. This means that between 8 and 18 kg of water are added to the mix for every 50 kg of cement used, to make the mix 'workable'.

Less water means:

1 the cement paste will be stronger (in flexure, bond, compression and tension);
2 the tendency to bleed (form laitence) is reduced, leading to increase of surface hardness;
3 the mix will be less liable to segregation;
4 durability will be increased, since the voids caused by evaporation of water will be fewer;
5 shrinkage and cracking are less liable to occur;
6 increased density, resulting from the reduction in unit water content, produces a less permeable concrete.

From the roadworker's point of view, air-entrainment additives are thought to be the most important.

10.27.1 Air-entrainment Admixtures

Air entrainment is important for all concrete members exposed to wetting and drying, freezing and thawing, or to the destructive action of chemicals. It is now a standard requirement for concrete roads, runways and, to some extent, for paved areas.

The main reason for the air entrainment is the significant improvement in durability obtained by the incorporation of the correct amount of entrained air in the mix.

Entrained air should not be confused with *entrapped* air (this latter is in the voids in the concrete, due generally to insufficient compaction). Entrained air in concrete is ideally in the form of microscopic air cells properly distributed throughout the concrete.

Between mixing and final compacting, concrete is usually roughly handled – transported, dumped, spread, vibrated and surface finished. At each of these operations, the air content is likely to be reduced. The larger the size of the 'bubbles' of air, the more likely they are to be 'knocked out' of the concrete in the handling processes. Hence the entraining agent which produces the more microscopic air cells is needed if the necessary amount of air is to remain in the concrete.

With certain types of air-entraining agents, the entrained air is produced by friction during the mixing process – a physical interaction between the solids in the additive liquid and the sand in the mix. (Even without the cement, the effect can be demonstrated by adding a few drops of the additive to a sample of damp sand. Vigorous stirring will transform the harsh unworkable sand into a smooth flowing material.)

The following changes occur in concrete due to the addition of an air-entraining agent:

(*a*) 'rich' mixes may lose some strength;
(*b*) 'average' mixes tend not to change in strength;
(*c*) 'weak' or 'lean' mixes should gain in strength.

(A rich mix can be defined as having at least 360 kg cement/m³; an average mix can be defined as having 240–360 kg cement/m³; and a lean mix can be defined as having less than 240 kg cement/m³.)

Factors Affecting the amount of Entrained Air

The volume of air produced from a given dosage of air-entraining agent varies according to a number of circumstances, the main ones being the:

1 particular air entraining agent being used;
2 type of mix and proportion of materials;
3 type of mixer;
4 length of mixing time;
5 consistency of the mix;
6 use of additions to the mix.

The exact amount of air-entraining agent is generally determined by trial and error, using the materials and equipment available on the specific job.

10.27.2 Water-Reducing Additive

This may also be referred to as a plasticizer and may be supplied in liquid or powder form. The liquid is often a solution of vinsol resin salt and appears as a dark brown watery liquid. It is either fed into the mixer directly or into the gauging water tank. Dosage is about 250–300 ml per 50 kg of cement.

The powder form of plasticizer is added either to the water supply tank or directly into the mixer with a dosage of 15–30 g per 50 kg of cement. The addition of this material enables a reduction of water in the mix, whilst still achieving the same slump value.

10.27.3 Retarder

This additive is often similar in colour to the plasticizer, i.e. it is brown, but rather more viscous than water, hence it is a 'thicker' liquid than the plasticizer.

It is best to add at the mixer stage, at the same time as the water is added to the mix. The undiluted mixture must not come into direct contact with dry cement.

Rates of feed depend upon the temperature; the higher the air temperature, the more additive is required. Rates vary from 0.15 l per 50 kg cement at about 20–25°C to 0.30 l per 50 kg cement at 32°C or over.

10.27.4 Accelerator

This additive is often used when the air temperature is low and falling, in order to make the setting time shorter. The composition includes calcium chloride in a water-reducing agent, with a catalyst to promote the chemical reaction. Dosage of this watery liquid is from 850 ml per 50 kg of cement, up to double this quantity.

It must not be used with cements which already contain calcium chloride or with high alumina elements.

10.27.5 Waterproofing Additive

An emulsion of an organic ester, combined with a lignin-based water-reducing agent is often used. In this form, it is a light brown liquid slightly thicker than water. A dosage of about 550 ml per 50 kg of cement is normal.

It is completely insoluble in water and becomes a permanent constituent of the hardened concrete. It also reduces the amount of water required for the mix.

10.28 Consistence of a Concrete Mix

Each batch mix on a particular job should be kept to the same consistency or workability. To check this, the slump test and compacting factor test can be used. *See* below and Sections 10.32 and 10.34.

(*a*) The slump test is often the only available site testing facility, though the compacting factor test is the more reliable test and should be undertaken in preference to the slump test, whenever possible.

(*b*) The use of the compacting factor test is usually confined to large jobs where there is a site laboratory. It is more accurate than the slump test and is better for concrete of low workability (i.e. where the slump is less than 25 mm).

(*c*) The compacting factor test must be used with machine laid concrete slabs.

10.29 Lightweight Concrete using Pulverized Fuel Ash (PFA)

PFA has been incorporated in concrete in two ways – as a partial replacement for cement (commonly 20–30% by weight of cement by an equivalent weight of PFA), or as a partial replacement for both sand and cement (on a weight or volume basis). Although satisfactory results have been obtained by these methods, research has now shown that a rational mix design method, taking into account the source and type of PFAs and their differing water requirements, can achieve specified strengths from three days onwards. Such a mix will not only satisfy strength specifications and meet workability and economy requirements, it will also contribute to reduction of bleeding and of temperature rise. Other benefits are increased resistance to chemical attack, reduced permeability, and reduction of the cement/aggregate reaction. The use of PFA can therefore undoubtedly produce concrete which is both cheaper and superior in many important characteristics to ordinary Portland cement concrete, but it is emphasized that it is the correct use of the most suitable PFAs that results in the most significant advantages.

PFA is usually of a similar size spectrum to cement, and should achieve the same degree of void filling. Its surface area permits greater reaction with other ingredients, and its particle shape contributes to workability. It has been found, however, that the most significant property determining the suitability of PFA for use in concrete is its water-absorbing capacity. In order to meet most design mix requirements, a PFA should be chosen that has about the same water requirements for standard consistency as an equivalent weight of cement. This class includes most PFAs produced.

In the rationalized mix design method developed for use with PFA concretes, the aim has been to achieve workability and strengths equivalent to ordinary Portland cement concretes of the same age. Thus if the PFA and Portland cement concretes have equivalent early strengths, then the additional lime–PFA interaction which continues after the concrete has hardened, produces an eventual superiority of strength in the PFA concretes.

10.29.1 PFA in Grouting

On both technical and economic grounds, PFA has been shown to be of value in all forms of grouting. It can be used neat, in a slurried form for void filling and as an aggregate with cement in high and low pressure work for foundation and structural grouting.

The chemical properties of PFA are quite significant in grouting. The alkaline nature of the material is such that it presents little danger of chemical attack on pipes or pumping plant. The sulphate resistance of cements is increased by the addition of PFA, which reacts with the free lime liberated by the cement on hydration. The small particle size improves pumping conditions by keeping the grout in suspension, thus reducing sedimentation and allows penetration of small interstices that could not be plugged by a cement/sand grout. The shape of the particles also gives a lubricating action, ensuring better flow qualities.

PFA is used in barrier grouting behind dams and around foundations, stabilization of embankments and railway tracks, bridge abutments, and void filling in abandoned tunnels where ground subsidence has been a problem. PFA grout is also being used in the repair of older masonry structures such as bridge piers, church towers, and masonry-lined railway tunnels.

Testing: Rigid Pavements

New concrete standards have recently been introduced, which are the European Standard (BS EN 206 – Part 1: Specification, performance, production and conformity) and a complimentary British Standard (BS 8500 Concrete). It should be noted that the term 'workability' has been replaced by the word 'consistence' and that there are four methods of assessing or measuring consistence. These are the Slump test, the Vebe test, the Flow Table test and the Degree of Compactability test (not the same as the Compacting Factor test). Details of the Slump test are given in Section 10.32.

10.30   The Silt Test

For this a 250 ml measuring cylinder is required and the procedure is as follows.

1 Fill the cylinder to the 50 ml mark with salt water (1 teaspoonful to ¹/₂ litre).
2 Pour in sand until the level of *sand is up* to the 100 ml mark.
3 Add more salt water until the 150 ml level is reached.
4 Shake the cylinder well, taking care not to spill any of the contents.
5 Stand the cylinder on a level surface and tap it gently until the surface of the sand is level horizontally.
6 Leave it to stand for 3 hours.
7 Measure the thickness of the silt layer (this will have formed on the top of the sand).

This layer should not be more than 8 ml thick or 8% of the sand volume if this aggregate is to be acceptable.

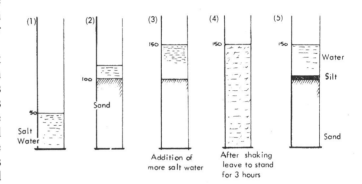

Figure 10.11   Diagrammatic representation of silt test procedure

10.31   Taking a Sample from a Concrete Mixer (either site or truck)

A proper sample consists of scoopfuls taken from four different parts of the load. To get a true sample use a sampling plan as follows:

1 Imagine the load divided into six parts (*see* Figure 10.12).
2 As the load is discharged:
  Let part 1 go
  Take 2 scoopfuls from part 2
  Take 2 scoopfuls from part 3
  Take 2 scoopfuls from part 4
  and 2 scoopfuls from part 5
  Let part 6 go.

N.B.: Scoopfuls *must* be taken from the moving stream of concrete (Figure 10.13).

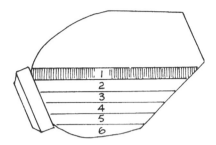

Figure 10.12   Sampling of concrete from a mix

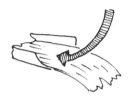

Figure 10.13   Sampling of concrete from a mix

10.32   The Slump Test

1 Mix the sample you have already taken, on the mixing plate, and put it back in the bucket.

2 Place the slump cone on the *levelled* mixing plate and stand with feet on the foot pieces (Figure 10.14).

3 Fill the slump cone in four equal layers. Rod each layer with *exactly* 25 strokes making sure the rod penetrates to the layer below (Figure 10.15).

4 Smooth off when the cone is full, using the float, and wipe away waste. Then slowly lift the cone straight up and off.

5 Lay the slump rod across the upturned slump cone.

6 Measure the distance between the underside of the rod and the highest point of the concrete. Record this distance (in millimetres) (Figure 10.16), this is the 'slump'.

7 If you do not get a 'normal' slump first time, repeat the test. If the second slump also is not 'normal', report the matter to the engineer or supervisor. There are three kinds of slump (*see* Figure 10.17):

normal
collapse
shear.

Figure 10.14   Slump test cone and base plate

Figure 10.15   Filling the slump test cone

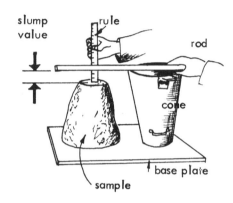

Figure 10.16   Measuring the slump

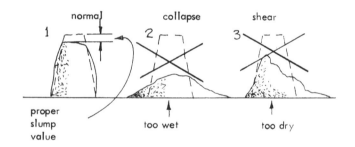

Figure 10.17   Types of slump

10.33   Making Cubes

After doing the slump test, use the rest of the sample to make cubes.

10.33.1   Procedure for Making 150 mm Cubes

1 Mix the concrete thoroughly on the mixing plate.

2 Check that the mould is fixed firmly to its baseplate and that all nuts are tight.

3 Fill to about one-third full.

4 Tamp all over, especially in the corners. *Tamp at least thirty-five times.*

5 Fill to about two-thirds full, tamp, as before.

6 Fill to overflowing and tamp again.

7 Remove surplus concrete. Smooth over with a float.
8 *Mark the cubes whilst still wet for identification later on.*

9 Cover the mould with a damp cloth and a polythene sheet and store inside at normal room temperature: 20°C.

N.B.: If making the smaller 100 mm cubes, fill the mould in *two layers.*

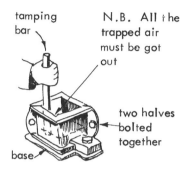

Figure 10.18   Concrete test cube mould

If the concrete is of low slump and, therefore, rather stiff, it may be preferable to vibrate the mix in the test cube. A pneumatic hammer with a steel plate to fit the cube is required, the procedure being similar to that for hand tamping.

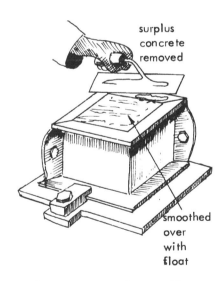

Figure 10.19   Concrete test cube mould

The cubes should be taken from the mould the day after making, numbered, and put into a curing tank – for this they may have to be taken to the depot or laboratory. To remove the cube from the mould:

1 Remove surplus concrete by wire brushing (Figure 10.20).
2 Slacken off all nuts with the spanner.

3 Part the sides of the mould, tapping gently with the hammer, and lift off (Figure 10.21).
4 Do not forget to number the cubes before putting them in the curing tank (Figure 10.22).

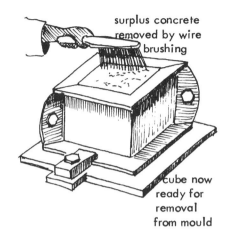

Figure 10.20   Removing cube from mould

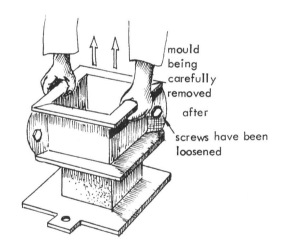

Figure 10.21   Removing cube from mould

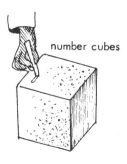

Figure 10.22   Numbering test cube

5 Check that the thermostat is set at 20°C and that the cubes are completely covered by the water (Figure 10.23).

Handle cubes with care.
Keep them warm and wet.
New cubes are fairly soft and could easily be damaged by rough handling.

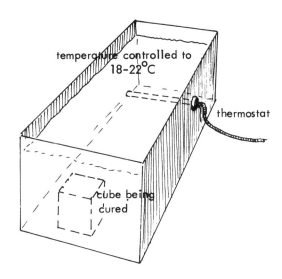

Figure 10.23   Curing the test cube

10.33.2   Compacting Cubes by Vibration

An electric or a pneumatic hammer must be used. Never try to compact cubes by using a poker vibrator. Under vibration, the moulds will tend to move about a good deal on the base plates and so it is preferable to use moulds which are firmly bolted to the base plates.

Fill the moulds with the same number of layers as for hand tamping.

When using an electric hammer, it is best to stand the mould on a piece of level hard wood rather than a hard surface and to vibrate it by holding the foot of the hammer against another piece of timber placed over the top of the mould.

Continue vibration until it is impossible to place any more concrete into the mould and the concrete is fully compacted. Full compaction has been achieved when the air bubbles no longer appear frequently on the surface. Finally, trowel the surface level with the top of the mould.

An identifying mark or number is required on the wet concrete: lightly scratch the surface with a matchstick to make the identification mark.

10.34   Compacting Factor Test

The apparatus for this test is shown diagrammatically in Figure 10.24.

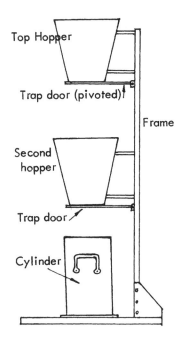

Figure 10.24   Compacting factor apparatus

The procedure for this test is as follows:

1 Check that the apparatus is clean and free from hardened concrete.
2 Weigh the empty cylinder (this must be to an accuracy of ±10 g).
3 Cover the cylinder (normally by means of two steel floats).
4 Using a hand scoop, gently and loosely fill the top hopper with concrete.
5 Open the trap door at the base of the top hopper, thus allowing the concrete to fall into the second hopper.
6 Remove the cover from the cylinder and immediately open the trapdoor in the base of the second hopper so that the concrete falls into the cylinder.

N.B.: If the concrete sticks to either of the two hoppers, free it by gently pushing the tamping rod vertically into the concrete from above.

7 Remove surplus concrete from above the top of the cylinder by holding one trowel in each hand and moving them towards each other, cutting off level with the top of the cylinder.
8 Wipe the outside of the cylinder clean and weigh it.
9 By subtracting the weight of the clean empty cylinder from its weight when full of concrete, the weight of 'partially compacted concrete' can be found.
10 Empty the cylinder and refill it in 50 mm layers, hand ramming each layer (or using vibration) to obtain full compaction.

11 Smooth the top of the concrete with a trowel to make sure that the cylinder is exactly full.

12 Wipe the outside clean and weigh it to find the weight of 'fully compacted concrete'.

13 The compacting factor can now be calculated by dividing the partially compacted concrete weight by that of the fully compacted concrete, giving

$$\text{compacting factor} = \frac{\text{weight of partially compacted concrete}}{\text{weight of fully compacted concrete}}$$

The higher the value (up to a maximum of 1) the more workable is the mix.

It is recommended that the compacting factor test should be carried out about 10 minutes after mixing (thus, allowing for the normal time between mixing and placing on site).

N.B.: The testing of concrete cubes is not included as it does not form part of the work of a roadworker. It is a laboratory test in which the cube (either 7 or 28 days old) is compressed until it breaks – thus giving the compressive strength (N/mm^2) of the concrete.

Tests on Soils

10.35 California Bearing Ratio (CBR) Test for Soils

The procedure for this test consists of finding what load is required to maintain a penetration rate of 1.08 mm/ minute via a circular plunger of 1935 mm^2 cross-sectional area.

The load required to maintain this rate of penetration is registered on a suitable load measuring ring (proof ring) and the results of the tests carried out on either natural or (more usually) re-compacted soils can then be compared with standard test result curves to evaluate the bearing capacity of the soil. This information is required for flexible pavement designs.

The test may be carried out either in the laboratory to which the soil sample is sent, or on site by means of the *in situ* CBR apparatus which can be mounted on a rolled steel joist cantilevered from the rear of a suitable vehicle or fitted to a mobile frame. The *in situ* apparatus enables very rapid test results to be available on site as work progresses.

The laboratory apparatus is shown in Figure 10.25 and the *in situ* equipment in Figure 10.26.

For the laboratory apparatus, the sample must be prepared in a standard mould of 152 mm inside diameter and 127 mm high to which is fitted a base plate and extension collar.

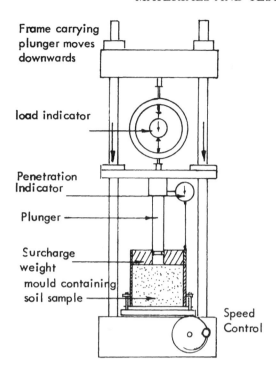

Figure 10.25   CBR apparatus

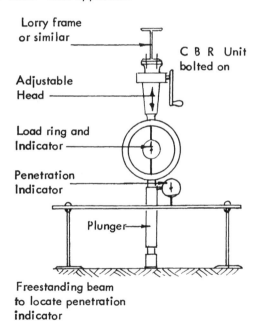

Figure 10.26   Portable CBR unit

The sample must be compacted to the expected density prior to the test, preferably by static compaction, though, in practice, it is sometimes easier to use a dynamic method.

10.35.1 Static Compaction

For static compaction the required weight of wet soil is prepared. This is a special calculation requiring knowledge of moisture content and dry density of the soil and is

$$m_1 = 23.05(100 + W)\rho_d$$

where

m_1 = mass of wet soil in ground, required for one specimen
W = moisture content of the soil sample (%)
ρ_d = dry density of the soil (Mg/m^3)

The static compaction may be carried out by one of the two following methods:

(1) Assemble the mould with collar and baseplate, the baseplate being covered by a filter paper (*see* Figure 10.27).

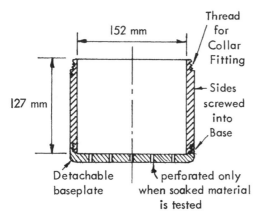

Figure 10.27 *Mould for soil*

The weighed soil is poured slowly into the mould, whilst being tamped continuously, using a 12 mm diameter 300 mm long steel rod. When all the soil has been added, the level should be 5 to 10 mm above the top of the mould.

Place a filter paper on the top of the soil with a 50 mm thick metal plug on the filter paper. The soil is then compressed in the compression-testing machine until the top of the plug is level with the top of the collar. The load is held constant for at least 30 seconds and then released.

If the soil rebounds further loading must be applied. The plug and filter paper are then removed.

The collar is also removed from the mould. If the sample is not to be tested immediately, the top plate of the mould must be screwed on to prevent evaporation.

(2) Assemble the mould, collar and baseplate covered with a filter paper. Divide the soil into three equal parts, each sealed in a container to prevent loss of moisture. Pour one-third of the soil into the mould.

Place three plugs in the mould and compress the soil (by machine) until the soil is compressed to one-third of the depth of the mould. Add the second and third layers of soil in a similar manner. Finally, the soil, covered by one plug only, is compressed until the top of this remaining plug is level with the top of the collar.

10.35.2 To Perform the CBR Test

1 The mould containing the compacted specimen soil and with the baseplate fitted, is placed on the lower platen of the testing machine (*see* Figure 10.25).

2 Surcharge masses are then placed on the specimen (50 N is the normal and is used for CBR values up to 30%).

3 The plunger is then seated and the motor started, to give a rate of penetration of the specimen of 1 mm/minute.

4 Readings of the force are taken at intervals of penetration of 0.25 mm up to a total of not more than 7.5 mm penetration.

5 After testing, small samples (350 g) of soil are removed from below the penetrated part of the mould and checked for moisture content.

6 The readings of force/penetration distance are plotted on a graph and a smooth curve is drawn through the plotted points. A typical set of curves is shown in Figure 10.28.

7 The CBR value is calculated at penetrations of 2.5 mm and 5 mm and the higher value is taken. Usually, it is most convenient to plot the force/penetration curve on a diagram on which curves corresponding to various CBR figures have been printed. The CBR value for the sample can then be read directly from this diagram.

Details are shown in Figures 10.29. *See also* Table 10.4.

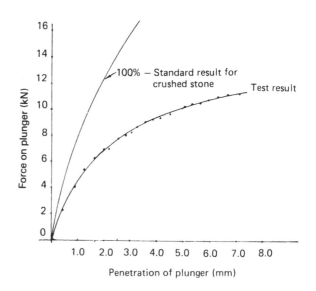

Figure 10.28 *Typical CBR test*

10.36 Optimum Moisture Content

Densities of soils which are achieved by compaction are normally expressed as 'dry densities', generally in kg/m^3, but occasionally in g/ml.

The moisture content at which maximum dry density is obtained for a given amount of compaction is known as the 'optimum moisture content'. This can be determined by means of the British Standard Compaction Test.

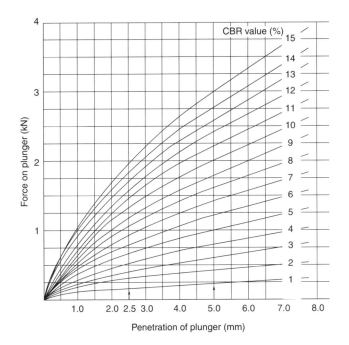

Figure 10.29 CBR plotting chart: this transparent overlay can be laid on top of a graph of a CBR test, and the CBR value then read off directly

Type of soil	Plasticity index	CBR (%)
Heavy clay	70	1–2
	60	1.5–2
	50	2–2.5
	40	2–3
Silty clay	30	3–5
Sandy clay	20	4–6
	10	5–7
Sand (poorly graded)	non plastic	10–20
(well graded)	non plastic	15–40
Sandy gravel	non plastic	20–60

Table 10.4

10.36.1 Standard Compaction test: BS 1377 Part 4

The soil sample to be tested, usually air dried, is first passed through a 20 mm sieve and the amount of 'gravel' retained by the sieve is noted. If this quantity is large, a correction must be applied to the test results.

2.5 kg of the soil passing the sieve is then thoroughly mixed with water to give a fairly low moisture content, about 6%, unless the natural moisture content of the soil is known, in which case it is that value less 5%.

The soil is placed in an airtight container for 2–3 hours for the moisture to 'migrate' through it. It is then compacted in a 101.6 mm internal diameter mould, by means of a 2.5 kg hammer with a 50.8 mm diameter head. This is allowed to fall freely from a height of 305 mm above the top of the soil. Compaction is effected in three approximately equal layers, each layer being dealt 25 blows by the standard hammer.

Compaction is considered to be satisfactory when the soil in the mould is not more than 6 mm above the top of the mould – an extension section is fitted initially to contain the soil sample. If it is more than this height, the test result is not valid and must be repeated.

The soil is then trimmed off level with the mould top and the mould and soil are weighed.

The test is then repeated with 2% more water content and continued with 2% increments in water content until the weight of the wet soil passes a maximum value. It is this maximum value which determines the optimum moisture content of the particular soil. *See* Figure 10.30 for details of the apparatus. For full information on carrying out the test, BS 1377 should be consulted.

When the values of dry density and moisture content are plotted on a graph, the resulting curve has a peak value, the optimum moisture content (*see* Figure 10.31). The reason for this 'peaking' is that a low moisture content results in the soil being stiff and difficult to compact, with a low dry density and a high void ratio.

As the moisture is increased, the water lubricates the soil, increasing workability and producing a high dry density with a low void ratio. Beyond this optimum value, the soil tends to become saturated, with the excess water keeping the soil particles apart and thus increasing the void ratio once more.

Where a correction has to be applied because of the large proportion of stone retained on the sieve, the percentage of the stone/gravel can be obtained by weighing the amount retained, provided that the soil is oven dried. Usually the particle specific density of the gravel will be higher than that of the finer soil particles, so the gravel excluded from the test would occupy less space than the soil which replaced it. Again details may be obtained from the British Standard.

The main disadvantage of the Standard Compaction Test is that, even with the correction applied, the results are not truly representative for a soil with a large percentage of particles larger than 20 mm. The test has been, therefore, adapted for a larger (152.4 mm diameter) mould, with soil passed through a 30 mm sieve. The test procedure is similar to that described above, using the same hammer, but increasing the compaction to three sets of 55 blows, with the soil compacted to a height of 127 mm, which gives a volume of 2.32×10^{-3} m³.

A similar type of test has been developed in America. This is known as the AASHO test (the letters stand for American Association of State Highway Officials). For this test, a hammer weighing 4.55 kg and falling freely from a height of 457 mm is used and compaction is carried out in five layers.

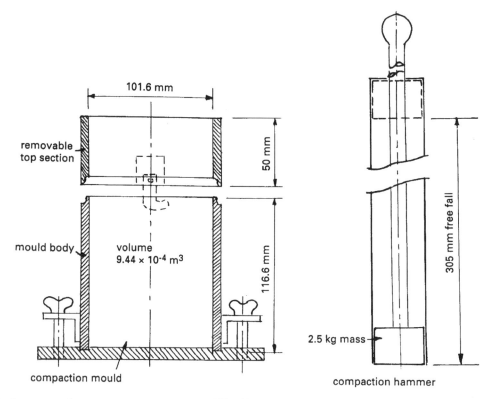

Figure 10.30   Optimum moisture content apparatus, BS 1377

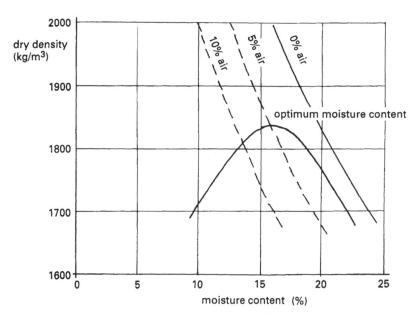

Figure 10.31   Dry density plotted against moisture content

Revision Questions

1   Name the three types of Cationic emulsions and their properties?
2   How might you determine the skid resistance of a surfacing material in the laboratory?

3   When a premixed concrete vehicle arrives on site which type of test would be most appropriate to determine the immediate suitability of the mix?
4   What test might you use on the formation level of a construction site to determine the bearing capacity of a new road or structure?

Note: See page 305 for answers to these revision questions

11

Highway plant

11.1 Introduction

This chapter deals with some of the plant which roadworkers are likely to meet in their daily work. Some of these machines are considered briefly, the descriptions being limited to the principles of operation and the main features of that machine. Other plant, generally that which roadworkers may themselves have to operate, is covered in greater detail with the inclusion of use, care and maintenance points. These notes cannot be regarded as an operator's manual for the equipment concerned and should be read in conjunction with the manufacturers' operating instructions.

The Provision and Use at Work Equipment Regulations 1992 (PUWER) have been in force in the United Kingdom since 1993. The aim of these regulations is to standardize the basic laws throughout the European Union. These regulations require all work equipment to be intrinsically safe to use and also to ensure that no person should be allowed to operate any equipment unless they have been informed, instructed and trained in its use.

Heavy Plant and Equipment

11.2 Excavators (Wheeled)

This type of machine is one of the most versatile and commonly used items of roadwork plant. It consists, basically, of a tractor unit (four wheeled) to provide the power to operate trench excavating equipment and loading equipment.

Excavator/Loaders

Trench excavating equipment

This is a hydraulically controlled dipper arm and bucket which is mounted on a strong rectangular frame on the rear of the machine and can be traversed across this frame for ease of lining up the trenching arm with a line of excavation. The frame also carries jacks for levelling and stabilizing the machine whilst excavating. Maximum excavation depth will depend upon the size of machine and is usually at least 3.5 m. Larger machines can excavate to 5.0 m depth. Rotation of the arm is limited to about 180°.

Loading equipment

This consists of a large bucket which is mounted on the front of the tractor frame by means of pivot arms (one each side) and raised or lowered by hydraulic rams. This bucket can be used for loading, dumping, dozing, grading, site stripping, snow clearing, etc. A self-levelling device enables the machine to traverse bumpy ground without appreciable spillage of the contents of the bucket.

The bucket capacity is usually about 0.75 m^3 and the weight of the loaded bucket in the raised (travelling) position is counter-balanced by wheel weights attached to the large diameter rear wheels and a ballast weight at the rear of the tractor (when the dipper arm and bucket are fitted, these act as counter-balance weights).

The operating pressure for the hydraulic systems is of the order of 7 N/mm^2 upwards which is often referred to as 0.7 hectobar (1 bar = 1 atmosphere).

The tractor-type excavator/loader is used on the majority of small scale roadworks. It has a considerable advantage over most other plant in that it can travel on the road under its own power at speeds up to about 20 mph (*see* Figure 11.1).

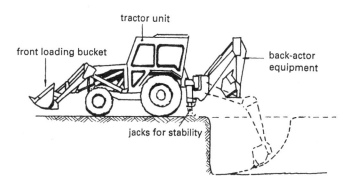

Figure 11.1 Wheeled excavator/loader

11.3 Excavators (Tracked)

For major trenching work and bulk excavation, tracked excavators are used with a variety of digging equipment for the different operations required.

The excavator is a crawler tracked chassis unit to enable the machine to travel over the ground, with a cab mounted on a large circular ring so that it can rotate through the full 360° circle. The motor and control gear are fitted in the cab from which the various digging units are mounted.

The machine can be used to provide itself with the level working platform which is essential to efficient and safe operation (*see* Figure 11.2).

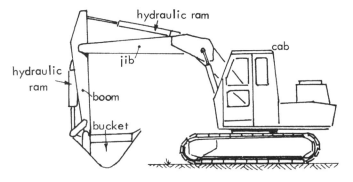

Figure 11.2   Tracked excavator (hydraulic)

A recent development of this machine is the long reach excavator/nibbler (*see* Figure 11.3) which is sometimes hired on bridgework sites for demolition purposes.

This type of plant has to be transported from site to site. Its speed across country is limited to about 2 mph.

A later version of the tracked excavator is a machine (*see* Figure 11.4) which has a parallel link mechanism and offset cylinder capable of moving the boom and bucket parallel to, and to either side of the machine superstructure, without the need to slew. This feature allows the excavator to dig ditches and trenches down to 2.5 m, directly alongside any obstacle such as a wall, fence or kerbline.

11.4   Mini Excavators and Skid Steer Loaders

Some small and versatile machines have now entered the range of highway plant and are common on many small maintenance roadwork sites.

These include the following:

(a) ¾ tonne tracked mini excavator (Compact) 0.68 m wide
(b) 1 tonne tracked mini excavator, 1 m wide
(c) 3 tonne tracked mini excavator, 1.56 m wide
(d) Skid steer loader with attachments.

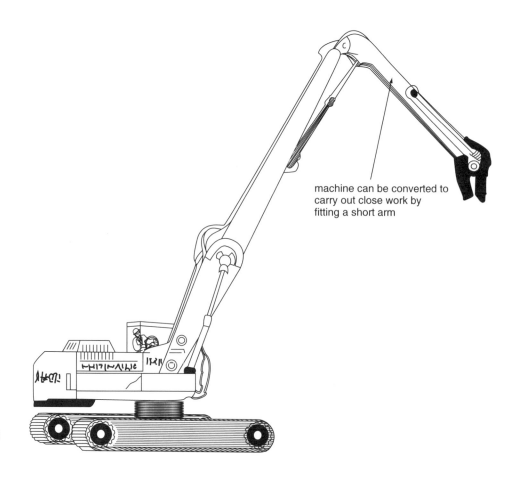

machine can be converted to carry out close work by fitting a short arm

Figure 11.3   Tracked long reach excavator/nibbler

Figure 11.4   Tracked excavator with parallel link

All these machines are capable of turning in their own length and have a mechanism for locking the track or wheels on either the near or off side in order to achieve a tight turning circle.

Mini excavators can be fitted with a variety of different bucket sizes (*see* Figure 11.5), hydraulic breakers or post hole borers with 150 mm, 225 mm and 300 mm augers and blades.

Skid steer loaders are multi-functional and are capable of use with 4 in 1 buckets, 1.5 m wide (*see* Figure 11.6), post hole borer (Figure 11.7), scarifier, forks, or concrete breaker and blade.

Figure 11.5   Mini-tracked excavator

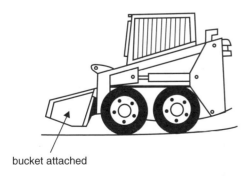

bucket attached

Figure 11.6   Skid steer loader (bucket attached)

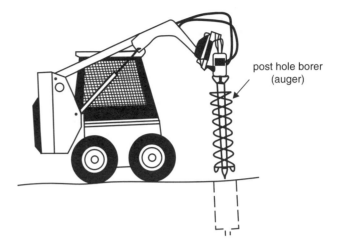

post hole borer (auger)

Figure 11.7   Skid steer loader (post hole borer attached)

Small Trenching Units

Various types of small walk-along trenchers are now in use for shallow trenching. A typical trencher weighs around 700 kg and is capable of digging a 100–250 mm wide trench, 1.25 m deep, using a heavy-duty rock and frost digging boom. Steerable drive wheels enable such a machine to be manoeuvred into tight places.

11.5   Dumper Trucks

Dumpers come in many different designs and configurations. They may be either hydraulically or gravity tipped. Two or four wheel drive options are available, with a great variety of skip sizes, varying from 15 cwt capacity (with 2 wheel drive and gravity skip operation) to 6 tonne capacity, having four wheel drive and hydraulic tipping.

Figure 11.8   Dumper truck

In addition, there is a mini dumper, which, although only 1 m wide, has a hydraulically controlled skip.

Other variations include fixed forward-facing skips or, alternatively, turntable mounted skips, able to rotate through an arc of 90° in either direction.

These machines, when operated by a competent driver, are very versatile all terrain vehicles and are ideally suited both for on site haulage and for use in confined spaces.

11.6 Lifting Appliances

This term covers a range of equipment such as a crab, winch, pulley block or gin wheel used for raising or lowering and also various items of plant such as a hoist, sheerlegs, crane, excavator/loader and tracked excavator.

Lifting appliances are subject to certain requirements made under the Lifting Operations and Lifting Equipment Regulations (LOLER).

When mobile cranes, excavator/loaders and tracked excavators are used on highway sites, weekly inspections have to be made and recorded by the operator. Mobile cranes are usually hired with a driver conversant with the Regulations but excavator/loaders are frequently handled by operatives unfamiliar with the legal requirements. An excavator/loader or tracked excavator may be used as a crane for the purpose of lifting pipes, etc., in connection with the excavation, but a Certificate of Exemption (from the full requirements for a crane) is necessary for certain machines. When lifting a load, this must be securely attached to the part of the machine designed for this purpose and the load must be correctly slung using properly attested chains, slings and hooks.

11.7 Lorry Loaders

Lorry loaders are increasingly being used on highway sites and within depots. The hydraulic loader is normally mounted behind the cab as shown in Figure 11.9 and a fairly rigid platform is achieved by using stabilizers each side of the vehicle. The stabilizers are load-supporting legs extended outwards from the side of the vehicle to give a wider and more stable base across the width of the chassis. The leg can be hydraulically extended vertically to reach the ground, and timber skids placed between the stabilizer and ground to assist in load distribution.

The maximum radius for loaders in the 6–10 tonnes metre (tm) range is from 5 to 13 metres, but larger loaders can have radii over 20 m (a capacity of 8 tm means 2 tonnes could be lifted at a radius of 4 m or 1 tonne at 8 m). When siting a lorry loader, particular attention should be given to the proximity of potential hazards such as overhead electric lines, nearby structures or cranes, etc. Nor should the danger to or from underground services or culverts be overlooked when placing the stabilizer feet on the ground. Strong packing

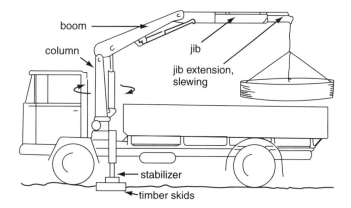

Figure 11.9 Typical lorry loader

should be placed under the feet to spread the pressure on the ground. Working on sloping ground should be avoided as the permissible loads at given radii are provided by the manufacturer for a vehicle standing on firm level ground. Audible and visual warning devices should be fitted to the vehicle to prevent safe working loads being exceeded.

11.8 Paving Machines

11.8.1 Floating Screed Pavers

The floating screed paver is the type with which most roadworkers will be familiar, and the majority of bituminous material mats are laid by paving machines which use this principle. It is useful to summarize the major functions and design parameters of such a paver. These are as follows:

(*a*) To receive hot or cold materials, coated or uncoated, from a very wide range of truck designs without spillage or damage to the vehicles in the shortest possible time.

(*b*) To translate and spread this material to a variety of widths and depths without segregation to form a hard level surface.

(*c*) To give the mat the maximum amount of compaction possible under varying conditions and still retain a satisfactory surface finish.

(*d*) To give very accurate control to the levels of the mat both longitudinally and transversely.

(*e*) To provide satisfactory flotation and traction over a wide range of operating conditions.

(*f*) To be capable of simple and highly reliable operation, with good accessibility to facilitate maintenance.

Although the basic design of pavers has changed little over the last twenty years, in detail, of course, many changes have been made. These were all intended to make the task of the operator and/or maintenance fitter easier, to lay material to a finer specification and to carry out this task more economically.

To provide for varying conditions, there is a range of size of paving machine available. These cover the following operating widths (typical figures are quoted):

1–3 m
2–4.25 m
2.5–5.25 m
2.5–6 m
2.5–8 m
7.5–12 m (twin, side by side machines).

These machines may be either wheeled or tracked.
The larger wheeled machines have:

1 Two pairs of driving wheel's (pneumatic tyred).
2 Two conveyors.
3 A variety of screeds (interchangeable):
  (*a*) tamping screed,
  (*b*) tamping/vibrating screed,
  (*c*) vibrating screed,
  (*d*) heavy duty concrete screed.
4 A speed of travel when laying which can be varied between 0.6 m/minute and 30.5 m/minute.
5 A travelling speed when not laying material of up to 16 km/hour (10 mph).
6 A hopper capacity of 9 to 13 tonnes.

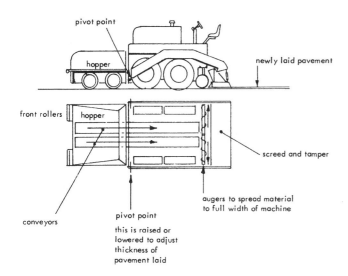

Figure 11.10 Flexible paving machine

11.8.2 Basic Principles of Free Floating Screeds

The basic principle of a free floating screed is that the thickness of the mat can be altered by adjustment whilst the machine is travelling (*see* Figure 11.11). The operating principle is as follows:

The screed is attached to the paver by the side arms and is free to pivot up and down about this point. When the screed is pulled into the paving material that has been spread in front of it by the augers, it will move up or down until the screed base plate is parallel to the paving base. To adjust the thickness of the mat produced, it is necessary

Figure 11.11(a) Increase in mat thickness: screed angle of attack (A–B) is increased by raising pivot point (P)

Figure 11.11(b) Increase in mat thickness: as the paver moves forward the screed rises until (A–B) becomes parallel to the base (C–D) and will remain parallel to the base until the position of the pivot point (P) is changed

Figure 11.11(c) Decrease in mat thickness: screed angle of attack (A–B) is decreased by lowering pivot point (P)

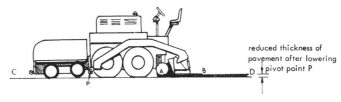

Figure 11.11(d) Decrease in mat thickness: as the paver moves forward the screed will lower until (A–B) becomes parallel to the base (C–D) and will remain parallel to the base until the position of pivot point (P) is changed

to alter the angle of the screed base plate which is termed 'the angle of attack' and this can be done in two ways:

(*a*) to change the angle of attack of the screed at the screed itself; or
(*b*) to change the angle of attack at the forward end of the screed towing arms which gives a more gradual change in the level.

An increase in the angle of attack of the screed will increase the thickness of the mat whilst a decrease will have the opposite effect.

Although method (a) may have an advantage when laying materials which are not being received in consistent conditions, for example, in the case of dry lean concrete, dry loads followed by wet loads, etc., or in the case of black top materials when major temperature changes are apparent, in the main, method (b) is favoured as this method gives a more gradual change in levels and, therefore, on major road works, airports, etc., when precision finish in levels is required, it is less likely to cause malformation in the mat.

It has been found by experience that it takes approximately 3 m travel from any change in height of the tow-point for the mat to reach its required thickness change. Hence, the change must be made early enough for the proper thickness of mat to be obtained where it is required.

In the case of the screed change (a), although theory may state that natural paving length can still be valid, in practice it is found that mat thickness is changed much more rapidly.

One very important point, however, when considering the paving action of a free floating screed, is that it is a dynamic and not a static problem. In other words, the screed remains in equilibrium only whilst moving and every time the paver is allowed to stop a malformation of the mat results. For precision paving it cannot be stressed too strongly that the paver should not be allowed to stop during the whole day's work if this can be accomplished, as even small pavement blemishes will readily be seen.

It is, of course, appreciated, that to keep a machine working from start to finish of the day and moving forward the whole time without stopping is virtually impossible, as any untoward incident such as loading vehicles breaking down, out of phase running of vehicles due to traffic lights, or any such rhythm breaking may necessitate the paver coming to a halt because of lack of material. However, the complete laying operation should be planned so that the machine laying speed is governed to the supply of materials available.

The figure of 3 m referred to previously is the natural paving length (distance from any change in height of the tow-point for the mat thickness to reach its required change) and will, of course, vary between individual pavers being dependent on screed design and its relationship to the base unit of the paver, and also on the type of material being handled.

When machine laying flexible materials, the following points are important:

1 Make sure that the bottom of the screed is clean (this should be cleaned at the end of each period of working and checked before restarting work).

2 Check the tamper for adjustment and wear. When it is at the bottom of its stroke, it should extend slightly below the bottom of the screed plate.

3 Heat the screed plate before starting laying operations.

4 Make sure that the material in lorries waiting to unload is kept covered with tarpaulins.

5 Do not start laying operations until at least two loads of material are on site.

6 Operate the paver with the minimum number of stops.

7 Keep a steady feed to the screed, with sufficient material to cover the shaft of the transverse distributing augers.

8 Shovel material from the sides of the hopper to the centre (to avoid it becoming chilled before laying).

9 Do not leave material in the hopper when laying is interrupted.

10 If narrow strips are laid by hand alongside the machine, they should be left high enough to allow for extra compaction and rolled immediately and before the material can cool.

11 When butting up to previously laid and compacted material, lay slightly 'proud', so that, after rolling, the joint is uniform and fully compacted. This joint should be 'pinched' using the heavy wheel of the roller.

11.9 Road Sweepers

Consist of a road vehicle fitted with a tank (for water and rubbish) and two brushes:

for carriageway sweeping (main brush)
or gutter sweeping (vertically mounted).

The system requires a suction pump and water pump. Water must be sprayed ahead of the brush. Trouble may be caused if the gutter brush hits the gutter; the weight is carried on the bristles and it can jam against the kerb, hence care must be taken when starting to sweep.

11.9.1 Main Operating Features

1 The gutter brush is on a fully floating suspension and will retain contact with the face of the kerb even if the vehicle deviates from the kerb line by 300 mm.

2 The main brush normally sweeps a width of 1.1 m, but can be increased to sweep the full width of the vehicle (2.25 m approximately) by means of an extension brush.

3 The suction unit is fitted with a nozzle which normally collects debris from behind the gutter brush. Debris collected by the main brush is swept across to the gutter end by the deflector blade.

4 The raising and lowering of the brush is usually by hydraulic power.

5 The sweeping equipment is powered by an auxiliary engine which is controlled from the cab.

6 The water and debris is discharged by tipping the tank with the rear door open.

7 The machine may be used as a gully emptier using the 200 mm hose from the rear of the vehicle.

11.9.2 Compact Sweepers

Smaller sweepers have been developed for use in shopping precincts and car parks and along footways. These have increased cab space, much better visibility and

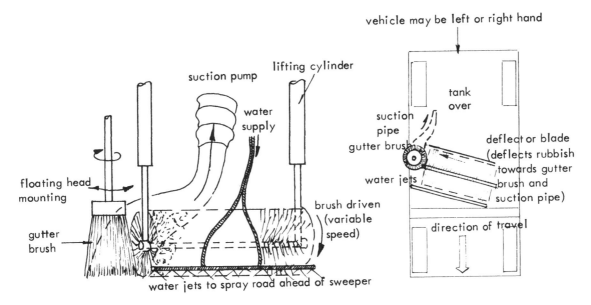

Figure 11.12 Diagrammatic layout of road sweeper

can operate in restricted spaces. They can be fitted with various attachments, such as a boom wander hose, snow plough and trailer gritter.

One type has an articulated chassis with a 100 mm diameter suction hose, and this can empty litter bins, clean flower beds and use a third feeder broom to get debris out of corners.

11.10 Gully Emptiers and Jetting Equipment

Gully emptiers and jetting tankers/trailers may be supplied separately or combined as a gully emptier having jetting equipment at the rear of the vehicle.

11.10.1 Gully Emptier

This consists of a two-part tank, mounted on a vehicle chassis so that it is inclined slightly to the rear. The tank is usually of 6 mm steel plate.

The front portion of the tank is for clean water, the rear compartment being for sludge. This latter compartment is fitted with a large door at the rear locked by means of steel clamps and sealed by a neoprene rubber joint. The water tank is usually about half the size of the sludge tank. Total tank capacities are 3600, 4500, 5400 and 6800 litres.

The filling and emptying of the tanks is by means of an exhauster–compressor pump unit. This enables the tank to be filled by suction and discharged by pressure, without the tank contents actually passing through the pump, i.e. it is a compressed air/vacuum system, the exhauster–compressor being an air pump.

A combined vacuum and pressure gauge is fitted to the tank where it can easily be seen by the operator. The level of the contents of each tank compartment is checked by a sight gauge.

The two compartments are inter-connected by a 100 mm full bore valve, mainly to facilitate cleaning. An air pressure equalizing valve is also fitted. The clean water compartment can be filled from a hydrant, via a hose and pipe. This incorporates a 'break' in the feed line and so complies with hygiene regulations.

The exhauster–compressor pump is usually of the rotary type. A change-over valve within the pump allows it to be used either:

(*a*) as an exhauster to create a vacuum; or
(*b*) as a compressor to create pressure.

A typical pump has a capacity for exhausting about 3 m³/minute at 1100 rev/min and is driven from the power take-off by direct drive, since its speed is compatible with that of the vehicle engine.

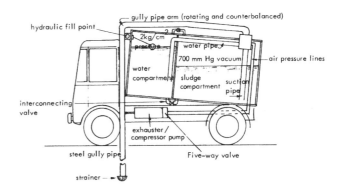

Figure 11.13 Gully emptier (see Section 9.5.11 for suitable vehicle signing in type C mobile work operations)

11.10.2 Operation

1 A vehicle equipped with two tanks: one containing clean water, the other to receive the material sucked out of the gully.

2 A pump which can be operated:

  (*a*) as a suction pump to empty the gully;

  (*b*) as a delivery pump to inject the flushing water under pressure.

3 After lifting the gully grid, the suction pipe and strainer are inserted.

4 The contents of the gully are sucked out into the receiving tank on the vehicle.

5 Clean water is injected into the gully under pressure to stir-up contents.

6 Gully is again sucked out.

7 Gully is refilled with clean water.

8 Suction pipe removed, grid replaced on gully.

When full, the receiving tank can be emptied at a convenient point.

The rear door is opened and a ram pushes out the contents.

11.10.3 Gully Emptier with Jetting Equipment

A high-pressure hose reel is positioned at the rear of the vehicle and a pump capable of a working pressure of 1400 psi and a flow rate of 30 gpm is mounted on the chassis at the side of the vehicle. A rigid lance can be used for cleansing operations or a jet nozzle can be fitted to the hose. The jet nozzle is introduced directly into a drainage pipe or sewer and the jet action of the water from the rear-facing jets propels the nozzle up the sewer, pulling the hose behind it. A forward centre hole in the nozzle breaks up the accumulation in front of the hose to give an easier passage (*see* Figure 11.14). When the hose is withdrawn, the water jets flush out the pipe.

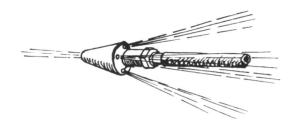

Figure 11.14   Jet nozzle

11.10.4 Jetting Tankers and Trailers

These are specialist machines having high-pressure jets well over 2000 psi. Some 4-wheel trailer jetting pumps operate up to 5000 psi at 60 gpm. The high-pressure pumps are often described as horizontal in-line positive displacement plunger pumps, having 3 cylinders (sometimes 5). A basic unit consists of pump and drive, close-coupled via an epicyclic gear box, all mounted on a

common frame. The hose reel can be hydraulically driven in both directions.

Larger machines from 80 to 150 hp can give water pressures up to 12 500 psi, having lances or guns capable of cleansing steel surfaces or cutting through concrete. These must be used with extreme caution.

11.11   Winter Services Vehicle (Principle of Operation)

This consists usually of a long hopper-type body with a conveyor belt forming the floor of the body. At the rear the conveyor terminates in a vertical rectangular funnel down which the sand/salt is fed to the spinner which is a horizontal rotating disc with small blades; this throws the falling material out horizontally in circular motion.

Calibration of the salt spreader is most important. Salts from different sources have differing flow properties and tests must be made with the same salt as will be used operationally.

The test site must be a paved strip, long enough for the machines to attain working speed and about 3 m wider than the roads on which they are to be used. The surface should be close textured so that salt is not lost in crevices.

The test surface is marked with 1 m squares and the vehicle driven over it fully loaded at normal working speed. The salt on each square metre is then swept, collected and weighed and a chart plotted of the distribution pattern.

Squares at the extremities of the test area which have less than 5 g of salt on them are disregarded. Modern machines however, can give reasonably even spreads over the full working width at rates of up to 40 g/m^2.

Some machines may give an uneven spread, in which case the only course of action is to set the spreader controls to

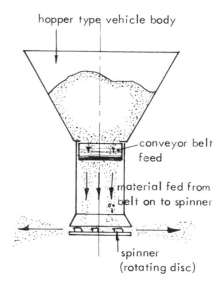

Figure 11.15   Gritting spreader (diagrammatic)

give the correct average over the effective width. Settings must be adjusted and test runs repeated until the correct values are obtained. These settings, etc. are then noted for that particular vehicle.

Spreaders fall into three main categories:

1 Those with spinners driven off the vehicle so that both the spread rate and width of spread varies with the road speed.

2 Those machines whose mechanisms are driven such that at all road speeds the spread rate and spread width remain constant. (This usually consists of a mechanical or hydraulic connection between the road wheels and spreader drive.)

3 Those with independently driven spinners. These can be set to give a constant width of spread at all speeds, but variation of road speed will give a different spread rate.

The corrosive action of road salt shortens the life of metal gritter bodies. Thermoset polymer panels have been developed to give a long life road gritter. Damaged panels can easily be removed and replaced.

11.12 Snow Plough

A snow plough is often fitted on the front of a winter maintenance vehicle so that it can perform the tasks of clearing snow and also treating the road behind the vehicle.

In areas where snowfalls are expected to be considerable and long lasting, the dozer type of snow plough has tended to be replaced by snow blowers. These are lorries fitted with a very powerful fan which is used to blow the snow to the side of the road.

Snow ploughs are not dealt with in detail due to the infrequency with which they are needed in the greater part of Britain.

11.13 Mowers and Tractor Cutting Attachments

A wide range of machines is marketed for grass cutting, but there are basically three categories in use:

(a) pedestrian-controlled mowers
(b) ride-on or mini-tractor mowers
(c) large tractor-mounted attachments.

There are basically three types of cutting mechanism used in each category and these are cylindrical knives, rotary blades and flails.

Cylinder mowers give the best finish but normally need regular specialist sharpening and usually cannot cope with grass above 125 mm in height. Some later cylindrical models have larger diameter cylinders with fewer knives and these can deal with growth up to 200 mm; however,

debris or protrusions in a verge, e.g. water stopcock covers, etc., can damage cylinder knives.

Rotary machines generally allow fewer cuts to be made within a season as they can handle taller growth, often cutting a swath of 2 m. One of the major problems with a rotary is that it discharges cuttings onto the grass, these being long-bladed cuttings which cannot be regarded as mulch and can yellow and weaken a lawn, encouraging growth of weeds, However, some machines have facilities for attaching grass-collection bags and rear-discharge machines leave cuttings in rows for collection.

The flail cutting method overcomes the problem of uncollected cuttings since the flail really does mulch-cut grass. This method was developed for cutting long grass often in inaccessible areas but can also cope with debris and protrusions in verges. The ride-on versions have a 800 mm cutting head and can cope with growth up to 300 mm in height.

Generally, self-propelled ride-on mowers are most effective when cutting verges, being more manoeuvrable than a tractor-drawn gang mower and covering ground much more quickly than a pedestrian-controlled mower.

The tractor-mounted or towed unit is versatile and can cover large areas of ground quickly, a typical machine trailing five large cylindrical gang mowers giving a cutting width of around 6 m. Some versions can hydraulically raise their side arms and can fold certain of the gang mowers tightly against the tractor body to enable the tractor to pass through gateways and also be used in confined areas (*see* Figure 11.16).

Flail hedge-trimmers are a common tractor-mounted unit and are frequently used in highway work. The operator can very easily trim a hedge, but must keep the flails spinning at the correct speed as centrifugal force is a vital factor to ensure correct cutting and to avoid damaging the attachment. To get the best out of the flail cutter, small cuts must be taken of, say, 450 mm at each pass with the flail head being held at an angle as shown in Figure 11.17.

Figure 11.16 Tractor and gang mower

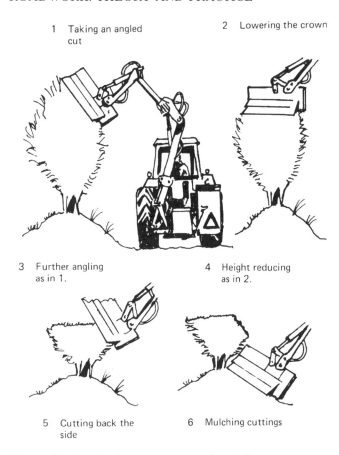

1 Taking an angled cut
2 Lowering the crown
3 Further angling as in 1.
4 Height reducing as in 2.
5 Cutting back the side
6 Mulching cuttings

Figure 11.17 *Cutting sequence using flail-type equipment*

11.14 Road Rollers

Compaction of soils and road material can be achieved by any one of three main methods:

1 pressure – compaction by dead weight;
2 vibration – compaction by means of both dead weight and excitation of the soil or road material;
3 impact – compaction by rammer, etc.

11.14.1 Dead Weight Steel-Wheeled Rollers

These are of the 6–10 tonne range with provision for ballasting to increase the total weight if necessary.

The layout of these machines is of a single (steerable) front roll with twin rear wheels, each of which is about half the width of the front roll. A large main wheel diameter is required in order to eliminate the tendency to 'over-penetrate' the surface, thus creating bow-waves and increasing rolling resistance.

The roller wheels are usually capable of being ballasted with water or sand (i.e. additional weight which may easily be added or removed).

Figure 11.18 *Dead weight steel-wheeled roller*

It is a requirement of the roller that it should be able to change the direction of travel, or start off, smoothly and without 'dwell' particularly when rolling such materials as hot-rolled asphalt. Speed ranges from 1.5–6 mph are usual. Hydraulic drive systems are used.

When rolling bituminous materials, the optimum (best) speed has been found to be 1.5–3 mph and for almost all bituminous materials a water sprinkler system (fitted as standard) needs to be used.

Tandem rollers may be used in place of the three-wheeled variety.

11.14.2 The Tandem Roller

This roller is often used and may be fitted with a chipping spreader for adding coated chippings to a new bituminous wearing course. Tandem rollers are more difficult to use for edge rolling than the three-wheeled roller.

Figure 11.19 *The tandem roller*

11.14.3 Pneumatic-tyred Rollers

These are used normally for compacting surface dressings and are of 8–10 tonnes in weight. The wheels are contained on two axles with 5 wheels on the leading axle and 4 on the trailing one, which is also the steering axle. Hydraulic steering is used. The wheels are independently mounted to give 5° to 6° of pivoting on the front axle.

11.14.5 Vibratory Compaction

Vibratory rollers fall into four main categories:

1(a) Small, pedestrian-controlled single drum type (Figure 11.22(a)) used mainly for footpath work.

1(b) Pedestrian-controlled double drum roller used for footpath, hard shoulder and carriageway work (Figure 11.22(b)).

Figure 11.22(a)
Single drum

Figure 11.22(b)
Twin drum

2(a) Medium-size tandem type, usually with one roll being vibrated (mainly 1–4 tonnes weight) (Figure 11.23(a)). Later types are articulated and are more versatile, as shown in Figure 11.23(b).

Figure 11.20   Pneumatic-tyred roller

Figure 11.23(a)   Tandem vibrating roller

11.14.4 Sheepsfoot Roller

This term embraces a very wide range of compaction equipment, covering anything other than a smooth roll.

The 'sheepsfoot' roller usually has a series of roughly rectangular-shaped feet and should be referred to as a 'tamping compactor'. Its use is on earthworks where its action reduces the amount of air voids, thus giving increased soil strength and the protuberances (feet) tie the layers of deposited material together.

Figure 11.21   5 tonne towed vibrating sheepsfoot roller

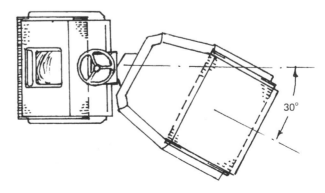

Figure 11.23(b)   Articulated tandem vibrating roller

2(b) Larger tandem machines (see Figure 11.23) having one or both rollers vibrated. Some have the facilities of 'offsetting the drums' to assist in radiused work. Vibration automatically cuts out and in when changing direction.

3 Medium to heavy towed or trailer type. 5–10 tonnes weight with single roll 1.2–1.5 m diameter and 1.5–2.5 m wide (Figure 11.24).

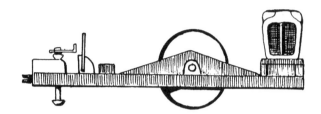

Figure 11.24   Trailer type vibrating roller

4 Self-propelled heavy rollers. These usually have a pair of driven pneumatic tyres and centre pivot style steering (allowing 45° angular movement on each side (Figure 11.25)). Certain later types, in which the smooth drum roller is replaced by a modular design pad roller, are equipped with a compaction meter which allows the operator to monitor the compaction process.

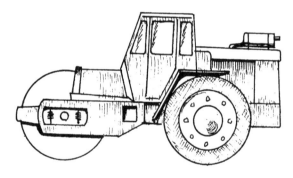

Figure 11.25   Heavy self-propelled vibrating roller

The small pedestrian-controlled type is the most widely used, and is changing from single to double drum construction. The drums are about 600–1000 mm wide and 450–700 mm diameter, mounted flexibly in a frame which carries the engine and transmission. The vibratory mechanism usually incorporates an eccentric weight mounted inside the roll or drum and rotating at 3000–5000 cycles/minute. The centrifugal force so produced amounts to about four times the dead weight of the machine (e.g. a one tonne vibratory roller could have an effective weight of four tonnes but the manufacturers' tables should be consulted).

In the larger vibratory rollers, the vibratory frequency tends to be lower, ranging from 1000–5000 cycles/minute

with a figure of about 2300 cycles/minute being found, by practice, to give good results for all materials.

Under certain conditions, such as:

(a) very thick lifts or layers;
(b) material with a high moisture content; and
(c) uniformly graded granular material;

a vibratory roller can experience traction difficulties. An additional wheel motor mounted directly on the roll is the usual method employed for overcoming this difficulty, but the cost of this modification is considerable. This additional drive improves traction by about 30%.

11.15   Concrete Mixers

For most roadworkers, there is an increasing tendency for on-site mixing of concrete to be replaced by ready-mixed truck delivered concrete. This is generally more convenient and the mix quality is more easily controlled.

Many types of mixer may still be in use. Details of these are included in this section.

Concrete mixers can be divided into three main types, which are:

(a) horizontal axis rotating drum mixers;
(b) tilting drum mixers;
(c) pan mixers.

11.15.1   Horizontal Drum Concrete Mixers

1 The principle of this type of machine is a mixing cylinder set with its axis of rotation horizontal and with a system of curved blades fixed to the inner face of the drum.

2 By rotating the drum, the cement and aggregates are mixed together by a tumbling action.

3 The rotation of the drum is provided by an engine fitted alongside the drum.

4 Aggregates are fed into the mixer drum by a hopper and water is added via a pipe leading into the drum from a water tank and water metering device.

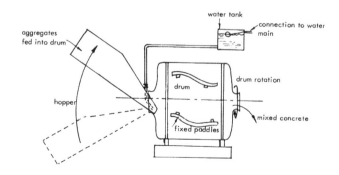

Figure 11.26   Horizontal drum concrete mixer

5 The operation of these horizontal drum mixers varies, the two main variations being:

(*a*) where the drum has both front and rear openings, is fed with materials from a rear hopper and delivers the mixed concrete from the front opening. The rotation of the drum is in one direction only.

(*b*) where the drum has a front opening only, the materials are fed in and the drum rotated to mix the contents. When thoroughly mixed, the direction of rotation is reversed and the concrete is fed out of the front opening by the discharge blades.

The location and angle of the blades is critical for this reversible action. This type of mixer is the one used on ready-mixed concrete vehicles.

11.15.2 Tilting Drum Concrete Mixers

1 In this type, the rotating drum is mounted on a cradle which can itself be rotated.

2 In operation, the drum is tilted upwards to receive a batch of aggregates to be mixed.

3 Mixing is carried out with the drum inclined upwards.

4 The drum cradle is then rotated so that the drum is tilted downwards to discharge the freshly mixed concrete.

5 The drum itself is normally shaped like a cylinder with a tapered nose section.

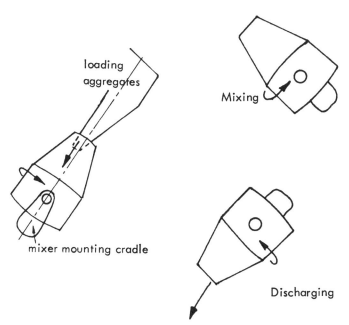

Figure 11.27 Tilting drum concrete mixer

11.15.3 Pan Type Concrete Mixers

The drum of this type of mixer may be either fixed or rotating with its circular end plate horizontal. The mixing of the materials is done by paddles rotating within the pan

and the mixed material is discharged by opening a panel in the floor of the drum. The rotating paddles then push the concrete to the hole through which it falls.

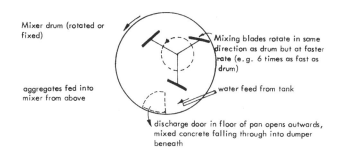

Figure 11.28 Pan type concrete mixer

11.15.4 Notes on the Mixing of Concrete

1 The mixer must have a water tank and gauge and the water supply must be clean and pure.

2 On some mixers, the cement can be added separately, in others, it is added with the aggregates.

Care should always be taken that cement is not fed in first, as it will then adhere to the inside of the drum and the blades.

3 Always wash out a mixer drum thoroughly, immediately on completion of mixing.

4 Do not, on any account, hammer the drum to 'crack off' hardening cement/concrete film, as this will damage the drum shape and reduce the efficiency of mixing.

5 Clean the outside of the mixer by rubbing down with an oily rag.

6 For proper operation, the mixer must be set up level. This helps to mix the concrete properly and keeps the water gauge accurate.

7 During a break in mixing, put coarse aggregate and water in the drum and leave the mixer running.

8 Check regularly that the mixer is running at its correct speed of rotation (this speed is set by the manufacturer) and that the water gauge is accurate.

9 Carry out the following daily checks/maintenance:

(*a*) grease all nipples;

(*b*) clean and inspect all moving parts for wear;

(*c*) clean and inspect all wire ropes;

(*d*) check and top up oil and fuel (engine);

(*e*) check that blades are not bent, broken or working loose;

(*f*) check that the mixer is set up level, re-levelling it if necessary.

11.15.5 Batching Plant

For the production of consistent mixes and for increasing output a batching plant or batch measuring hopper is useful (being preferred to the use of gauge boxes).

The increased output is attained by feeding the materials into the hopper in the correct proportions (by weight) whilst the previous batch is being mixed.

Water Measuring Tanks

These are of two main types:

1 open cistern, fitted with ball valve and float;
2 syphon tank.

Both tanks are provided with measuring devices; visual operation is used with the open cistern type, the syphon type operation is automatic.

N.B.: The smaller tilting drum concrete mixers have been included in the heavy plant, with the larger mixers, but are more correctly considered as light plant.

Light Plant and Equipment

11.16   Compressors

In roadwork, compressors are used mainly to provide power to operate road breakers.

11.16.1   Power Unit

This power is provided by means of compressed air, produced on demand by means of an air pump driven by a petrol or diesel engine. A receiver tank is usually fitted to a compressor, its purpose being to separate oil particles from the compressed air. Later designs omit the receiver vessel as the oil separation unit is integral with the pump.

Many of these air pumps are of the rotary vane type (*see* Figure 11.29), which gives smooth operation with an even (pulsation-free) flow of air, normally at an operating pressure of 7 kg/cm² or 7 bar. Reciprocating type pumps are also used.

At this pressure 1.5–2.0 m³/min of compressed air is required to operate each breaker. 1 m³/min is equivalent to 1000 litre/min or 16.6 litre/second.

For most roadwork applications the smaller two-breaker machines are adequate and are in trailer-mounted, self-propelled lorry or tractor-mounted form. There is a limit

Air capacity of machine cubic metres per minute	Operating pressure	No. of breakers
2.5–4.5 m³/min	7 kg/cm²	2
5.0–7.5 m³/min	7 kg/cm²	4
10.0 m³/min	7 kg/cm²	6
		etc.

Table 11.1

to the size of compressor capable of being mounted on a tractor (usually not larger than 5.5 m³/min capacity).

A typical trailer mounted compressor is shown in Figure 11.30. The unit is mounted on a heavy chassis within a tough canopy, weighing under one tonne. Noise reduction is important and some models achieve a level of 69 dBA at 7 m from the compressor (normal conversation at 1 m is 60 dBA).

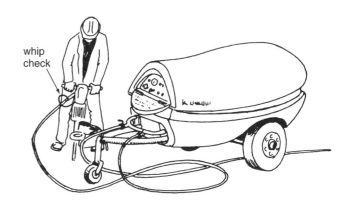

Figure 11.30(a)   Trailer mounted air compressor

Hose joints must have 'whip checks' (*see* Figure 11.30(*a*) and (*b*)) to prevent loose hoses from flailing if becoming disconnected.

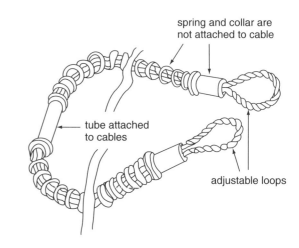

Figure 11.30(b)   Whip check

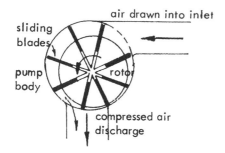

Figure 11.29   Rotary vane pump

11.16.2 *Tools for Compressors*

Some are breaking tools such as the:

pick (peg point; narrow chisel or wedge chisel);
spade (digging tool, standard spade and wide flat spade);
asphalt cutter with a slightly curved cutting edge

and others are specialist tools such as:

tampers;
vibratory air pokers;
grinders, etc.

The breaker consists of a heavy forged steel body which contains a throttle valve, valve block and piston. Compressed air produces a rapid oscillating motion by the piston (by diverting the air from top to underside of the piston) which in turn forces an anvil or tappit on to the tool holder. This, together with the weight of the breaker, drives the tool into the material being broken.

A muffler or jacket may be used to reduce the noise from a breaker tool. This encloses the breaker body and is basically a bag made from reinforced terylene, lined with foam rubber and strapped to the handle (*see* Figure 11.31).

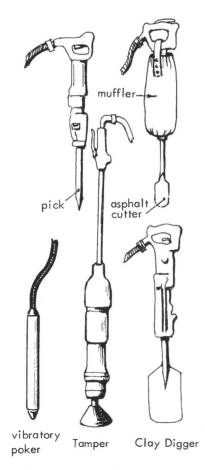

Figure 11.31   Compressor tools

11.17   Hydraulic Breaker Pack

This is a more recent development in powering breaker tools, etc. The hydraulic pack consists of a portable 500 ml petrol engine which drives a hydraulic pump. The unit will take one or two breakers which are hydraulically operated by supply and return hoses. The unit is compact and easily transported in the back of a van (*see* Figure 11.32).

Figure 11.32   Hydraulic breaker

11.18   Hand-Operated Power Breakers

These breakers are either independently powered by petrol engines or are electrically driven. They can be fitted with a variety of tools and may be used for breaking up road surfaces, drilling stone and driving in posts. As the petrol-powered machine is independent of a generator or compressor, it can be used in awkward situations (*see* Figure 11.33(*a*)).

The version with an electric motor as shown in Figure 11.33(*b*) can be used directly from a power source of 200/380 V or off a transformed site voltage of 110 V. All these breakers have a percussion rate of 1200 to 1500 blows per minute.

A drilling device may be fitted to the machine which gives an alternating drilling and hammering action. This is capable of drilling 26 to 100 mm diameter holes to a depth of 1 m. *See* Figure 11.33(*c*).

By mounting an additional attachment, these machines can be quickly converted into a pile and sheet driving unit.

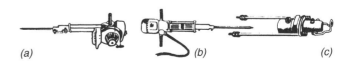

Figure 11.33(a)–(c)   Hand-operated power breakers

11.19 Hand-Operated Tampers and Vibrators

11.19.1 Auto-tampers

As with the breakers, two main types are in use, one powered by a petrol engine and the other electrically driven. The auto-tamper delivers a series of blows giving a percussion rate of 450 to 650 blows per minute. Because it can be easily transported and handled, it is used on many construction sites for a wide variety of compaction work.

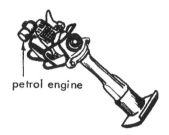

Figure 11.34   Auto tamper (petrol)

Figure 11.35   Auto tamper (electric)

11.19.2 Plate Compactor

A range of hand operated 'vibration plates' (*see* Figure 11.36) is available for the final compaction of bedding concrete and sub-grade materials. These machines use a vibratory action through a plate and with self-assembled screed boards, can be used as a vibratory unit. They can also be used in the laying of block paving (*see* Section 7.8).

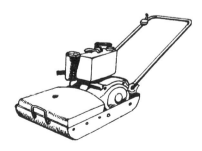

Figure 11.36   Vibration plate

11.19.3 Internal Vibratory Pokers

For vibratory work when placing concrete, the usual poker operating from a compressor may be replaced by an independent hand-operated unit powered by its own portable petrol engine as shown in Figure 11.37. One type has a flexible drive shaft from the motor to the poker, which rotates a 'pendulum' inside the poker head. The driven pendulum runs inside the outer housing at around 2800 rev/min, hence different outer housing sizes will give different vibration rates. *See* Figure 11.38. Another version is the electrically-powered high-frequency vibrator which has either a flexible drive shaft or a power connection to an electric motor incorporated in the vibrator head.

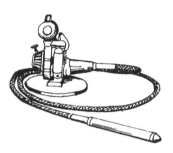

Figure 11.37   Poker vibrator

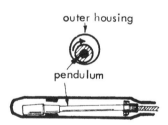

Figure 11.38   Detail of poker vibrator

11.20 Pumps

Many types of pump are available for highway work and these may be powered by petrol, diesel or electric motors. Pumps are essential for handling water or sewage, disposal of storm water, removing water from excavations, etc. Often a pump will need to handle water containing solids, which may range from sand to large stones, without the risk of choking. Another asset is the ability to 'pick up' without interruption should the input water temporarily fall, which might force the pump to handle air or a mixture of air and water. Self priming or the ability to get going without an injection of water is a vital factor in a pump.

Early mechanical pumps were of the piston type and this form still remains together with single or double diaphragms and rotary patterns (including centrifugal types). Some common versions are illustrated.

11.20.1 Diaphragm Pump (Figure 11.39)

This is a simple type of pump capable of pumping water containing grit and sand. This material would rapidly damage a piston type pump.

Depressing the handle raises the diaphragm (of leather or rubber), causing suction on the underside of the diaphragm, thus drawing in liquid. On lifting the handle, the diaphragm is forced down, closing the flap valve and causing the non-return valve to open, allowing the water to flow through. When the handle is again lowered, the valve closes and the rising diaphragm causes the water to flow out through the spout.

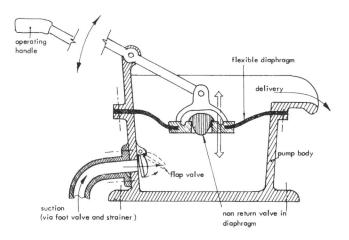

Figure 11.39 Diaphragm pump

11.20.2 Double Diaphragm Pumps

These may be of the open or closed type but both can operate in sludge, etc., without damage to the working parts and can deal with small, intermittent flows.

(a) Open type

Figure 11.40 shows, by this method of coupling, two pumps, working 180° out of phase, so that one pump is on the suction stroke when the other is on the delivery stroke, and continuous suction and delivery can be maintained.

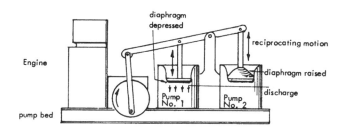

Figure 11.40 Double diaphragm pump (open type)

(b) Enclosed type

This is a more modern type of pump and is shown in Figure 11.41.

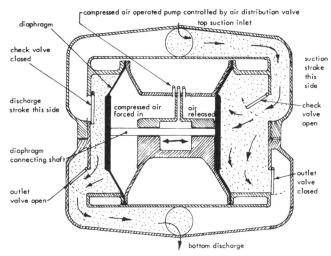

Figure 11.41 Double diaphragm pump (closed type)

11.21   Wheeled Cutters (concrete/asphalt saws)

Because of the increased use of concrete in many forms of construction, machines have been developed to remove laitance with rotary planes, scrabble the surface with a series of multiple heads and cut with abrasive or diamond-edged circular saw blades.

A wheeled cutter is capable of cutting considerable thicknesses of concrete, even if reinforced with steel bars. Known by roadworkers as a 'floor saw' it is also used for work in asphalt. A typical machine is either hand or self-propelled and is powered usually by a petrol or diesel engine (see Figure 11.42). This often allows a 356 to 508 mm diameter abrasive or diamond-edged blade to be mounted on one or either side of the machine.

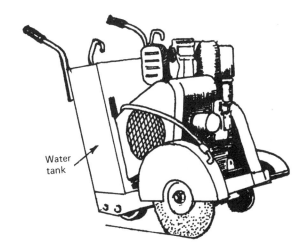

Figure 11.42   Floor saw (wheeled)

The 3 mm wide cut can be made to a depth of 100/120 mm to 175 mm, with the blade being cooled by water from a 40 litre tank. Larger versions can mount a 700 mm blade cutting to 280 mm in depth.

This machine can extract the dust and debris away from the operator and, when cutting expansion joints in the carriageway, the blade can be replaced with a rotary brush to clean out the joint. Large ride-on machines are available to groove carriageway surfaces and these will produce lateral grooving in patterns designed to prevent aqua-planing, etc.

11.22 Chainsaws

Chainsaws have many uses in the forestry, arboriculture and farming industries, but are also widely used in roadwork operations for cutting back overgrowth, removing fallen trees and overhanging branches.

Chainsaws are potentially very dangerous machines and need to be handled with the utmost care.

Training of operatives, together with regular updates, is vital for ensuring safe operation and is, of course, a legal requirement.

The following gives the list of most of the tasks for which specific training is necessary:

(a) Aerial rescue
(b) Chainsaw maintenance
(c) Chainsaw work in the proximity of overhead lines
(d) Crosscutting and stacking
(e) Windblown trees (individual and multiple)
(f) Hung up trees
(g) Use of ropes and harnesses
(h) Using chainsaws from ropes and harnesses.

The most obvious hazard associated with chainsaws is contact with the moving chain, due to the limited amount of guarding (*see* Figure 11.43). Other associated hazards include being struck by timber, together with noise and vibration. Adequate personal protection must be worn in order to mitigate against injury. Figure 11.44 shows the minimum necessary protective dress standard for operators of chainsaws.

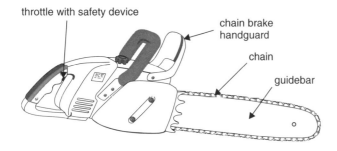

Figure 11.43   Chainsaw

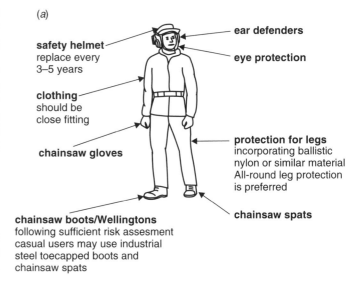

Figure 11.44(a)   Personal protective equipment

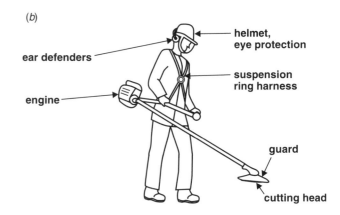

Figure 11.44(b)   Brush cutter/strimmer

Chainsaws range in guidebar length from 225 mm to 600 mm, the length being dependent upon the size of tree to be cut. For instance, large girth trees being cut near to ground level would need a greater length of saw than if working higher up the tree, where a more compact and easier to use saw would be used.

11.23   Brushcutters and Strimmers

Both brushcutters and strimmers are used to cut back undergrowth from roadside verges.

The brushcutter normally is fitted with a steel cutting head (*see* Figure 11.45) which is designed to cut through dense scrub in areas where there is open terrain.

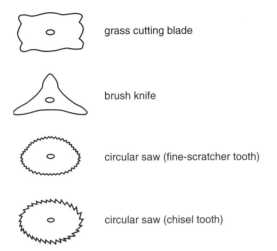

Figure 11.45 Brushcutter tools

grass cutting blade

brush knife

circular saw (fine-scratcher tooth)

circular saw (chisel tooth)

The strimmer uses nylon line, on a reel, or small plastic blades which fly outwards when the throttle is opened (*see* Figure 11.46). This allows it to be used for cutting around obstacles, such as trees, boundary walls fences, etc., without causing damage either to the machine or any of the obstacles.

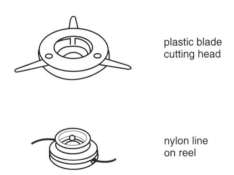

plastic blade cutting head

nylon line on reel

Figure 11.46 Strimmer tools

Operators must be trained in the use of these machines and wear protective equipment (*see* Figure 11.44(*a*)). Care must be taken when using these machines as there is considerable risk of thrown debris (such as stones) and even parts of the machine. A first aid kit must be available.

When working close to the carriageway, adequate signing must be set up.

All brushcutters/strimmers must have a clearly marked, positive ON/OFF switch, a power cut-off handle, serviceable anti-vibration mountings, adjustable suspension ring harness and handles to provide an ergonomic working position (*see* Figure 11.44(*b*)). A guard over the cutting mechanism is provided both for transportation and for when in use, together with an adequate tool kit.

When in use, the cutting edge must be kept clear of the ground. If any person or animal is in close proximity, cutting must be stopped.

When using a blade type tool, there will be a degree of blade thrust. Always use the most appropriate blade, flexible line or other tool for the task.

If the cutter becomes jammed, stop the engine before attempting to remove the obstruction.

11.24 Vehicle Actuated Portable Traffic Signals

Temporary traffic signals should always be vehicle actuated (VA) except where otherwise instructed by the Highway Authority.

Modes other than vehicle actuated are provided on the controller (Figure 11.47) but these should only be used to relieve short-term difficulties, e.g. traffic build-up in one direction or heavy plant crossing the carriageway. The object of using VA is to reduce delay to traffic by ensuring that the time for which the green signal is shown in one direction is automatically adjusted to suit the amount of traffic flow using the shuttle lane through, or past, the obstruction.

A microwave vehicle detector, or small 'radar', is fixed to each signal head and this is pointed in the general direction of the approaching traffic. When the detector registers an approaching vehicle up to 70 m away, the signal head facing in that direction changes to green and will remain so until all detected vehicles have passed.

Detectors cannot work properly if:

1 they are pointing at the sky or over a hedge
2 they do not face the oncoming traffic
3 they are roughly treated, e.g. thrown on to the backs of vehicles.

To ensure efficient use of vehicle actuated signals, only approved equipment, which complies with the DoT specification, should be used.

The signal equipment should be tested before being set up and the control box correctly adjusted.

Setting Up the Equipment

Where temporary traffic signals are needed, the signal head must be set up at each end of the obstruction, ensuring clear visibility at up to 70 m distance.

Adjustment of the time of the 'all red' setting on the control box allows the shuttle lane to be cleared from one direction before the signal head at the other end changes to green.

Adjusting the 'maximum green' ensures that, if a continuous flow from one end is registered by the MVD, the signals will change after the pre-set maximum period.

Electrical Supply

The electrical supply is 110 V a.c., which is taken either from a 230/250 V a.c. mains supply (possibly a convenient

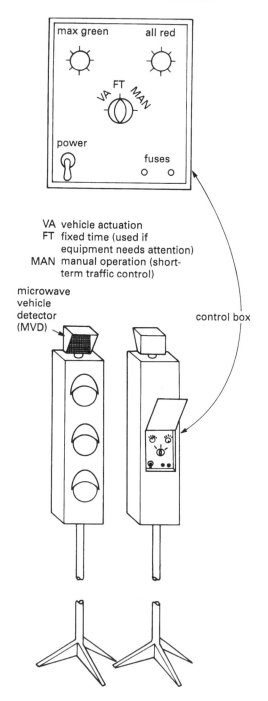

VA vehicle actuation
FT fixed time (used if
      equipment needs attention)
MAN manual operation (short-
       term traffic control)

Figure 11.47   Vehicle actuated portable traffic signal. The control box is located in the rear of the signal head, or may be an independent unit

lamp-post) being transformed with a mains adapter unit, or from a portable generator. The generator used could be a four stroke, single cylinder vertical air cooled diesel engine driving an alternator at 110 V a.c. and would be mounted on a two wheeled chassis. Most portable generators are handle starting and have a fuel tank capable of some 16 hours running time.

Microwave vehicle detectors work best if positioned on the near side of the road at each end of the obstruction. In this case, the control cables must cross the shuttle lane and must, therefore, be protected from crossing traffic by a cable crossing protector (Figure 11.48). A 'ramp' sign must be positioned at this location.

More recently, combined trailer units have become available. These house the generator, control box, cables, lighting heads and stands.

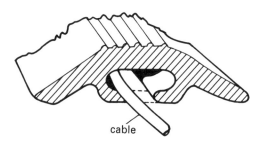

Figure 11.48   Flexible rubber cable protector, supplied in various sizes in 2 m or 3 m lengths. The largest size has a solid base (shown dotted)

11.25   Remote Control Temporary 'Stop/Go' Boards

Where temporary traffic control is needed at roadworks, this system greatly increases safety, both for the worker operating the control unit and for traffic on the road at the site of the work.

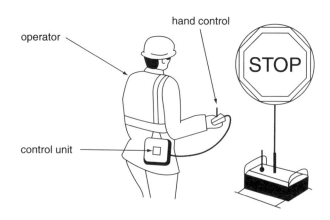

Figure 11.49   Remote control Stop/Go system

The Stop/Go boards are controlled from a hand held radio remote control unit, linked to a transmitter attached to the operator's belt. A single press on the control button rotates both signs simultaneously on their 'radio' bases.

The single operator can work from a safe vantage point out of the direct line of the approaching traffic.

The system incorporates a fail safe programme, by which, in the event of a fault, both ends of the work zone display the 'Stop' sign.

11.26 Cable/Pipe Detectors

There are several types of electronic detector tools and most models are simple to operate and easy to handle. They are usually sensitive to around 3 metres in depth and will receive signals through dense materials such as concrete and rocks. These tools are designed for tracking 'live' sources such as current-carrying cables and also signals from 'dead' cables, metal pipes and rods through non-metallic pipes, either by re-radiated radio signals or by trace signals induced by a signal generator.

The detector tool can be switched to three modes: power, radio and generator mode (*see* Figure 11.52).

(*a*) *Power mode*: Switched to this mode the tool detects loaded power cables (radiating 60 Hz signals). The device gives an audible response when held directly over the hidden cable and the position of the cable is determined by finding the point where the signal is at its strongest.

(*b*) *Radio mode*: This enables the tool to detect long conductors even if they do not carry any energy of their own. These conductors include live, but not loaded, power cables, main telephone cables and some continuous metal pipes. The signals originate from distant VLF radio transmitters and these signals are re-radiated by the buried conductors in sufficient strength to be detected by the tool.

(*c*) *Generator mode*: A special-purpose generator can be used to apply a signal to a conductor either by being placed in line over the buried conductor (Figure 11.50), or by connecting to one end of a conductor at a chamber, etc. (Figure 11.51). The detector tool when switched to

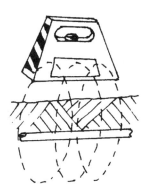

Figure 11.50

'generator' can then be used with great accuracy to trace the route of a specific conductor.

The power and radio modes are both used initially when sweeping an area in the highway and between them they will detect main cables and most other buried conductors (*see* Figure 11.52).

Figure 11.51

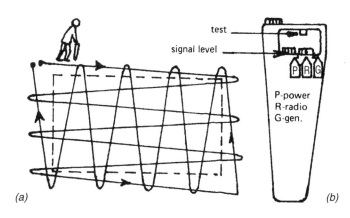

Figure 11.52 (a) Sweeping an area; (b) detector tool

Revision Questions

1 What is the aim of the provision and use of work equipment regulations which were introduced into the UK in 1993?
2 What is the maximum radius for loaders in the 6–10 tonnes metre (tm) range of lorry loaders?
3 The high pressure hose reel on a gully emptier with jetting equipment is capable of operating at what psi?
4 What are the three main types of concrete mixers?

Note: See page 305 for answers to these revision questions

The term 'street furniture' embraces all highway fixtures or markings that have to be provided to assist pedestrians and road users in the safe usage of the highway. Street furniture has to be considered carefully when the difficult task of 'designing the street' is undertaken by planners. To try to create pleasant but effective street scenes, through the use of well designed and carefully sited street furniture, entails many problems. The publication *Street Furniture from Design Index* by the Design Council gives information on this aspect of street design. In 'designing the street', consideration is given to:

the selection and siting of co-ordinated street furniture (including signs, markings and bollards);
the colour and texture of the ground;
the use of trees;
architectural conservation;
poster displays and outdoor advertising signs;
amenity lighting and lighting columns; and
miscellaneous items such as guard rails, litter bins, bus shelters and outdoor seating, etc.

For the purpose of this section, some of the more important items that concern the Highway Engineer will be examined. These are:

permanent signs and road markings;
traffic activated signals;
vehicle crash barriers and safety fences;
pedestrian guard rails; and
street lighting.

12.1 Permanent Signs

Clear and uniform signing is an essential part of highway and traffic engineering. Road users depend upon this for information and guidance and highway authorities use signing to ensure efficient traffic control together with the enforcement of traffic regulations. Although 'signing' is usually thought of as signs on posts, it also covers carriageway markings, beacons, studs, bollards, traffic signals and other devices.

Signs must not only give road users a clear message, but this must be given at the correct moment in time. For the message to be of maximum use, it must be given not too soon for the information to be forgotten before it is needed and not too late for the safe performance of consequent manoeuvres. Hence the positioning of signs is just as important as the shape, colouring and lettering of signs.

Uniformity in the design of signs is essential to allow quick recognition and the types of signs and carriageway markings allowed are prescribed by the Traffic Signs Regulations and General Directions Act.

To obtain the fullest benefits of uniformity, there must not only be uniformity of signs but also uniformity in their use, in their siting and in their illumination. The *Traffic Signals Manual* (Department of Transport) sets out the codes to be followed in the use, siting and illumination of signs and should be referred to for detailed information.

12.1.1 Classification of Signs

Apart from temporary signs (*see* Chapter 9) and carriageway markings there are three main classes of road signs. Each class has its basic shape and the use of certain colours is restricted to particular classes of signs.

These three classes are as follows.

Regulatory Signs

These include all, signs which give notice of requirements, prohibitions or restrictions. They may be either mandatory or prohibitory. All regulatory signs are circular, with the exception of the 'STOP' and the 'GIVE WAY' signs shown in Figures 12.1(*a*) and (*b*).

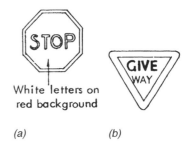

(a)   *(b)*

Figure 12.1(a) and (b)

Warning Signs

These give warning of a hazard ahead and are triangular as in Figure 12.7(*b*). There are one or two exceptions in shape such as the inverted triangles used in the advance warning signs prior to the 'STOP' and 'GIVE WAY' signs (*see* Figures 12.9(*b*) and 12.10(*b*)).

N.B.: Rectangular plates may be used below most signs in the regulatory and warning sign categories in order to give additional information. An example is shown in Figure 12.6.

Informatory Signs

These normally give road users information about their route or places of interest. Most signs are rectangular but direction signs usually have one end pointed.

12.1.2 Positioning of Signs

The four factors considered in positioning a traffic sign are its siting, placement, mounting height, and orientation.

Siting

Its siting along the road in relation to the junction, hazard or other feature to which it applies, is mainly determined by the speed of road traffic. Sufficient distance must be provided to allow a driver adequate time in which to react to a sign message. The driver's line of sight to the sign should not be obstructed over this distance. Road users are accustomed to signs on their nearside and hence this should be the normal position for signs. An exception is on the approach to a left-hand bend, or where doubling up of signs is required on the approach to a hazard.

Placement

The placement of a sign is largely determined by the following:

(*a*) it should normally be set a minimum distance of 0.5 m from the edge of the carriageway to the edge of the sign.

(*b*) This distance to be a minimum of 0.6 m where there is severe camber or crossfall or where signs are mounted on the central reserve of a dual carriageway.

(*c*) The distance to be a minimum of 1.25 m on high-speed dual carriageways (where there is a hardened verge the nearest edge of the sign is to be 0.6 m from the verge).

Mounting Height

Mounting heights should ensure that the lower edge of a sign is between 1 and 1.5 m above the highest point of the adjacent carriageway.

Where signs are erected across and above footways, the lower edge must be a minimum of 2.1 m in height.

Orientation

The orientation of signs should be such that they are set approximately at right-angles, usually 95°, to the line of travel of traffic, to prevent too much reflected light being directed back to a driver; with the exception of direction signs and also plates detailing the hours of waiting restrictions, which should be parallel to the kerb.

Signs must be positioned so that they are clearly visible. Overhanging trees and hedges should be cut back and, if necessary, bus stops moved. Bends, hill crests, narrow verges, buildings, etc. will require special consideration when siting signs and it is preferable to increase the standard distance between the sign and the site to which it relates, rather than decrease it.

Signs may lose their effectiveness because of a distracting background. Where it is impossible to avoid such a background, a plain black, grey or yellow backing board should be erected behind the sign.

On roundabouts, vertical chevron warning signs (white chevrons on a black background), or the inverse configuration on a sloping side in black and white paving blocks, are now commonplace. However, there are concerns over the use of chevron blocks without any signage, as the blocks can be overgrown with vegetation or be obscured by snow.

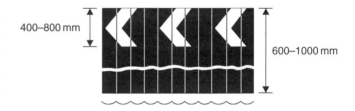

Figure 12.2 Sharp deviation of route at a roundabout

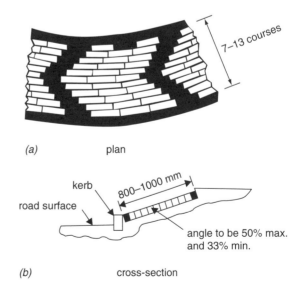

Figure 12.3(a) and (b) Black and white paving block signing on a roundabout

12.1.3 Mounting Signs

Wherever possible, signs should be either fixed to existing posts, such as street lighting columns (*see* Figure 12.4) or attached to walls, adjacent to the footway (if footway is under 2 m in width). Alternatively, the top of the post may be angled (*see* Figure 12.5).

Certain signs with small letters, such as 'no loading' plates, must be mounted close to the edge of the carriageway.

When mounting signs on one or more purpose made tubular metal posts, the posts should not project above the sign or lighting unit (*see* Figure 12.6). The sign plate should be fixed level across the posts.

It is sometimes possible to mount two or more signs on one post but care is necessary to ensure that clarity is not lost. Some signs such as 'STOP' and 'GIVE WAY' must be mounted by themselves on the post.

Posts used to support pedestrian crossing beacons, hazard markers, some traffic control signals and load gauges are striped black and white. All other posts, backs of signs, bracings and clips should be grey in colour, but where concrete supports are used, these may be left their natural colour. Tape is used with clips and the clips should

Figure 12.6   Positioning of signs on posts

be tightened before the weight of the sign is allowed on them.

Posts should be set in the ground to a depth of between 0.6 and 1.5 m (about one quarter of their total length) and should have a concrete surround.

Dimensioned sketches of the mounting and positioning of the following signs are shown in Figure 12.7(*a*), (*b*) and (*c*):

(*a*) a regulatory sign placed in a narrow footway which is adjacent to a high wall;
(*b*) a large warning sign mounted on two tubular metal posts fixed in a wide grass verge;
(*c*) a 'no loading' plate fixed to a tubular metal post located in a footway.

12.1.4   Types and Details of Regulatory Signs

Regulatory signs give instructions to road users and disregard of these instructions makes an offender liable to a penalty on conviction, under the Road Traffic Act.

These signs are usually the means of putting into effect Traffic Regulation Orders made under the Acts and regulatory signs are either the *mandatory* type or the *prohibitory* type.

(a)   Mandatory signs

Mandatory signs give instructions to drivers as to what action they should take. Examples of these are 'STOP' (Figure 12.1(*a*)) and 'GIVE WAY' (Figure 12.1(*b*)) signs. All other mandatory signs are circular with white symbols on a blue background such as the 'KEEP LEFT' sign (Figure 12.8).

Details of the more important mandatory signs are as follows.

The Stop sign and Advance Warning sign
The 'STOP' sign requires every vehicle to stop at the transverse lines before entering the major road. Often visibility is so restricted for a driver entering the major road, it is essential for the driver to stop. The junction layout is shown in Figure 12.9 and sizes of sign and visibility distances are given in Table 12.1.

Where the stop sign is not clearly visible to an approaching driver, an advance warning sign with a

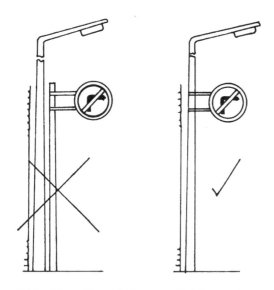

Figure 12.4   Mounting of signs on lighting columns

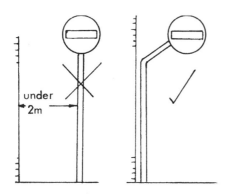

Figure 12.5   Mounting of signs on posts in footway

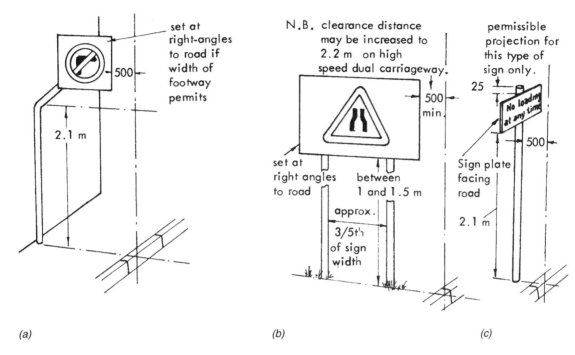

Figure 12.7(a–c)

Figure 12.8

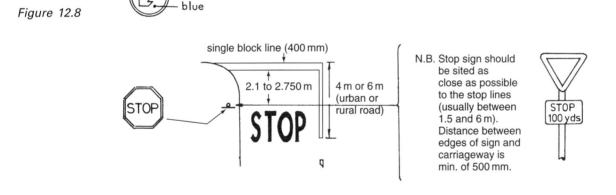

Figure 12.9(a) and (b)

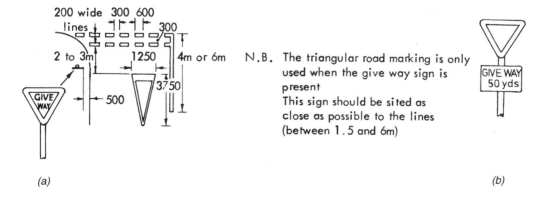

Figure 12.10(a) and (b)

Approach Speeds of Vehicles (typical cases)	Examples of Typical Roads Shown in 7/75	Stop and Give Way Signs		Triangular Warning Signs		Regulatory Signs (Dia.)
		Diameter (mm)	Min Visibility* (m)	Sign Height (mm)	Distance from Hazard (m)	
Between 20/30 mph	Urban and other rural roads of a local nature	750 (stop) 600 (G.W.)	45 45	600	45	600
30/40	Urban and rural single carriageway 2 lane roads	750	45	750	45–110	750
40/50	Urban Motorways and high standard 2 or 3 lane rural roads	900	90	900	110–180	750

*Distance below which an advance warning sign is necessary

Table 12.1   Regulatory and warning sign sizes and siting distances (information from Circular: Roads 7/75)

distance plate is required (Figure 12.9(*b*)). Table 12.1 shows the circumstances in which this sign is justified.

The Give Way sign and Advance Warning sign
The 'GIVE WAY' sign requires that a vehicle will not proceed past the transverse lines in such a manner as is likely to cause danger to the driver of any other vehicle on the main road.

The junction layout is shown in Figure 12.10 and sizes of sign and visibility distances are given in Table 12.1. As before, an 'ADVANCE WARNING' sign is used where the 'GIVE WAY' sign is not clearly visible to an approaching driver.

Keep Left (or Right) (see Figure 12.8)
The sign is most frequently used in bollards on islands and refuges in the centre of the carriageway and at the beginning of central reserves of dual carriageways.

The normal size of the sign is 600 mm diameter but when used in a bollard, it is 270 mm diameter. Where it is necessary to supplement the sign at the beginning of dual carriageways a plate is used.

N.B.: The 'KEEP RIGHT' sign is generally only used for roadworks (*see* Chapter 9).

Turn Left (or Right), Ahead Only and Turn Left (or Right) Ahead
These 'TURN LEFT (or RIGHT)' signs (Figure 12.11(*a*)) are usually used at roundabouts on the central island, surmounting the chevron warning sign. These, together

with the sign 'AHEAD ONLY' (Figure 12.11(*b*)) are normally found at road junctions.

The signs 'TURN LEFT AHEAD (or RIGHT)' as shown in Figure 12.11(*c*), are for use in advance of junctions, a reasonable distance being approximately 50 m.

Figure 12.11(a)–(d)

Where any of these signs are used in one way streets, it is usual to erect them on both sides. The supplementary plate 'ONE WAY' is used (as shown in Figure 12.11(*d*)) with signs in this instance.

(b)   Prohibitory Signs
Prohibitory signs, of which there are many different types, give instructions to drivers as to what they must not do.

Signs banning turning (*see* Figure 12.12) or entry, or imposing speed restrictions are examples of those in this category. All are circular with a red border and details of some of the more important are given.

Figure 12.12

Figure 12.13

Signs banning turns

'NO LEFT TURN (or RIGHT)' signs, as shown in Figure 12.13 are used to prohibit left or right turns or are seen in association with one-way streets. When mounted on signals, they should be alongside the green light.

The 'NO U TURNS' sign (*see* Figure 12.12) may be used where this sort of turn is prohibited at a junction or along a length of road.

When used for a length of road, supplementary plates are needed; one to indicate the affected length at the beginning and an 'END' plate to show the point where the restriction is lifted.

Signs banning vehicles

The 'NO ENTRY' sign is shown in Figure 12.5.

A sign 'prohibiting all vehicles in both directions' is used for play streets and must always be accompanied by a supplementary plate, as shown in Figure 12.14.

Figure 12.14

Other signs prohibiting all motor vehicles or certain classes of vehicles, such as cycles, lorries or buses and coaches, have pictograms to this effect (*see* Figure 12.16).

Weight, height, and width restrictions

Signs banning all vehicles exceeding a certain weight may be used (*see* Figure 12.15), or alternatively, goods vehicles over a certain specified unladen weight (*see* Figure 12.16).

Figure 12.15

Figure 12.16

Similar signs are those restricting the width or height of vehicles.

Waiting restrictions

The 'NO WAITING' sign shown in Figure 12.17 is now only erected at, or near, the main points of entry to any area in which there are temporary waiting restrictions.

They are erected on both sides of the road to face traffic entering the area. Only one size (200 mm diameter) is erected, but small replicas of this sign appear on all time plates indicating the precise details of the time of the restrictions (*see* Figure 12.18).

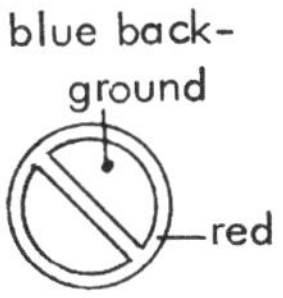

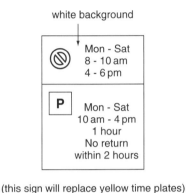

Figure 12.17   *Figure 12.18*

White plates are used (without the replica sign) to give details of loading restrictions (Figure 12.19).

Figure 12.19   No loading at any time

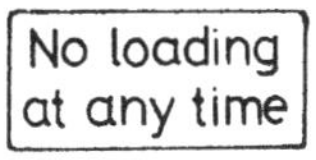

Yellow line markings on the road and kerb indicate the types of restrictions and where they apply. See Table 12.2. and Figure 12.20.

Controlled parking zones

The system of road markings already described is also for use in controlled parking zones, a broken white line being

Marks kerb	Marks on road	Figure 12.20	Restrictions
Two marks	Single yellow continuous line	See A	Restrictions in force as per plate
Three marks	Double yellow continuous lines	See B	Restrictions in force as for A, plus any given period
Single mark	Single broken white box	See C	Any other restrictions, i.e., peak hours only

N.B. Other combinations of these kerb and road marks will give different restrictions between waiting and loading.

Table 12.2   Controlled parking zones

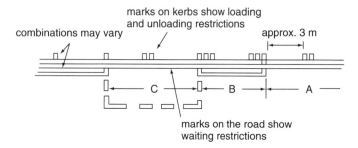

Figure 12.20 Waiting restrictions: road and kerb markings

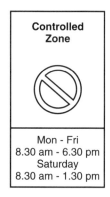

Figure 12.21

used throughout the zone outside the parking places. Signs, as in Figure 12.21 are used (having meter disc or controlled zone wording) together with a plate giving the periods of control. The parking bays should be marked out with white lines.

Clearways
On clearways, all stopping on the carriageway is prohibited with certain exceptions, such as public service vehicles. Waiting bays are usually provided for other vehicles to stop or unload.

The 'CLEARWAY' sign has a red cross on a blue background erected at the beginning of a clearway and is available in 4 sizes, 300, 450, 600 and 900 mm (the smallest being a repeater sign).

The signs should be erected on both sides of the carriageway, with repeaters at mile intervals and a sign used with an end plate at the end of the restriction.

Priority signs
These are described in Chapter 9.

12.1.5 Types and Details of Warning Signs

To be most effective, these must be used sparingly, as frequent use of them to warn of conditions which are obvious tends to cause disregard of signs generally.

Five sizes of sign are available and information on their size and siting is given in Table 12.1.

Some of the more important signs are as follows.

Junction signs
These are not generally used on primary routes but rather on major roads where a driver cannot easily see a side turning, etc. Their use on minor roads is usually confined to places where there is not a 'GIVE WAY' or 'STOP' sign, or traffic signals.

Roundabout and merging traffic
The 'ROUNDABOUT' sign should only be used for true roundabouts and may have a 'REDUCE SPEED NOW' plate on high speed roads (*see* Figure 12.22). Elsewhere, the sign may be used without a plate. The plate requires site authorization from the Department for Transport.

Figure 12.22

A sign 'MERGING TRAFFIC' is erected to give warning where two physically separated streams of traffic join the same section of carriageway.

Bend signs
These are to give drivers advance warning of a curve that may be difficult to negotiate without slowing down.

'DOUBLE BEND' signs should only be placed where two severe bends occur close together (*see* Figure 12.23).

Figure 12.23   *Figure 12.24*

'FREQUENT BENDS' signs (*see* Figure 12.24) are still to be seen but are not now erected. The double bend sign is used, together with a plate indicating the length of the road having frequent bends. If this combination is used, then additional signs are not necessary for individual bends.

Miscellaneous hazards
There are numerous warning signs, some of which are:

(*a*) A 'ROAD NARROWS' sign is necessary where there is a sudden reduction in road width (*see* Figure 12.25).

Signs indicating that the road narrows on one side are usually seen at roadworks (*see* Chapter 9).

Figure 12.25

Figure 12.26

(*b*) Steep hills are usually shown, as in Figure 12.26, which indicates a steep hill of 1 in 10 gradient, sloping downwards ahead. Supplementary plates may be attached if the gradient is severe and they could instruct a vehicle to 'keep in low gear', etc.

(*c*) Many signs deal with hazards at points such as 'hump bridges', 'animals', etc. and reference should be made to the *Traffic Signs Manual* for their correct use.

12.1.6 Types and Details of Informatory Signs

Most of these are directional, giving drivers essential information to enable them to find their way to their destination. Other signs give either information of practical importance such as parking facilities or information of general interest, such as a town name.

The most important types of directional signs are:

(*a*) Advance direction signs which give a driver information about the route ahead before a road junction is reached.

(*b*) Direction signs which give route information at a junction (*see* Figure 12.27).

(*c*) Route confirmatory signs which give confirmation and often additional information about the route ahead after a road junction.

The colours used on a sign indicate the type of route. Signs concerning motorways have white lettering on a blue background.

Those concerning primary routes (these are the most important traffic routes such as trunk roads and certain A roads) have a green background with white lettering and yellow numerals.

Signs concerning non-primary roads have a white background with black lettering.

Local direction signs on primary and non-primary roads have a white background with a blue border.

Tourist attraction signs have white letters, sometimes with a symbol such as a castle or railway engine, etc., on a brown background.

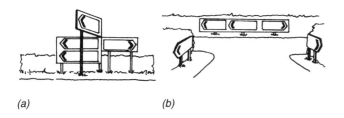

(a)   *(b)*

Figure 12.27(a) and (b) *Arrangement of directional signs*

Arrangements of directional signs tend to become visually confusing if care is not taken in minimizing the cluster.

Obtrusive and complicated arrangement in Figure 12.27(*a*) is improved by the horizontal arrangement in Figure 12.27(*b*).

To reduce the size of a sign two destinations may be placed on one line and often the sign may also be placed on a wall.

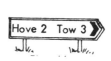

Figure 12.28(a)
Multi-destination sign

Figure 12.28(b) Positioning of multi-destination signs

Street nameplates may be mentioned under informatory signs. Figure 12.29 shows a typical street nameplate in a residential street and indicates the dimensions to which it should conform in relation to its height and location.

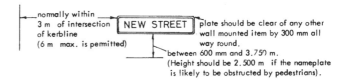

Figure 12.29   Street nameplate

If possible, the nameplate should be fixed to the back edge of the footway and attached to a permanent structure, e.g. wall or building.

The nameplate may be mounted on post(s) if considered necessary.

12.1.7 Some Special Cases of Signing

As might be expected, adequately to sign a particular situation, a mixture of regulatory, warning and informatory signs is required. Some examples are given.

Signs for One Way Streets

These require signs such as:

'NO ENTRY' (*see* Figure 12.5). This is a prohibitory sign and is normally 750 mm diameter but 270 mm when on a bollard.

'NO LEFT TURN (or RIGHT)'. Again on a prohibitory sign (*see* Figure 12.4) and should be used on a road wherever it crosses or joins a one-way street, unless a turn left (or right) sign, shown in Figure 12.11(*a*), is used instead.

'PASS EITHER SIDE' This is a mandatory 270 mm diameter sign for use in bollards erected in the centre of a one-way street.

'ONE-WAY TRAFFIC' (*see* Figure 12.30(*a*)). This is really an informatory sign used in conjunction with regulatory signs. If there are no central refuges and, thus, no bollards, then this 300 × 450 mm sign is used along the street.

'ONE-WAY STREET' (*see* Figure 12.30(*b*)). Again, this is really an informatory sign used mainly to help pedestrians to cross a one-way street, where traffic approaches from their left. The arrow indicates the direction of the traffic – not the direction in which pedestrians should look.

'TWO-WAY TRAFFIC' (*see* Figure 12.30(*c*)) This is really a warning sign and is erected after a section of one-way working.

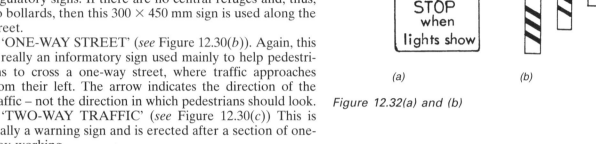

Figure 12.32(a) and (b)

Figure 12.30(a)–(c)

Figure 12.33 Level crossing (open signal controlled)   *Figure 12.34 Level crossing (no signal control)*

Level Crossings

These may be:

 gated or full barriered;
 automatic half barriered;
 open with traffic signals;
 or open without traffic signals.

(*a*) A gated or full barriered level crossing should have a sign, as in Figure 12.31(*a*). If equipped with twin flashing red signals, the plate shown in Figure 12.31(*b*) is used with the sign.

N.B.: The sign shown in Figure 12.35 is used at crossings where a barrier is not present.

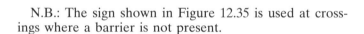

Figure 12.35 Level crossing without barrier

12.2 Road Markings

These are used to control, warn and guide road users. They supplement kerbside signs and may be seen when the signs are obscured. Equally, when markings are obliterated by snow or when very dirty, the signs then become more necessary. Unlike signs, they provide a continuing message to the moving driver.

Figure 12.31(a) and (b)

(*b*) The automatic half barriered crossing is always equipped with twin flashing red signals and has a supplementary plate, as shown in Figure 12.32(*a*). Where the crossing is difficult to see or is on a high speed road, then count down posts are provided on the nearside approaches to the crossing. These are white posts with red bands (Figure 12.32(*b*)).

(*c*) At open level crossings, controlled by traffic signals Figure 12.33 is used, together with plate (Figure 12.31(*b*)).

(*d*) When the crossing is without traffic signals, the sign Figure 12.34 is used.

12.2.1 Classes of Marking

They may be classified as follows:

(*a*) Transverse white markings which are at right-angles (or thereabouts) to the centre line of the carriageway.

(*b*) Longitudinal white markings, which are generally parallel to the centre line of the carriageway.

(*c*) Yellow or white markings for waiting restrictions.

(*d*) Worded and box markings (e.g. parking bays, etc.).

(*e*) Junction markings.

(*f*) Crossings.

Transverse Markings

The two main cases are 'STOP' lines and 'GIVE WAY' lines.

(*a*) Stop lines are of the single line type, shown in Figure 12.36.

(*b*) Give way lines are used for all junctions other than those controlled by stop lines.

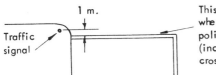

This indicates the position where a driver must stop by police or by traffic signals (including signals at level crossings).

Figure 12.36

Minor junctions have only the give way lines, whereas the more important junctions have, in addition, triangular give way road markings and give way signs (*see* Figure 12.10(*a*)).

If, at a minor junction, the transverse marking is used alone, then the 'SLOW' marking on the road may be used to give advance warning of the junction.

Longitudinal Markings

These are for the guidance of traffic and comprise:

(*a*) Double lines – which are centrally placed to prohibit traffic from either direction from overtaking (*see* Figures 12.37(*a*) and (*b*)).

(*b*) Warning lines – the broken single line is used simply to define the centre of the highway, as in Figure 12.38(*a*) or to warn drivers when approaching or negotiating hazards (*see* Figure 12.38(*b*) for alternative methods). A continuous line is used parallel with a broken line to indicate in which direction overtaking is prohibited (*see* Figures 12.39(*a*) and (*b*)).

(*c*) Lane lines contribute greatly to the safety of the traffic. Alternative methods of marking are shown in Figures 12.40 and 12.41.

(*d*) Edge of carriageway markings are laid to assist a driver to see the limits of the available carriageway. Some

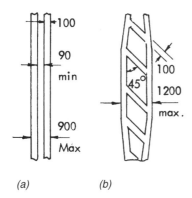

Figure 12.37(a) and (b)

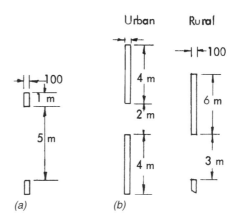

Figure 12.38(a) and (b)

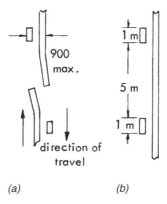

Figure 12.39(a) and (b)

Figure 12.40 Urban method of lane marking

Figure 12.41 Rural method of lane marking

instances are given below and reference is also made to lined edge details on p. 226.

The line shown in Figure 12.42(*a*) indicates the edge of the carriageway at a bend or similar hazard. The line used elsewhere is shown in Figure 12.42(*b*). Where a hard shoulder or strip is provided, the line shown in Figure 12.42(*c*) may be used.

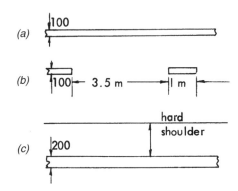

Figure 12.42(a)–(c)

Yellow or White Markings for Waiting Restrictions

These are detailed in Table 12.2.

Worded and Box Markings

Markings under this group are white and yellow. Some give instructions or warnings (as with STOP and SLOW) and others indicate areas of carriageway used for a special purpose.

Examples of this type of marking are:

(*a*) for a bus stop – shown in Figure 12.43;
(*b*) for a school entrance – shown in Figure 12.44.

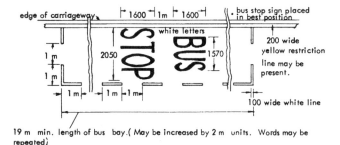

Figure 12.43   Bus stop

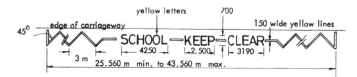

Figure 12.44   School entrance

Figure 12.45 shows the method of indicating the limits of places at the side of the road where hackney carriages may stand in accordance with the Traffic Signs Regulations and General Directions.

The sketch in Figure 12.46 indicates the marking of parking bays within a meter-controlled parking zone.

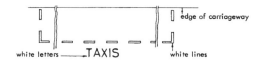

Figure 12.45   Taxi rank (white letters and lines)

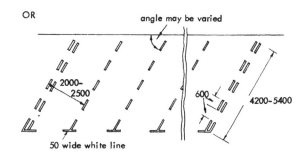

Figure 12.46   Parking in echelon (white lines)

Junction Markings

This means the marking of signal controlled and other junctions. As many variations are possible due to the layout of the junction and the number of lanes, reference should be made to the *Traffic Signs Manual*.

However, a typical traffic signal controlled road junction in an urban area, with two approach lanes, is shown in Figure 12.47 and indicates:

(*a*) the siting of a pedestrian crossing and width of the crossing;
(*b*) the siting of pedestrian refuges;
(*c*) the siting of the two primary traffic signals;
(*d*) the position of the detector pads in relation to the stop line;
(*e*) the location and dimensions of the stop and lane line markings.

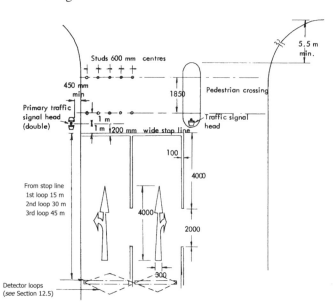

Figure 12.47   Typical junction markings

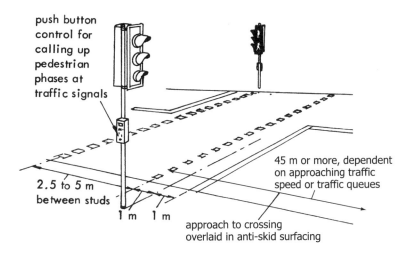

push button control for calling up pedestrian phases at traffic signals

2.5 to 5 m between studs

1 m 1 m

approach to crossing overlaid in anti-skid surfacing

45 m or more, dependent on approaching traffic speed or traffic queues

Figure 12.48 Controlled pelican crossing

Pedestrian Crossings

Many types of pedestrian crossings are in use, the main ones being the Pelican, Puffin, Toucan and Zebra. Details of two types are shown in Figures 12.48 and 12.49.

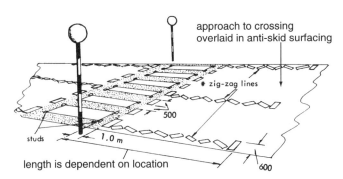

approach to crossing overlaid in anti-skid surfacing

zig-zag lines

500

studs

1.0 m

length is dependent on location

600

*Single zig-zags are placed in the road centre when the road width exceeds 6 m and the double zig-zags are used where a central refuge exists

Figure 12.49 Zebra crossing

12.2.2 Road Marking Materials

These are available in many forms. The main types in use in the UK are:

1 Thermoplastic (applied hot by hand or machine)
2 Paint (applied as a wet film)
3 Pre-formed (in tape or sheet form).

Additionally, the markings may be of:

4 inset concrete blocks
5 inset rubber blocks
6 steel studs and plates.

Thermoplastic road marking material consists of a combination of plasticized resin, aggregate and pigment. It is supplied in either block or powder form, being heated in a boiler to between 150° and 180°C before being applied to the road surface. Application can be by hand, machine-drawn screed box, extrusion or spray equipment. The material is wholly solid when cold and cools to road temperature within one minute of laying.

Paint is a pigmented liquid which is converted to an opaque film on drying. Composition is of pigments dispersed in a binder with solvent. It is applied wet, to a thickness of about 4 mm, normally drying within 5 to 15 minutes.

Pre-formed materials may be supplied either as tape or in sheet form. They are of metal foil or polymer based and can be self-adhesive or fixed with glue applied on site. These materials are most useful for temporary markings and for panels for pedestrian crossings.

Surface Dressed Lines

A narrow strip of binder (bitumen) is sprayed on the road and covered with white chippings. Calcined flint is the normal material used.

This method of lining is cheap and durable, but the appearance is rather ragged.

Chemically Set Resinous Markings

These are based on polyester, epoxy and other similar resins. The liquid resins are hardened by the addition of a chemical curing agent and form a tough 'horny' mass. Polyester resins may be reinforced with glass fibre.

Temporary Markings

These may be of the following types:

1 white self-adhesive plastic type – this is very slippery when wet;
2 a weak paint (e.g. distemper) – usually the most convenient and economical temporary lining material.

Marking Materials for City Streets

Steel plates/studs

These are almost indestructible but very slippery when wet (*see* Section 12.2.3 for types of road stud).

Inset thermoplastic material

This is commonly known as 'white mastic asphalt' and is laid about 12 mm thick in prepared insets in the road surface. Its composition is white pigment, filler and binder and it is laid by wooden float. It is widely used for the white strips in pedestrian crossings having a life of at least 4 years even under very heavy traffic conditions.

Inset rubber blocks

These are used for some stop-lines and centre lines. These usually consist of black rubber blocks with a white top face and fluted sides to help in keying them into the asphalt into which they are set. A pattern may be cut into the rubber surface to improve skid resistance.

Prefabricated Stick-on Markers/Lines

These may be of the following materials.

Thin PVC sheeting

It adheres well to the road, but becomes slippery or yellows with age.

PVC material with a high filler content to give a gritty texture

It retains its non-skid quality but only bonds well to smooth surfaces (e.g. mastic asphalt or cold asphalt).

Rubber markers

Generally patterned rubber studs can be used in place of steel studs: a synthetic rubber is used which is resistant to oil-staining and which bonds well.

Quality of Markings

The main criteria to bear in mind when selecting materials for marking a specific site are:

(*a*) the degree of durability required (which must be balanced against cost);
(*b*) the light reflection properties at night (depend upon the illumination, e.g. street-lighting or headlights);
(*c*) slipperiness;
(*d*) local conditions (i.e. standing vehicles, etc.).

Durability

The expensive inset materials are the most durable, but tend to be used only in heavily trafficked city streets.

Non-urban roads have superimposed thermoplastic or paint lines as these roads are often surface dressed and the life of the marking is limited to the intervals between dressings.

Road markings should be clearly visible by both day and night, skid resistant, durable and cost effective, as well as being safe to apply.

Visibility by day is achieved by contrast with the road surface, mainly through the use of pigments such as titanium dioxide in white materials and chromates in yellow materials. Additionally the use of high quality clear resins and clean white aggregate will help to achieve high luminance values.

Visibility by night relies almost entirely on the retro-reflectivity produced by the inclusion of tiny glass beads in the materials. The beads may be added during manufacture (commonly called 'in-mix' or 'pre-mix' beads) or added to the markings during application to the road surface.

Skid resistance depends mainly on the rugosity of the road surface and on the thickness of the marking material.

Cost effectiveness is dependent upon the anticipated life of the road surface, the traffic density and location of the marking, as well as the durability of the material itself.

Safety

The process of road marking is in itself a hazardous operation, but mechanization and fast drying times contribute to reducing the time for which the hazard exists.

All operatives must now be accredited as competent under a UK quality assurance scheme.

Inspection of Markings

The DfT has produced a Code of Practice for Routine Maintenance which recommends that visual inspection of road markings should be carried out every year for paint and every two years for thermoplastic.

Defects found during these inspections should be actioned under two categories:

1 Those which require prompt attention because they represent an immediate or imminent hazard (e.g. badly worn 'Stop' or 'Give way' double continuous white lines, etc.).
2 Less urgent repairs to worn or otherwise defective markings, which can be left for at least six months before repairs need to be effected.

Category 1 markings should be replaced within 28 days.

The following list contains suggested inspection intervals for various types of road and traffic density, expressed as vehicles per day (VPD) (average 24 hour traffic flow in both directions):

1 Single carriageway roads: 7.3 m wide
   (i) up to 10 000 VPD   paint: 1 year
                                    thermoplastic: 2 years
   (ii) over 10 000 VPD   paint: 6 months
                                    thermoplastic: 1 year
2 Single carriageway roads: 10.0 m wide
   (iii) up to 15 000 VPD   paint: 1 year
                                    thermoplastic: 2 years
   (iv) over 15 000 VPD   paint: 6 months
                                    thermoplastic: 1 year
3 Dual carriageway roads: 2 lanes
   (v) up to 20 000 VPD   paint: 1 year
                                    thermoplastic: 2 years
   (vi) over 20 000 VPD   paint: 6 months
                                    thermoplastic: 1 year
4 Dual carriageway roads: 3 lanes
   (vii) up to 30 000 WD   paint: 1 year
                                    thermoplastic: 2 years
   (viii) over 30 000 VPD   paint: 6 months
                                    thermoplastic: 1 year

Further information may be obtained from the 'Road Marking Industry Group' and from BS 3262 (for thermoplastic materials) and BS 6044 (paints). In addition to this, the use of high speed lining quality test apparatus, is now becoming widespread.

12.2.3 Road Studs

Road studs can be of various types and designs, but they are in two groups.

Non-reflecting Road Studs

These are in general use for marking the limits of pedestrian crossings and their approaches (circular or square in shape) and of parking bays (triangular).

They may be of stainless steel or types of plastic, but they are not allowed to project above the adjacent carriageway more than 16 mm at stud centre.

Reflecting Road Studs

Studs of this type have either a reflex lens, reflecting cube, or are solid with beads.

Solid reflecting studs can be circular or rectangular and are made of a hardened resin with ballotini. These are usually stuck to the road surface.

The reflex lens type is usually referred to as a 'cats-eye' and consists of a rubber insert carrying two reflex lenses set into a cast steel base. A vehicle wheel passing over the cats-eye depresses the rubber insert over the lenses, thus keeping the lenses clean.

The lens may be coloured white, red or green and the stud may be bi-directional or uni-directional. The 'short' cats-eye shown in Figure 12.50(a) is the old type set in road surfaces, but a 'long type' is available for use in roads carrying heavy traffic. This type has a longer base of 254 mm.

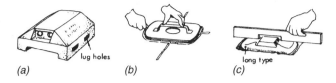

Figure 12.50(a)–(c)

Fitting and Replacing Road Studs by Hand

This is relatively simple. A template (*see* Figure 12.50(*b*)) is used to mark out the area to be cut and after excavation (usually 40 mm in depth), bedding material is placed in the base of the cavity and compacted by means of a hand rammer or vibrating rammer.

The recesses in the rear of the stud are filled with say, cold asphalt, and the stud is aligned in the cavity and pounded solidly to the bed. A template (*see* Figure 12.50(*c*)) is used to check the height at the centre of the stud, which should be no more than 25 mm. Hot bitumen is then grouted around the stud to the level of the road surface.

A special tool is used to lever out the pad if this should require changing. The pad is made of rubber reinforced with canvas and is held in the steel base by four lugs (*see* Figure 12.50(*a*)).

A more recent technique involves the use of polymer-modified rapid setting mortar when replacing studs in concrete road surfaces. Hand-mixed material can be applied by trowel and will withstand heavy traffic flow within 3 hours.

Road Stud Development

A reflecting road stud which replaces the twin reflex lenses with a reflective ring has the advantage of being non-directional and also provides a facility for the use of warning identification colours. The circular casing is equipped with a flange which, when impacted into a core-drilled hole, displaces the road building material and locks the device into position.

Road studs, some incorporating solar cells, are now in use in areas prone to fog, ice or other hazardous situations, such as in bends.

12.2.4 Traffic Calming/Restricting

There is a wide range of measures which are becoming increasingly used to control traffic, mainly for residents, but also for pedestrians and cyclists, by inhibiting the speed of passing traffic.

This is achieved, and traffic volumes reduced, by introducing such restricting measures as a type of chicane, road narrowing, footway widening and the creation of cul-de-sacs. Figure 12.51 shows one example of these techniques.

12.2.5 Vertical Control Measures

Road humps have proved to be an effective means of keeping down the speeds of vehicles. At low speeds, the

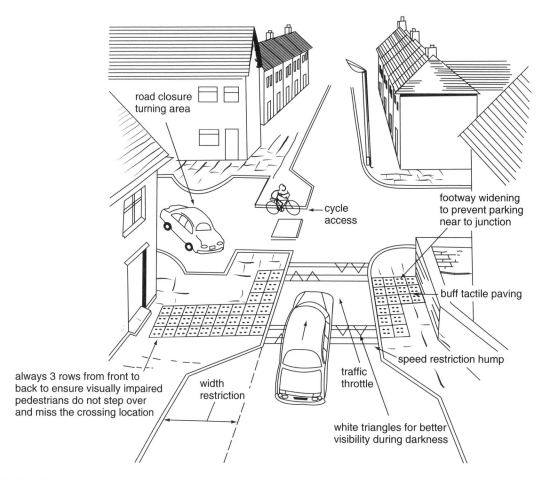

road closure
turning area

cycle
access

footway widening
to prevent parking
near to junction

buff tactile paving

always 3 rows from front to
back to ensure visually impaired
pedestrians do not step over
and miss the crossing location

width
restriction

traffic
throttle

speed restriction hump

white triangles for better
visibility during darkness

Figure 12.51   Traffic control measures

vertical features can be crossed without undue discomfort to passengers or damage to vehicles, but, as speeds increase, the humps become increasingly uncomfortable and possibly damaging to vehicles.

The Highways (Road Humps) Regulations permit considerable flexibility in the siting and shape of features. Flat top humps, cushions and speed tables are permitted and each may be tapered at the sides to allow a drainage channel to remain between the hump and the kerb. Heights can vary between 50 mm and 100 mm, but the usual practical minimum is 75 mm. Outline details of speed humps are shown in Figure 12.52

It is appropriate to use humps on both single and dual carriageways, provided that the road concerned has a 30 mph speed limit and is not a trunk, special or principal road. Unless the road has 20 mph limit, lighting must be provided to comply with the regulations.

Road humps should be preceded by a speed reducing feature, such as a junction, road markings, a bend or the end of a cul-de-sac. Traffic signs indicating the approach to humps are also needed except when within a 20 mph zone. There is no restriction to the number of humps in a series.

Where a road hump coincides with a pelican or zebra crossing, the flat type hump should be used extended to produce a speed table. Studs indicating the limits of a controlled crossing should be contained within the flat top of the hump.

Humps should not be constructed on a bridge or subway, or inside a tunnel, as there could be a risk of structural damage due to vehicle impact loads.

12.3   Pedestrian Guard Rails

The erection of pedestrian guard rails along the edge of a footway gives effective segregation of pedestrians from traffic, but may lead to access difficulties. Possibly the main function of guard rails is the guidance and protection of pedestrians at points of special danger such as busy road junctions.

The guard rails should be neat and simple in appearance, with their height and construction being such as to deter children from climbing over or through them.

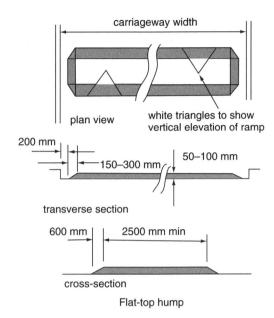

Figure 12.52 Road speed control humps

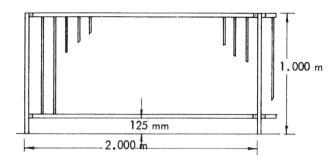

Figure 12.53(a)   Close panel type (square section rails)

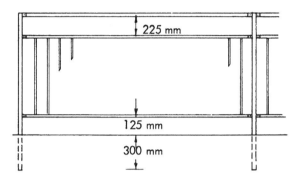

Figure 12.53(b)   Open panel rails to give improved sighting at T-junctions, pedestrian crossings, schools, etc. (square section rails)

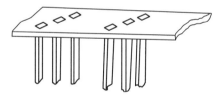

Figure 12.53(c)   View of visi-rail type fencing with slotted posts to aid see-through by children and approaching motorists

There are minimum clearances between the edge of the carriageway and any obstruction on the footway, etc. These clearances are:

450 mm (min.)	where the camber is 1 in 40 or less;
525 mm	where the camber is 1 in 40 to 1 in 24;
600 mm	where the camber exceeds 1 in 24.

Pedestrian guard rails are generally of three types as shown in Figures 12.53(*a*), (*b*) and (*c*).

Guard rails should, whenever possible, be set back 450 mm from the kerb face to give enough clearance for passing vehicles and also to leave sufficient room on the footway for two prams to pass. It may, therefore, be necessary to widen a footway before guard rails can be erected.

In providing guard rails, the following points need to be considered:

(*a*) Where loading/unloading is unavoidable, gates must be provided. Great care is needed to ensure that these gates are kept shut all times when not in use for loading/unloading.

(*b*) Gaps have to be left to accommodate trees, pillar boxes, electrical control boxes, etc., where these are near the kerb edge of the footway. The guard rails must be fixed close enough to these items to prevent small children from squeezing through the gap into the carriageway.

(*c*) Street furniture may have to be re-sited in locations where they would cause difficulties in fitting the guard rails.

(*d*) Suitable access must be provided at bus stops. This usually requires a second guard rail being erected to 'cover' the gap and so prevent direct access to the carriageway by pedestrians.

(*e*) On dual carriageways, the guard rails can be erected on the central reserve, instead of on both sides of the carriageway.

(*f*) In the case of a staggered pedestrian crossing (each half leading to the central reserve, or island, but not forming a straight path across the road) guard rails are necessary on the island to channel pedestrians from one part to the other.

(*g*) At particularly busy or dangerous sites, guard rails should be erected on *both* sides of the carriageway.

(*h*) At road junctions, guard rails should be erected/sited to prevent people from crossing the junction diagonally. They should channel pedestrians to the studded crossing provided at traffic lights, or, be long enough at unsignalled junctions to guide pedestrians away from turning traffic.

(*i*) Guard rails are essential at school entrances where children could otherwise run directly into the road. If there is a nearby pedestrian crossing or school crossing patrol, the guard rails should, if possible, extend to it.

(*j*) Care should be exercised that guard rails do not mask pedestrians from the view of approaching drivers. The visi-rail type fencing assists in this aspect.

The erection of pedestrian guard rails, which are to be fixed in a flagged footway fronting a school, is given as an example of the work involved. The work should preferably be done during a school holiday. The rails should be positioned parallel to the kerb line and centrally to the centre line of the school exit. The rails should be sufficiently behind the front face of the kerbs to permit vehicles to travel with their inside wheels close to the kerb, without striking the guard rails.

Procedure

(*a*) Set out for post holes (at 2 m intervals) 300 mm deep.

(*b*) Remove flags prior to excavation of post holes.

(*c*) Excavate post holes (for 50 mm diameter (nom.) or 50 mm square steel uprights).

(*d*) If the guard rail is of tubular steel prefabricated units, the whole rail must be erected upright in the post holes and adjusted for line (parallel to the kerbs) and level.

(*e*) If separate posts, with infill panels, the first two uprights are fixed with the first panel bolted between them. Then the third upright and second panel are fitted, aligned and fixed.

(*f*) When erected, the post holes are concreted, up to footway surface level, tamped and screeded to a fine finish to match the adjacent flags as nearly as possible and left to set.

(*g*) Alternatively, the flags are cut and relaid as necessary on top of the concrete in the post holes (this having been left low to permit the relaying of the flags).

(*h*) Warning signs and cones are required throughout the operation.

(*i*) All excavated material must be loaded and carted away.

(*j*) Care must be taken in excavation to avoid damage to any services in the footway and to the kerb backing.

(*k*) The rail will have to be supported in the upright position until the concrete has set.

(*l*) It may be necessary to lamp the work overnight if the concrete has not set at the end of the day's work.

(*m*) The rail must be set back at least 450 mm from face of kerb (more if camber exceeds 1 in 40).

12.4   Vehicle Safety Fences

In Britain there are several types of permanent safety fence in use. These are the untensioned open box beam type, tensioned corrugated beam, untensioned corrugated beam, tensioned wire rope, rectangular hollow section and double rectangular open box section.

The main objective in installing safety fencing is to restrain a vehicle within the confines of the carriageway on which it is travelling and prevent it from rebounding into that carriageway, thus causing an additional hazard. Safety fences are designed to 'give' when hit in order to absorb as much as is possible, of the energy of the impact and to redirect the vehicle along the line of the fence.

It is the policy of the Highways Agency to install fences along the central reserve of motorways, trunk roads and adjacent to all street furniture, e.g., road signs, lighting columns, bridge abutments and piers on these roads. They may also be provided on other roads where there are dangerous locations and obstructions.

The provision of safety fences has increased in recent years because of the extremely high cost of serious accidents, a cost to the community which is rising. The most dangerous form of accident occurs when a vehicle crosses the central reserve and collides with vehicles travelling in the opposite direction, where the closing speed of the vehicles may legally be as great as 225 km/hour or 140 mph.

The type of safety fence to be installed will be determined by the containment level and the working width. The containment level of each type of fence varies, being dependent on its design. The working width is the distance between the face of the fence and the obstruction. In central reserve situations where double sided fences are to be installed, care must be taken that the working width is maintained for both carriageways. The standard setback from the edge of the carriageway is 1.2 m, but at the edge of hardshoulders/hard strips it may be reduced to 600 mm.

12.4.1   Untensioned Corrugated Beam Safety Fence

This is used only on roads where vehicles are limited to speeds of less than 50 mph. The fence is carried on Z section steel posts (90 × 125 mm × 6 mm thickness) with steel blocking out pieces. These fences were previously installed using 150 × 150 mm × 1.8 m long wooden posts (oak or elm) with timber blocking out pieces. The posts are spaced at 3.2 m centres. These fences are inferior in strength to the other types. *See* Figures 12.54 and 12.55.

12.4.2   Tensioned Corrugated Beam Safety Fence

This is by far the most commonly used type of safety fence in Britain. It consists of 'W' section beams, used either single or double sided; it is attached to 'Z' section posts (49 × 110 mm) by M8 (8 mm) shear screws and is tensioned between anchorages. *See* Figure 12.56.

Tensioning is carried out by first of all ensuring the fence is sufficiently anchored at both ends. This is checked by applying 280/300 NM to the adjusting assembly nearest the anchor and then checking for movement in the anchor. If the anchor is firm, then the appropriate torque

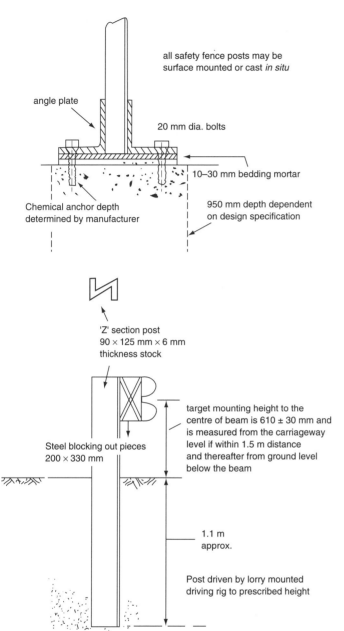

Figure 12.54   Untensioned corrugated beam

all safety fence posts may be surface mounted or cast *in situ*

angle plate

20 mm dia. bolts

10–30 mm bedding mortar

Chemical anchor depth determined by manufacturer

950 mm depth dependent on design specification

'Z' section post 90 × 125 mm × 6 mm thickness stock

target mounting height to the centre of beam is 610 ± 30 mm and is measured from the carriageway level if within 1.5 m distance and thereafter from ground level below the beam

Steel blocking out pieces 200 × 330 mm

1.1 m approx.

Post driven by lorry mounted driving rig to prescribed height

Torque (Nm)	Ambient temperature (°C)
0 (finger tight)	25
50	20
100	15
150	10
200	5
250	0
300	−5

NB 1. Interpolation between values is allowed.
2. Tolerances on torques is to be ± 5%.

Table 12.3   Relationship between torque and ambient temperature

is applied to the end assembly and all lock nuts tightened (*see* Table 12.3).

This type of fence may be used on central reserves and for long lengths on verges. Tensioned corrugated beam fences should not be used where the length of fence at full height is less than 45 m or on curves of less than 120 m radius.

On motorways and trunk roads, all street furniture in close proximity to the edge must be protected. This protection must extend from 30 m before the obstruction to 7.5 m after it, with an absolute minimum length overall of 45 m (Figure 12.57). When protecting a road sign, the 45 m minimum rule applies. When protecting a much longer obstacle, the '30 m in advance and 7.5 m after' rule applies.

Protection must be given to embankments which are over 6 m in height in the UK. Common practice in Europe, tends to consider protection at 2 m or more, in height.

12.4.3   Open Box Beam and Double Rail Open Box Beam Safety Fence

Where the safety fence needs to be of more rigid construction, it may be of the box beam type (Figure 12.58). For this fence, the posts are of similar size to tensioned corrugated fences (49 × 110 mm) and are placed at 2.4 m centres. The beams are normally 4.8 m in length, but 2.4 m beams are used where deflection space is limited or where the fence is alongside a curved length of road This type of fence should not he used on curves of less than 50 m radius.

The Double Rail Open Box Beam is installed where a higher level of containment is required and will normally be found at sign gantry legs, multiple message signs and railway features. It consists of a large steel post (90 × 125) cross-section, upon which open box beams are attached using offset brackets (*see* Figure 12.58(*a*)). It also incorporates a rubbing rail of 100 × 100 box steel, below the beam mounted at a 610 mm centre height by support straps.

12.4.4   Rectangular Hollow Section Safety Fence

There is no guidance on the situations where safety fences and barriers should be erected, or about the type to be used. However, an alternative to those already mentioned is the rectangular hollow section safety fence. This is not readily available from stock and is, therefore, only manufactured to order.

It is constructed from rectangular hollow section steel which is tensioned between anchorages. There are two sizes of section used, either 200 × 100 mm or 100 × 100 mm section. The 200 × 100 mm section is the more common in highway use. Figure 12.59 shows the general arrangement of this type of fence. It is essential that all end posts and angled posts are set in concrete.

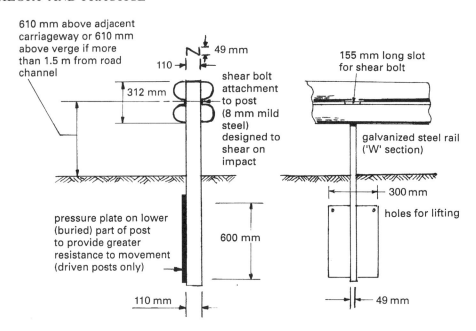

Figure 12.55 Double-sided corrugated beam safety fence. Posts are erected at 3.2 m centres or may be reduced to 1.6 m spacing where greater containment is required

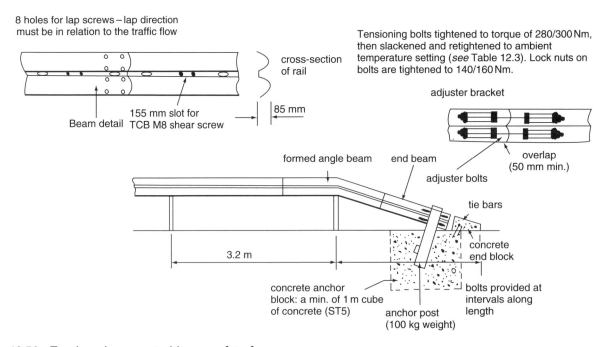

Figure 12.56 Tensioned corrugated beam safety fence

12.4.5 Tensioned Wire Safety Fence

Experiments have been carried out with the wire rope type of fence, which is used extensively on the continent of Europe. This type of fence has now been accepted in Britain, incorporating a three or four rope system.

Wire rope has been found unsuitable for 'soft' reserve or verge situations, as the height of the wire is critical; nor is it suitable for use around lighting columns or piers.

This system is designed with the intention of the wire 'digging into' the bodywork of the vehicle on impact.

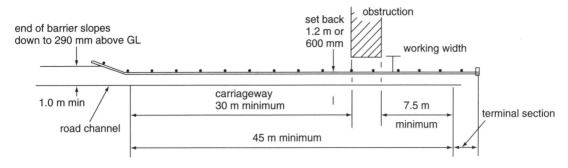

Figure 12.57   Plan view of safety fence. The distance between the line of the fence and the obstacle determines the type of fence to be used

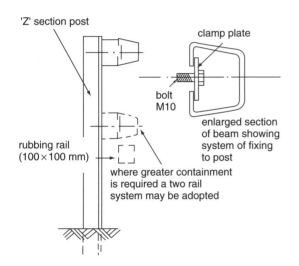

Figure 12.58   Open box beam safety fence and double rail open box beam variant

Figures 12.60 and 12.61 show layout and fixing post detail.

12.4.6   Erection and Maintenance of Safety Fences

In erecting safety fences, great care must be taken to drive the posts accurately to line and level. This is best done with a specially designed post-driving rig, which may be lorry mounted or portable, and which can be set to drive the post and maintain both vertical and horizontal alignment whilst so doing.

Repairing damaged lengths of safety fence is a job which must be done as soon as possible after the damage has occurred, and which must also be carried out very accurately as well as speedily. In order to restore the fence, posts should be replaced exactly into the position from which they have been removed on impact, ensuring first that the ground conditions are favourable. This is

determined by driving in a test post and applying to it a load of 6000 Nm at a point 600 mm above the ground. If the post deflects by more than 100 mm, then it has failed. In this case, either the post must be concreted into position or a longer post must be driven in. Alternatively, two additional posts at half spacing may be erected, to provide additional support.

It is possible to straighten, by re-rolling, the beam sections where they have not been too severely buckled by the impact. This decision must be made by the Site Engineer. Regalvanizing would also have to be considered.

If the special rig is not available for driving/extracting posts, a jacking frame can be of assistance in removing a damaged post (Figure 12.62).

12.4.7   Concrete Safety Barriers

Various shapes of reinforced concrete barrier have been trialed in a very limited way. The principle is different from that of the tensioned steel guardrail and beam, in that it is not designed to 'give', but still to contain the vehicle which crashes into it.

One type of barrier (Figure 12.63) is so shaped that small vehicles can climb partially on to it, but will not be overturned or thrown back across the carriageway. In tests, however, it has been found that small cars are redirected (bounced back) across three lane widths of road.

The British concrete barrier is a redesigned version of this type of containment barrier, the modification in design being carried out in cooperation with the Cement and Concrete Association. In this barrier, the face has been changed to be nearly vertical and in tests with small and medium-sized cars, as well as a 16 tonne heavy goods vehicle, it has been found that all were contained and redirected close to the barrier. (*See* Figures 12.64 and 12.65.)

12.4.8   Temporary Concrete Safety Barriers

A further addition to the range of concrete barriers is the Temporary Vertical Concrete Safety Barrier (TVCB).

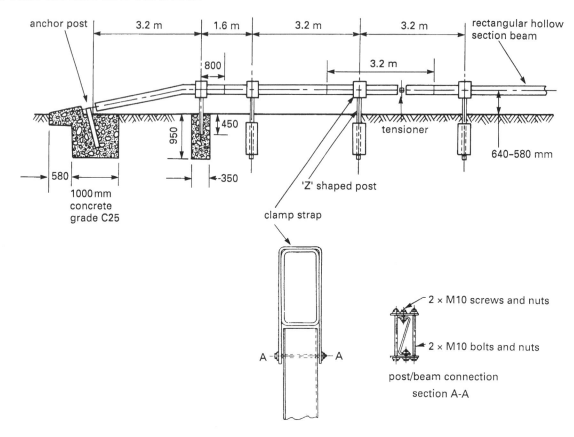

Figure 12.59   Rectangular hollow section safety fence. The beam centreline is 610 ± 30 mm above the road surface

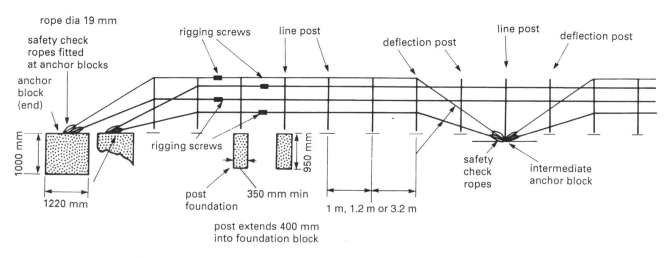

Figure 12.60   Layout of four-rope tensioned wire safety fence

These barriers provide added protection for operatives working on high speed roads and motorways.

The barrier consists of 3 m long concrete blocks having a base width of 450 mm and tapering to 400 mm at a height of 800 mm. (*See* Figure 12.64.)

12.5   Traffic Signals and Roundabouts

Much traffic congestion occurs at road junctions where streams of traffic heading for different destinations compete for limited road space. At certain junctions,

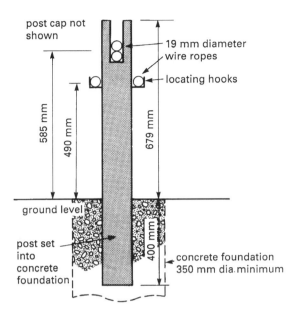

Figure 12.61   Detail of posts for tensioned wire rope safety fence

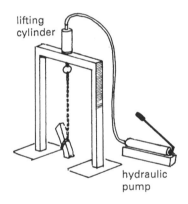

Figure 12.62   Jacking frame

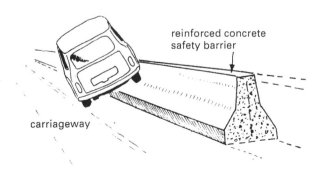

Figure 12.63   Reinforced concrete safety barrier (alternative designs are available)

queues of traffic can be prevented by building flyovers, but such a solution is too expensive for general use and a flyover can be unacceptable in an urban environment. Where sufficient space is available a roundabout or traffic island can work well (*see* Section 12.5.4), but traffic signals are generally used where the road space is limited and traffic flows are heavy.

Traffic signals were originally of the 'fixed time' type, which took no account of actual traffic conditions. They are now of the vehicle-activated type. The older versions of these operate by means of a pressure pad. The operation of these signals is initiated by a passing vehicle compressing the rubber tube in the tray of the pressure pad. This operates bellows which, in turn, operate a switch and alter the signal.

12.5.1   Loop Type Traffic Detector

A later and common type of traffic detector is the *loop type*. In this system, three loops of wire (the X, Y and Z loops) are set in the carriageway on the approach side of the signal lights and operate on a 'variation of magnetic field' basis. The cable consists of a single strand of wire with butyl rubber insulation and the loop operates at 40 V. The loops are set in the surfacing, at about 40 mm depth, and are connected to the controller (placed on the footway).

An approaching vehicle first passes over the X loop. This is the 'call' loop which initiates the change to 'green' for this direction. As it nears the signals and, therefore, the junction, the vehicle will pass over the Y and then the Z loop. These are 'extend' loops, designed to extend the 'green' period as traffic continues to arrive at the junction. Normally, each 'extend loop' will lengthen the period of 'green' by 1.5 seconds.

In addition, loops may be set in the junction to detect the presence of vehicles waiting to turn right. These are called 'presence loops', For a presence loop to operate, the waiting vehicle must be static for at least 3 seconds.

Although the loop system operates at 40 V the actual light operation is at 240 V. The light signal heads use tungsten halogen bulbs of 50 W and 100 W. The 100 W bulbs are used where greater light intensity is needed. An increasing number of installations in the UK now use Light Emitting Diode (LED) bulbs, which give a greater intensity of light. These are particularly useful in bright sunlight and also use less energy and have a longer life.

12.5.2   Computerized Traffic Signals (TRANSYT)

With the earlier traffic signal systems, timing changes are normally made without taking into account the traffic flow at adjacent signals at say, the other end of the street. The first generation of Urban Traffic Control

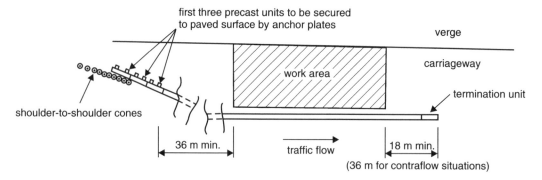

Figure 12.64 Diagrammatic layout of TVCB units at roadwork site

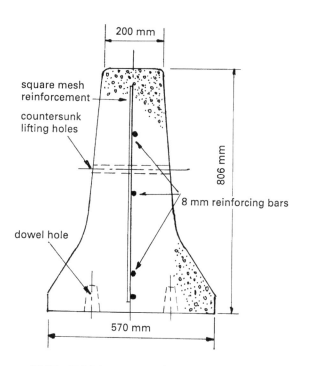

Figure 12.65 British concrete barrier – precast

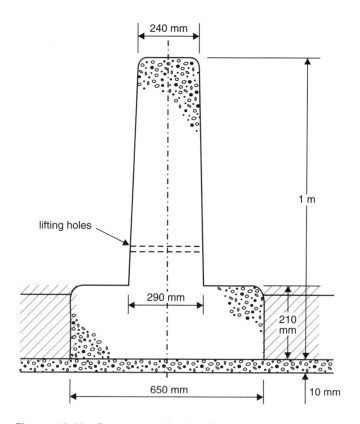

Figure 12.66 Permanent Vertical Concrete Barrier (precast) to BS 6579 Part 9

(UTC) systems linked adjacent sets of traffic signals to a central computer, which controlled the operation of all the traffic signals in a given network. This system, known as TRANSYT, gave great improvements to the traffic flow, particularly at peak periods, and relied on preset timings stored in the computer. The computer holds several different sets of timings (called plans) and the appropriate plan is selected on a time of day basis. However, if there is a change in the traffic conditions during a plan period, the system is unable to respond. Also, as the traffic patterns change in an area, the various signal timing plans need updating periodically to ensure that this system is effective.

12.5.3 Computerized Traffic Signals (SCOOT)

A second generation of UTC has been developed which measures the vehicle flow patterns using detector loops buried in the carriageway. This traffic flow information is sent to the central computer by British Telecom lines. The SCOOT (Split Cycletime and Offset Optimization Technique) program receives this information from all the detector loops in the network and determines the best coordination of signal timings for minimum traffic delays. Information to control the traffic lights is then sent back to the signal controllers. *See* Figure 12.67(*b*). The central computer continually monitors the traffic flow and adjusts the cycle time, altering the 'splits', i.e. the period of 'green' time to each signal installation, and adjusts the 'offsets', i.e. the amount of time by which adjacent signals are 'out of step'. The shortcomings of conventional UTC Systems are largely remedied by SCOOT.

12.5.4 Roundabouts

As already mentioned, where space permits and traffic flows are not too heavy, roundabouts can effectively deal with traffic congestion at road junctions. The most effective roundabout design for a given situation will depend upon approach speed, visibility, entry width, entry deflection and the circulatory carriageway.

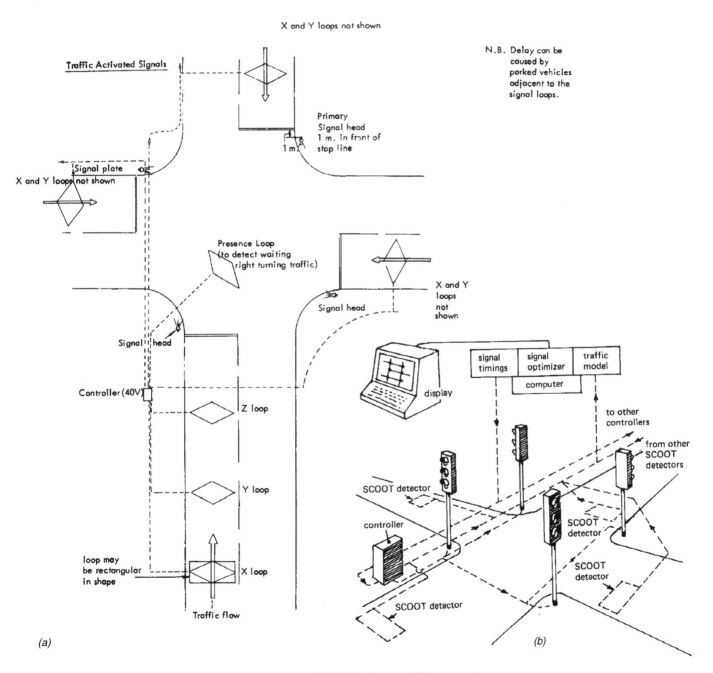

Figure 12.67 (a) Traffic activated signal system (b) SCOOT urban traffic control

Three main types of roundabout are in use in the UK:

(*a*) *Normal roundabout (Figure 12.68)*: a roundabout having a one-way circulatory carriageway around a kerbed central island 4 m or more in diameter and usually with flared approaches to allow multiple vehicle entry.

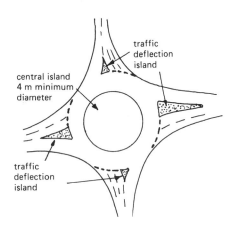

Figure 12.68   Normal roundabout

(*b*) *Mini roundabout (Figure 12.69)*: a roundabout having a one-way circulatory carriageway around a flush or slightly raised circular marking less than 4 m in diameter and with or without flared approaches.

(*c*) *Double roundabout*: an individual junction with two normal or mini roundabouts either contiguous, i.e. adjacent, or connected by a central link road or kerbed island.

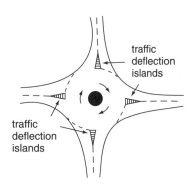

Figure 12.69   Mini roundabout

12.6   Street Lighting

These notes on street lighting are essentially very brief. They relate mainly to the needs of the roadworker and provide an appreciation of basic essentials of illuminating roads at night. The notes do not provide a complete background to the science of street illumination.

12.6.1   Lamps

These are of two main types; filament and discharge.

Filament type

These are very expensive to run, produce only a low than intensity lighting and are old fashioned.

Discharge type

N.B.: Discharge lamps need control gear for their operation and cannot be plugged directly into the mains.

(*a*) *Mercury or Metal haline lamp* – gives a bluish light but can be colour corrected to give white light emission (when coated for this purpose).
(*b*) *Low pressure sodium lamp* – gives a yellow light and is the most efficient lamp. It uses 135 W power and is a fire hazard when broken. It gives maximum light at minimum cost.
(*c*) *High pressure sodium lamp* – gives white light: rather 'cold' in appearance. The bulb is similar in appearance to mercury lamp.

12.6.2   Lighting Columns

Concrete columns: these are tending to become rather old fashioned and are used mainly on residential estates not on main roads. Concrete columns are obtainable up to 12 m in height.

Steel or aluminium tubular columns: these are commonly used for main road lighting and have a painted finish and, therefore, need regular repainting. They are made either in single or double arm form (*see* Figure 12.70).

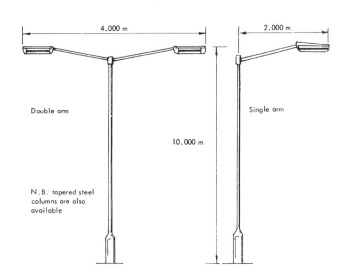

Figure 12.70   Aluminium tubular column

Glass fibre columns: these are very light and easy to erect and are normally used for estate roads only.

Heights of Lighting Columns

5 m (smallest standard size) – only extend 0.5 m below ground level;

10 m (for main road use) – 1.5 m below ground;

12 m (for wide roads: dual carriageways) – 1.5 m below ground.

There is also an 8 m intermediate size: this is now only used in exceptional circumstances.

12.6.3 Lanterns

Small Mercury Vapour

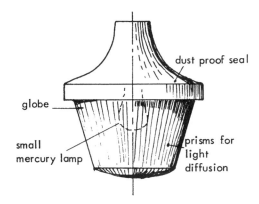

Figure 12.71 Small mercury vapour lantern – globe generally of glass.*

* Glass globes are easily broken. Can be polycarbonate or biturate globe (semi-unbreakable), but these globes are expensive.

Figure 12.72 Newer version of lantern top

Sodium Discharge Lantern

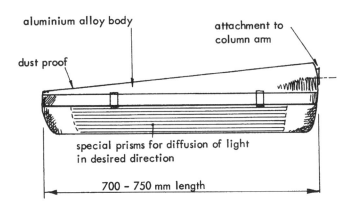

Figure 12.73 Sodium discharge lantern

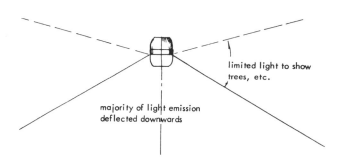

Figure 12.74 Semi cut-off type: allows the dispersion of a limited amount of light to the side and upwards

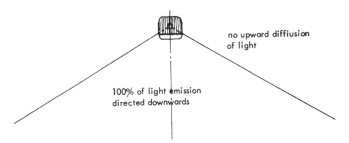

Figure 12.75 Complete cut-off type: no upwards diffusion

Diffusion of Light

The lantern can be either of the semi cut-off or of the 100% cut-off type.

12.6.4 Control Equipment

The equipment generally included is as follows.

Solar Dial System Time Switch

In this, the on/off contact cams vary day by day of the year to give correct lighting up/switching off times. Can also be set to switch on/off at predetermined times.

Photo-Electric Control Switch

A photo-cell is fitted on top of the lantern. As daylight fails the switch is operated and the light comes on. Light must be left on all night, only switching off when morning light reaches the necessary intensity. This type has the advantage of coming on early in bad light conditions.

Column controls must include a cut-out and fuse box.

12.6.5 Erection of Column

Brief details of the erection procedure are:

(*a*) Excavate minimum possible size of hole.
(*b*) Place small slab in bottom of hole (to support weight of column).
(*c*) Fit bracket on to column whilst on ground, and pull wires through.
(*d*) Hoist column into position on rope sling (check that bitumen coating of root of column is sound and intact).
(*e*) Lower column on to slab.
(*f*) Set column vertical with spirit level.
(*g*) Ram and compact concrete round base/root of column. N.B.: Care must be taken to leave access to hole in root.

The larger types of column may be up to 1000 kg in weight. In these cases the lantern may be fitted prior to erection, but is generally left to be fitted *after* erection of the column, due to handling difficulties.

The choice of position depends on:

column height
lamp type
road width
road conditions (e.g. junctions, bends, etc.).

12.6.6 Principle of Street Lighting

The underlying principle is to make the whole of the road bright.

BS 5489 contains tables for determining spacing of columns, heights, etc., i.e. it gives guidelines about the layout of lights in various situations.

Typical situations are:

(*a*) on straight roads (shown in Figure 12.77(*a*))
(*b*) on bends (shown in Figure 12.77(*b*)).

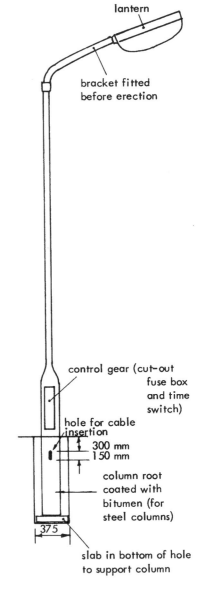

Figure 12.76 Erection of street lighting column

12.6.7 Basic Lighting Installation Design

Consider a stretch of road as shown in the plan (Figure 12.78). Critical positions for lighting are at the:

1 bend;
2 cross-roads;
3 junction.

When using lights with 100% cut-off, spacing is reduced compared with semi cut-off because there is no tail-back of light.

On dual carriageways lighting columns may be opposite each other, not staggered as on a single carriageway (Figure 12.79).

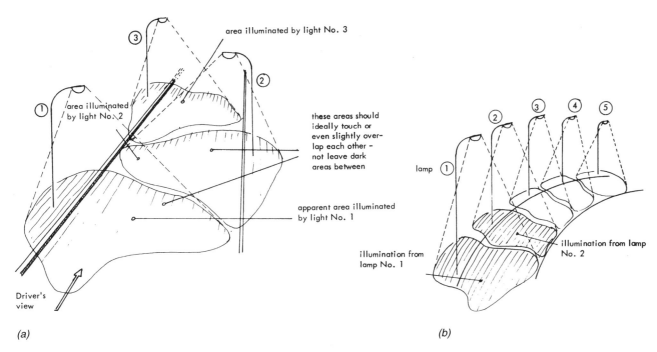

(a)   (b)

Figure 12.77   (a) Straight road illumination. For a main road lights are normally at 44 m (staggered); for a side road lights are normally at 33 m (staggered). (b) Illumination on bends. All lamps are positioned on the outside of the bend

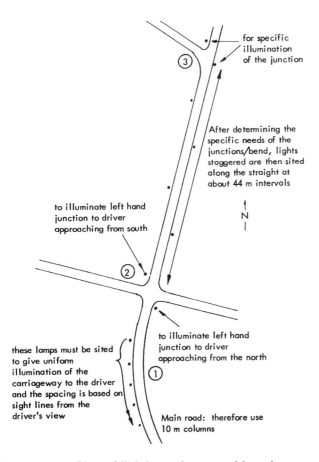

Figure 12.78   Plan of lighting column positions in varying circumstances

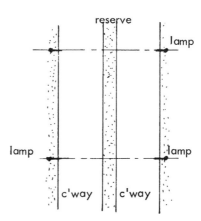

Figure 12.79   Plan of lighting column positions on dual carriageways

Revision Questions

1   What is the minimum clearance from the edge of carriageway (kerbline) to the edge of a sign?
2   Where signs are erected above a footway, what should be the minimum mounting height?
3   What is the practical road hump height used to provide maximum speed reduction whilst minimizing vehicle impact risk?
4   Which type of standard street lighting gives the 'whitest' and most bright light?

Note: See page 305 for answers to these revision questions

13

Bridgeworks: construction & maintenance

a limited number of suitable routes and can only be moved after giving due notice to the Highway and Police Authorities.

13.2 Design

In bridge design, certain considerations have to be taken into account, including answering the following:

What is the normal load to be carried?
What span is required?
Is the height of structure sufficient to clear flooding?
Siting of abutments, piers, etc.?
What is the navigation clearance (if required)?
Appearance of bridge? Must, if possible harmonize with its surroundings. If not, it must be a pleasing feature.
What type of subsoil? And what subsoil conditions exist?
What is the estimated cost?

When building over a canal or river or railway, centring (support formwork) cannot be used, due to its impedance to traffic on water or rail. Hence, precast construction must be considered.

13.1 Historical Notes

All roads have associated bridgeworks where they cross rivers, railways and other roads. Hence, bridges form a major feature of the road system of the country. All local authorities have the responsibility of maintaining the bridges in their area, but some authorities do not concern themselves with bridge design. However, it is necessary to consider the types and methods of construction of bridges in order that their maintenance may be adequately carried out. The method of repair must have prior approval of the DfT in the case of Ancient Monuments.

Many early bridges are now classified as ancient monuments, and must be repaired to match original materials, etc. Many early bridges were built and maintained by the Church; these are almost entirely of stone construction.

In the eighteenth century, the building of the canal systems caused many bridges to be built, in this case mainly in brickwork. In the 1830s and 1840s, the development of the railways caused another spate of bridge building, this time, mainly in cast iron.

In the present century, bridges have been designed and built purely as highway bridges. Due to increasing vehicle loads, considerable extra stress has been put on existing bridges, often well in excess of their designed load.

Normal maximum load is 44 tonnes. In extreme cases, loads of up to 300 tonnes have been transported by road and over bridges. Such loads are necessarily restricted to

13.3 Bridge Types

13.3.1 Arch

This is the oldest form of bridging and is shown in Figure 13.1(*a*), with a typical coping and parapet detail shown in Figure 13.1(*b*).

Materials used are: masonry, brick, or concrete (either mass or reinforced).

If there is risk of scouring of the water channel, an invert slab is placed under the bed of the watercourse.

When constructing a concrete arch bridge, or brickwork arch, the arch is usually built up about two-thirds of the height, allowed to set and cure and the centre section of the arch is subsequently added (*see* Figure 13.2). In the case of larger spans, the construction of the arch is divided into five stages.

13.3.2 Portal Type (Figure 13.3)

This is a form of distorted arch bridge. Usually in reinforced concrete construction (may be steel framed).

13.3.3 Rectangular Opening

Often with a deck of precast/prestressed beams supported on abutments of either (*a*) reinforced concrete (Figure 13.4(*a*)); or (*b*) mass concrete (Figure 13.4(*b*)). Alternatively, the abutment may be partially mass and partially reinforced concrete.

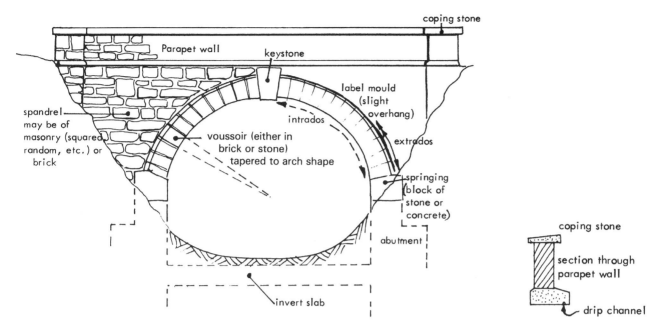

Figure 13.1(a)   Stone arch bridge

Figure 13.1(b)   Bridge parapet detail

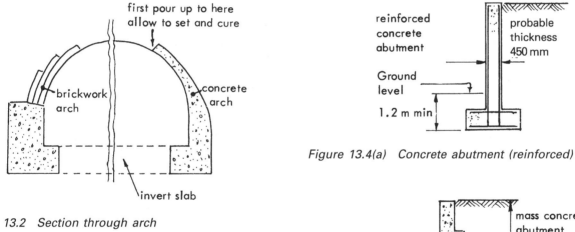

Figure 13.2   Section through arch

Figure 13.4(a)   Concrete abutment (reinforced)

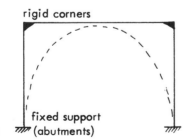

Figure 13.3   Portal frame

Figure 13.4(b)   Concrete abutment (mass concrete)

13.4   Abutments

For the construction of abutments, sub-soil conditions must be known, from site investigations, geological maps, etc. Abutments must be on sub-soil having uniform settlement conditions, and must be deep enough to be below frost penetration depth.

Where ground is unsatisfactory, piling may be used, Piles may be of:

(*a*) reinforced concrete (*see* Figure 13.5(*a*) and (*b*)); or
(*b*) steel (I section).

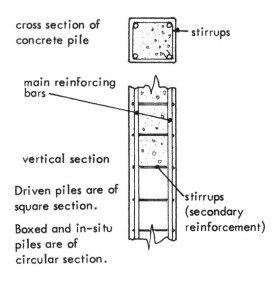

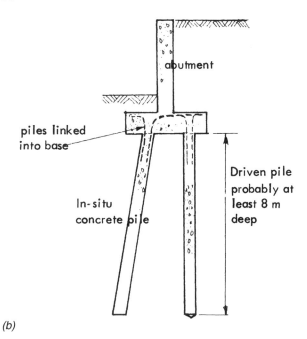

Figure 13.5(a) and (b)   Bridge support piles (concrete)

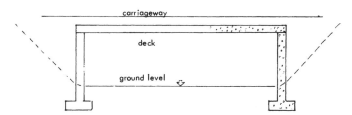

Figure 13.6   Rectangular concrete bridge structure

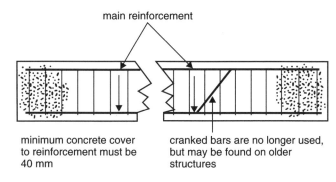

Figure 13.7   Concrete deck beam

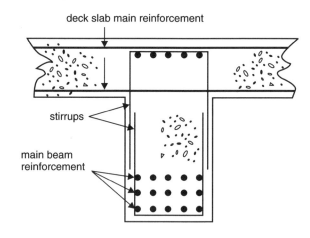

Figure 13.8   Transverse section of beam and slab construction

13.6   Composite Construction (Figure 13.9)

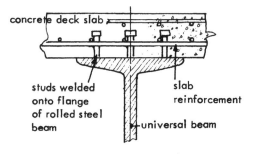

Figure 13.9   Composite construction

13.5   Superstructure (Figure 13.6)

This is the deck of the bridge which rests on the abutments.

The reinforcement of the deck beam is usually as shown in Figure 13.7. If the cover were to be less than 40 mm, moisture would eventually penetrate and corrode the reinforcement, which would then expand and burst the slab.

Many rectangular type bridges have a considerable length of span and thus require 'stiffening' of the deck. This can be achieved by 'beam and slab' construction (*see* cross-section in Figure 13.8).

The main problem with composite construction is the maintenance of the exposed steel beam, which requires painting every five to seven years; for new construction this may be extended to ten to fifteen years.

13.7 Prestressed Concrete Beams (Figure 13.10)

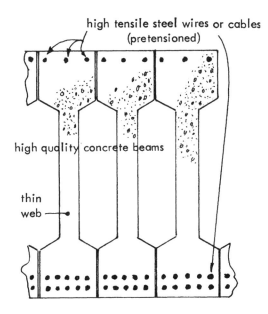

Figure 13.10   Prestressed beam construction

This type of construction is required for longer spans. The usual loading (prestressing) of the wires is about 1000 N/mm^2. The wires are tensioned prior to concreting by means of jacking systems fitted to the beam mould (shuttering). The concrete is then poured. When hardened the prestressed beam is removed from the mould. It tends to bend upwards, but when loaded, deflects to an approximately horizontal position (*see* Figure 13.11).

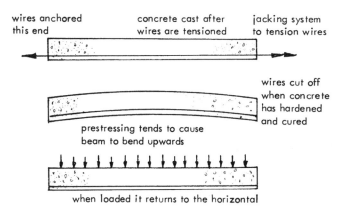

Figure 13.11   Principle of prestressing of concrete beams

A deck slab can be constructed of prestressed beams, placed side by side on the bridge abutments and concreted as shown in Figure 13.12.

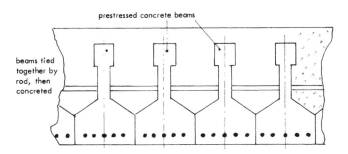

Figure 13.12   Transverse section of deck

13.8 Expansion

It is most important that allowance be made for expansion of the bridge deck, and for spans of more than 8–10 m, a bearing must be provided. One of the most simple forms is shown in Figures 13.13(*a*) rocker end (fixed end) and (*b*) roller end (free end).

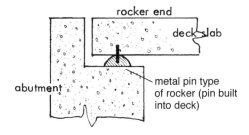

Figure 13.13(a)   Provision for deck expansion: rocker end

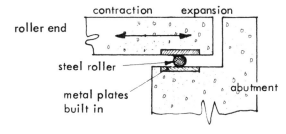

Figure 13.13(b)   Provision for deck expansion: roller end

13.8.1 Using a Neoprene 'Rubber' Block (Elastomeric)

This distorts under movement of the slab and is composed of a neoprene block with steel plates built in (Figure 13.14).

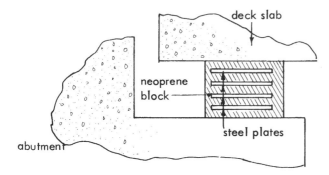

Figure 13.14   Expansion joint with neoprene bearing (Elastomeric)

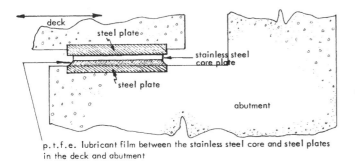

Figure 13.15   Epansion joint with PTFE film

In place of the roller, a stainless steel core plate with polytetrafluoroethylene (PTFE) lubricant can be used (*see* Figure 13.15).

13.8.2   Expansion Jointing

Without expansion jointing at the ends of the deck, movement of this slab will cause cracking of the surfacing material, letting in water with consequent deterioration of the expansion provision (*see* Figure 13.16).

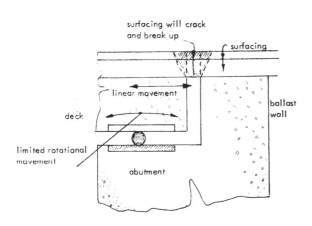

Figure 13.16   Surfacing over expansion joint position

(*a*) The simplest form of expansion jointing is shown in Figure 13.17.

(*b*) An alternative type of joint, as shown in Figure 13.18, uses a special flexible bridge jointing material over the joint at the end of the deck.

(*c*) Epoxy mortar nosing/polysulphide, etc., sealant (Figure 13.19).

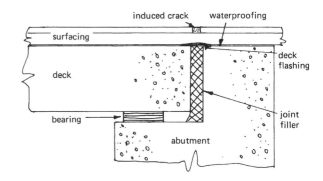

Figure 13.17   Fixed end buried joint

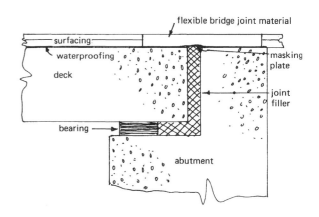

Figure 13.18   Alternative joint using flexible material

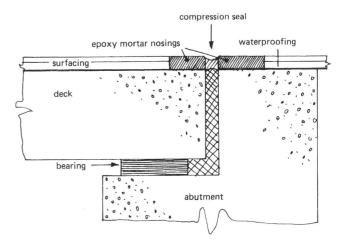

Figure 13.19   Use of epoxy mortar nosings

The nosings are bonded to the deck and the gap is filled with the sealant, which bonds to the sides of the deck/abutment joint.

On larger bridges steel plate nosings or proprietary manufacturers' joint systems may be used.

13.9 Waterproofing

This is laid on the top of the bridge deck or slab and was only introduced in the early 1950s. Prior to this only a bitumen spray was used to act as a tack coat for the surfacing.

The reasons for waterproofing are:

(*a*) Concrete tends to shrink on hardening, thus causing fine shrinkage cracks. These may allow water to penetrate and eventually to corrode the reinforcement and burst the deck/slab.

(*b*) Salt used for defrosting will penetrate asphalt and also will, in solution (i.e. as salt water) rapidly penetrate the shrinkage cracks in the deck, causing more rapid corrosion and breaking up of the concrete than clean water.

(*c*) Dangerous chemicals are often carried by road tanker and can leak on to the carriageway (particularly after an accident). If on a bridge, this chemical may rapidly penetrate and eat away the concrete, causing the bridge to fail.

13.9.1 Waterproofing materials

All these hazards must be prevented from affecting the concrete deck which provides the strength of the structure.

Waterproofing may be by mastic asphalt, membranes, 'two pack' materials or spray applied systems.

Mastic Asphalt

This is laid in two courses, each 10 mm thick. With this material, it is not really possible to produce a complete seal, particularly as it does not adhere fully to the concrete. Nor is it an economic process, since it is extremely labour intensive.

Membranes (of pitch polymer)

This material is supplied in black sheets (in rolls) of about 1 mm thickness, and 1 m wide.

The waterproofing process is as follows:

(*a*) prime deck surface with light bitumen sealer and allow to dry;
(*b*) apply hot bitumen (adhesive);
(*c*) roll out membrane on hot bitumen;

(*d*) press down and roll out air bubbles;
(*e*) lay 10 mm thick layer of sand asphalt (zero stone content)* to protect membrane from damage by site traffic and/or stone in the asphalt surfacing.

*N.B.: It is most important that there is no stone in the sand asphalt. It is also important to keep to about 10 mm maximum thickness for the sand asphalt layer in order to avoid the possibility of horizontal shear caused by the stress induced by braking and accelerating traffic, with consequent slipping and disintegration of the road surfacing.

'Two Pack' Materials

These materials may be polyester resin + hardener, or epoxy resin + hardener. In these cases, 6 mm thick board is laid on the 'two pack' material – which remains plastic indefinitely. The boarding protects this plastic sealant (as the sand asphalt protects the membrane).

13.10 Parapet Walls

13.10.1 Post and Three Rails (Figure 13.20)

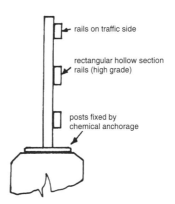

rails on traffic side

rectangular hollow section rails (high grade)

posts fixed by chemical anchorage

Figure 13.20 Fixing of rail-type parapet

This must be capable of withstanding impact by a $1\frac{1}{2}$ tonne vehicle at 113 km/h at 20° to the fencing (*for trunk roads*). For less important roads, the impact speed is 80 km/h (for a $1\frac{1}{2}$ tonne vehicle at 20° to the line of the wall).

13.10.2 Railway Bridges

In such cases, the Railway Authorities require that the parapet be a solid wall, brick faced (*see* Figure 13.21).

13.11 Large Bridges

These tend to be either suspension bridges or box girder bridges, and require special design teams, since they are

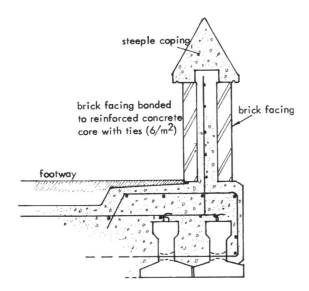

Figure 13.21   Detail of parapet of railway bridge

beyond the scope of a local authority Bridges Section. Examples of suspension bridges are:

(a) Severn suspension bridge ⎫
(b) Forth suspension bridge ⎬ identical basic designs: steel frame towers, high tensile steel cables, steel decks.
(c) Humber suspension bridge ⎭ reinforced concrete towers with steel cables and deck.

Box girder bridges have been used extensively on sections of elevated motorway in this country.

13.12   Bridge Maintenance

All bridges should be subject to two yearly bridge maintenance inspection, in accordance with official regulations regarding the 'Inspection of Highways – Bridges'. In addition, they should be checked visually every time a roadwork foreman/ganger passes over or under. Any apparent defects should be reported to the bridge engineer as soon as they are seen, so that immediate expert attention can be given to rectifying these defects.

13.12.1   Bridge Inspection Report (see page 254)

The Report form lists all the features of the many types of bridge; not all of these are applicable to any one particular bridge.

The inspection should cover the following:

(1) Invert – this may be of concrete or stone pitching, which must be maintained in good condition. Not all bridges have inverts.

(2) Aprons (Figure 13.22) – the extension of the invert beyond the limit of the bridge structure. Examine for signs of scouring at the end of the apron (join to normal ground invert).

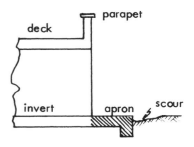

Figure 13.22   Aprons

(3) Foundations – often, the detail of bridge foundations is not known, particularly with older bridges. Check for any 'under-cut' of structure (Figure 13.23).

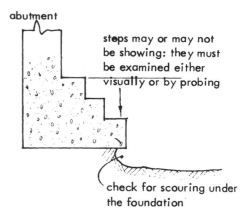

Figure 13.23   Foundations

Serious scouring must be remedied immediately by underpinning (Figure 13.24). If there is scouring, loose material must be cleaned out, and the hole filled with concrete.

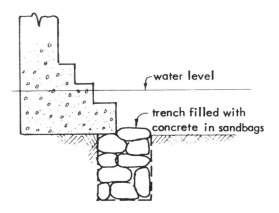

Figure 13.24   Underpinning foundations

If under water, dig a trench and fill with sandbags containing dry concrete. Divers may be required for this work.

(4) Cutwater (Figure 13.25) – this is the lower part of a pier in a multi-span bridge.

The cutwater suffers extremely high rate of wear from the water flowing past (particularly on the upstream side) due mainly to debris, grit, etc., carried by the water. It may also be subject to severe scouring, particularly if tree trunks, branches, etc., have lodged against the pier.

(5) Piers/columns – if concrete, check particularly for rusting on surface of concrete. This indicates penetration of water to reinforcing steel. Cracked areas must be cleaned to the steel, and the rusted steel cleaned. A rust inhibitor should be applied to the steel after cleaning. The damaged area is then re-concreted using a bonding agent to ensure good adhesion of the new material. If *brick* or *masonry*, check the state of the brickwork/stonework.

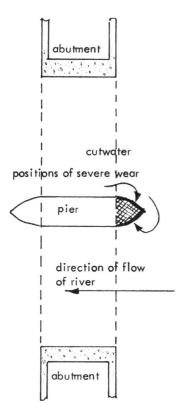

Figure 13.25 Cut water wear

(6) Abutments – check for cracks. Care must be taken to differentiate between shrinkage and movement cracks. Shrinkage cracks are *not* significant. Movement cracks must be corrected/checked. In general, shrinkage cracks are uniform in width, movement cracks tend to vary in width.

(7) Wing walls – to protect and prevent erosion from under wing walls, it is essential that gullies be provided at the ends of bridge decks, to collect surface water (Figures 13.26(*a*)). Proper outfall drains must be provided from these gullies to conduct the water away.

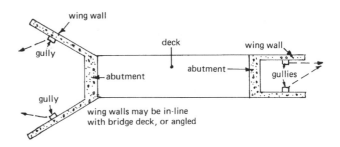

Figure 13.26(a) Surface water gully locations

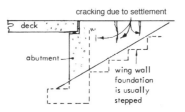

Figure 13.26(b) Settlement adjacent to abutments

(8) Embankments
  (*a*) settle behind the abutments.
  (*b*) need to be made up.
  (*c*) cause cracking of road structure (due to this settling).
  (*d*) the 'run on' for vehicles must be smooth and well aligned to prevent 'impacting' by vehicular traffic. (Any impact can easily double the live load or over-stress the deck.)

(9) Training walls – as wing walls.

(10) Drainage sub-structure – weep holes must be clear and operating properly. They can easily become buried and must be dug clear.

(11) Parapets – are they stable and safe? Joints may perish in brick and stonework. May be damaged (and not reported) by vehicular impacts. Vandals often aggravate initial minor faults. Steel parapets must be checked for rusting.

(12) Bearings – must be free to move. If seized up, they may be rectified by greasing. They must be 'free' as, otherwise, temperature variation can cause deck expansion which will then tilt the abutments and cause excessive stressing.

Neoprene – look for splits and distortion on the sides.

Generally, bearings can be replaced by jacking up the deck slab (by about 50 mm). Care must be taken if water mains, etc., are carried in the deck (should be diverted before jacking).

(13) Expansion joints – must be examined for condition – polysulphide sealant can crack, joints can become distorted, adhesion may be lost (particularly with polysulphide sealer), steel plate joints must be firmly fixed and not rattle as traffic passes over.

(14) Main beams – if of steel – is there any rusting? Do they need painting? If of concrete – is rust marking appearing on surface of concrete?

(15) Encased ends – ends of beams are built into an end 'cap' – the web of a steel beam can become distorted or 'bowed'. Beams (steel) can corrode badly where they fit into the concrete encasing.

(16) Steel troughing (Figure 13.27) – this type of construction was common in the 1930s and 1940s, but is no longer used. Must be checked for water leaks (particularly at joints between trough sections).

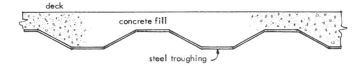

Figure 13.27   Steel troughing

(17) Jack arches – check for damage to bricks in arch. Outside (or end) RSJ may tend to tilt outwards thus not supporting the brick arch (this movement is due to failure of the tie bars). Common defect is rusting of bottom flange – even to the extent that the flange can fall off (Figure 13.28).

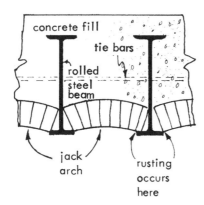

Figure 13.28   Jack arch rusting

(18) Transverse beams and diaphragms – these may be universal beams with steel diaphragms or concrete beams with reinforced concrete diaphragms (*see* Figure 13.29).

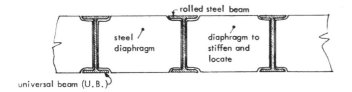

Figure 13.29

(19) Waterproofing – if underside of bridge deck is damp, this may be either leakage through the deck or condensation. Only the newer bridges are waterproofed.

A visual check cannot determine whether or not the waterproofing is adequate/effective.

(20) Drainage superstructure – are gullies in the deck functioning or are they blocked?

Drainage may be provided from the bearing area by a channel (as shown in Figure 13.30).

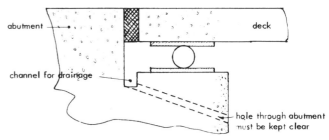

Figure 13.30   Drainage channel

(21) Concrete deck – check visually for cracks. There will generally be shrinkage cracks, particularly at joints between shuttering panels. Structural cracks must be reported at once.

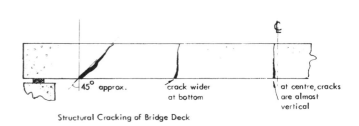

Figure 13.31   Structural cracking of bridge deck

(22) Arch springing – check condition.

(23) Arch ring – check that rings are not parting from one another. Is ring of true shape? (Even if apparently distorted, it may be quite sound and safe) (Figure 13.32).

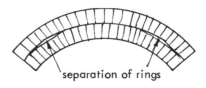

Figure 13.32   Separation of arch rings

Alternatively, excessive traffic loads can cause cracking as shown in Figure 13.33; with the result that the outer part of the arch tends to move outwards.

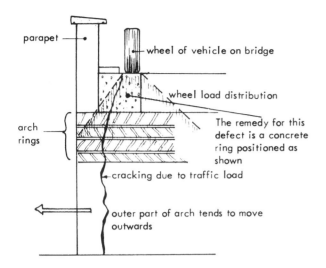

Figure 13.33   Circumferential damage to arch

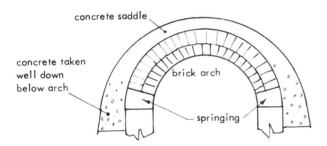

Figure 13.34   Concrete saddle to strengthen damaged arch

Where there is general deterioration of the arch, the remedy is to provide a concrete saddle (for strengthening) as shown in Figure 13.34.

(24) Voussoir/arch face – visually inspected for condition.

(25) Spandrel walls – are they starting to bulge?

(26) Tie rods – can rust away.

(27) Pointing – is generally covered by inspection of other items in report.

(28) Condition of masonry – are joints sound? Is stone face spalling away?

(29) Surfacing – vehicles must have a smooth run across the bridge. Some old bridges still have grass verges on them, which retain water. They should be removed and replaced with impervious material.

The bridge inspector is only there to record what he sees. It is the duty of the bridge engineer to decide what action is to be taken based on the report which is submitted to him. Simplified forms used for bridge inspection and culvert inspection are shown in Figure 13.35.

13.13   Footbridges

In addition to the main road bridges, the roadworker may, on occasions, be concerned with the construction of footbridges. The style and construction detail will vary, but in most cases the decking will be of timber with a standard form of handrailing. This handrailing may be on one side only of the deck, in the case of the more simple and small-scale bridge.

One example of a footbridge of a fairly complex nature is considered in detail below. Variations to suit different situations can then be made.

A footbridge is to consist of natural stone abutments and timber decking and is to be constructed over a stream approximately 2 m wide. Describe and sketch a suitable bridge, indicating timber sizes to be used and explain how the work would be carried out.

One method of doing this would be as follows: (*see* Figures 13.36(*a*) and (*b*)).

(a) Divert stream clear of site of footbridge.

(b) Excavate footings for abutments.

(c) Place concrete for footings, probably due to the small span, taking the footing across the whole width of the stream (but below stream invert level).

(d) Build first course of stone facing for the abutments on the footing at each side of the stream.

(e) Concrete behind the stone facings, tieing the stonework into the concrete (use a grade 20 or stronger concrete).

(f) Continue building up stonework and concreting behind it until abutments are completed.

(g) Set dummy end posts into concrete and withdraw these as the concrete begins to set or set fixing bolts into concrete ready for bolting on handrailing uprights and deck timbers.

(h) The timber deck is placed and fixed after the concrete of the abutments has cured.

The main beams are positioned as are the end posts. These are tied together by means of a tie bar and bearer (between the beams).

(i) Fix the decking (of transverse boards) to the beams.

(j) Fix handrails.

(k) Re-direct stream to its original course.

(l) Provide tarmacadam ramps to link the bridge deck level to the adjacent path.

N.B.: 1   All timber should be treated with a preservative, preferably pressure impregnated.

  2   At locations where children are present, i.e., play areas, parks and footpaths, it is essential that the side rails prevent climbing or falling incidents.

13.14   Timber Bridges

Wood is a material which, unlike steel or concrete (whose properties can be altered by changing their composition, etc.) has natural mechanical properties that cannot be changed. Laminated beams do, however, allow optimum strength and stiffness to be obtained.

Bridge Inspection Report

Bridge No......
Location........
Date of inspection.....
Inspected by.........

Item No	Description	Remarks
1	Invert	
2	Aprons	
3	Foundations	
4	Cutwaters	
5	Piers/Columns	
6	Abutments	
7	Wingwalls	
8	Embankments	
9	Training walls	
10	Drainage substructure	
11	Parapets	
12	Bearings	
13	Expansion joints	
14	Main beams	
15	Encased ends	
16	Troughing	
17	Jack arches	
18	Transverse beams and diaphragms	
19	Waterproofing	
20	Drainage super-structure	
21	Concrete deck	
22	Arch springing	
23	Arch ring	
24	Voussoirs/Archface	
25	Spandrel walls	
26	Tie rods	
27	Pointing	
28	Masonry condition	
29	Surfacing	

Action required..........

ACS Bridges...........
Date..................

Culvert Inspection Report

Culvert No.............
Location.............
Date of inspection.............
Inspected by...................

	Condition (Good, fair, bad)	Remarks
Brick arch, Concrete tube, Armco, Deck slab		
Abutments		
Wing walls, Head walls		
Parapets (Brick, Stone, Post & tube)		
Waterway Debris, Overgrowth etc.		
Training walls		
Foundations Scour etc.		
Carriageway (over culvert)		
Drainage		

Bridge Inspector..............

Divisional Surveyor..................

Figure 13.35 Bridge and culvert inspection report forms

Timber is a resilient material with good shock load resistance and its properties are only slightly affected by extremes of temperature. Its coefficient of expansion is lower than that of steel or concrete. Timber is only used for footbridges in the UK.

In use, the design of a timber bridge must guard against water leaks from the deck, which could result in decay and direct contact with the ground must be avoided.

Consideration must be given to preservative treatment to prolong the life of the timber.

13.14.1 Structural Forms of Timber Ridge

This applies to bridges over 10 m in length.

(1) Truss

The forms of truss most commonly used are the Howe, Warren and Bowstring (*see* Figure 13.37), bridges up to about 75 m in span being practically possible using glued laminated members (glulam).

(2) Web Girders

These make use of fabricated beams and have the advantage of taking up less height than the trusses. Construction is of two or more layers of boards running 'diagonally' and at right-angles to each other and joined, by nailing, to form the web of an I-beam with laminated flanges (*see* Figure 13.38).

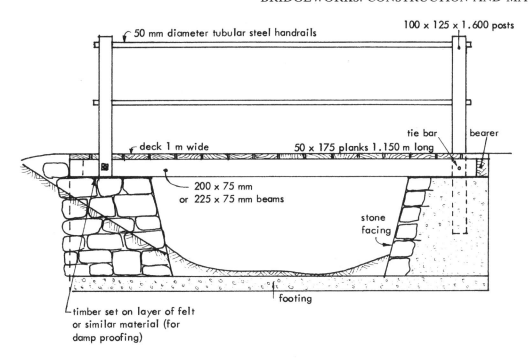

Figure 13.36(a)   Side elevation of footbridge (suitable in rural locations with limited child access)

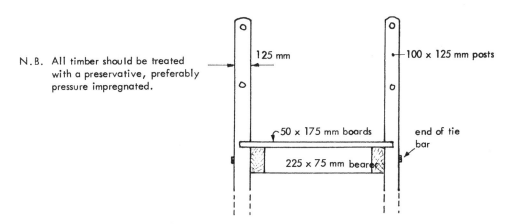

Figure 13.36(b)   Cross-section of footbridge

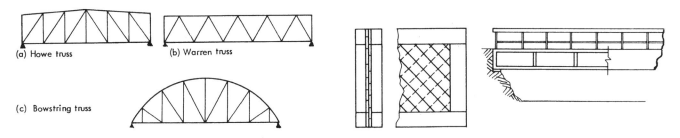

Figure 13.37   Forms of timber bridge

Figure 13.38   Web girders fabricated in timber

(3) Plywood Box Beams (Figure 13.39)

In these, the plywood is employed as a shear force carrying material, the flanges (top and bottom) being of high quality timber which has to be capable of carrying the main direct and bending stresses. It is essential that these box beams are made under controlled factory conditions.

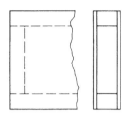

Figure 13.39 Box girders fabricated in timber

(4) Glulam Beams (Figure 13.40)

These are laminated with the stronger timber laminations being in the positions of maximum stress. In the glued/laminated beam, short lengths can be used (using scarfed or finger joints). It is essential that the adhesive is of adequate strength and weather resistant. Usually synthetic resins (resorcinol formaldehyde or phenol formaldehyde) are used.

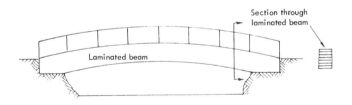

Figure 13.40 Laminated timber beam bridge

13.14.2 *Small Span Structures (suitable for footbridges)*

These may be of any of five types, as shown in Figure 13.41.

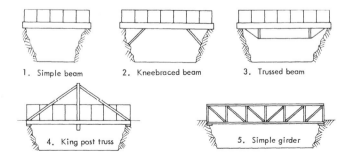

Figure 13.41 Types of small span timber bridges

13.14.3 Decking

Simple planked (longitudinal or transverse planking) is quite effective. More complex decking may consist of the nailed laminated or glued laminated type. The deck should be watertight and, thus, will provide excellent protection for the supporting beams.

13.15 Preservation of Timber

Any timber used in bridges should be either naturally resistant to decay or sufficiently permeable to allow adequate preservation treatment. To obtain maximum natural durability, sapwood must be excluded, but if preservatives are to be used, then sapwood is permissible.

Preservation treatment should be by pressure impregnation, the preservatives being of three main types:

(*a*) tar oil (e.g. creosote, now not used in current preservation techniques)

(*b*) organic solvents (e.g. copper napthenate penta-chlorophenol or tributyl-tin oxide;

(*c*) water borne (e.g. copper/chrome/arsenate solutions).

Where laminated timbers have been utilized, care must be taken to choose a preservative which will not affect the adhesive.

13.16 Repair of Corroded Concrete Bridge Structures

Concrete bridge structures may be subjected to a number of damaging substances, mainly in the form of chemical solutions, which attack the concrete, causing both deterioration of the strength of the concrete and corrosion of the embedded reinforcement. Some of these aggressive substances are airborne, others can occur in ground water and yet others result from accidental spillages on the bridge itself.

Deterioration of the structure may be as follows:

1 Simple leaching of free lime.

2 Reaction between attacking solutions and cement compounds, resulting in the formation of secondary compounds. These may either leach from the concrete or remain in a non-binding form causing gradual loss of strength.

3 Reactions similar to 2, but resulting in crystallization of secondary compounds, with associated expansion and disruption of the concrete.

4 Crystallization of salts directly from the attacking solution, again causing damage to the concrete.

5 Corrosion of the embedded steel reinforcement, causing spalling of the protective concrete cover.

More than one of these problems may occur at the same time.

Damaging chemicals include acids, sulphate solutions, chlorides (salts), nitrates, phenols and alkalis.

In particular, concrete bridge structures may frequently suffer from reinforcement corrosion induced by penetration of de-icing salt. Such damage can only be repaired by the removal and replacement of all the severely chloride-contaminated concrete. This is extremely difficult, often structurally undesirable and very costly.

One alternative is to repair only cracked and spalled concrete, leaving the less contaminated concrete in place and relying on the surface treatment to minimize further metal corrosion.

Any repair must allow water vapour to pass out of the concrete, thus allowing it to dry out, and prevent ingress of additional water and chlorides. Water-repellant surface impregnations are available. Silane is one such treatment.

A suitable repair operation may be carried out on a bridge structure as follows:

1 Cut out areas of cracked and spalled concrete.
2 Grit blast exposed reinforcement.
3 Repair small areas with acrylic repair mortar, larger areas with concrete.
4 Check drainage holes, spouts, etc., and areas where water may stand, thus allowing seepage into the structure.
5 Steam clean all concrete surfaces.
6 Apply 'Silane' or similar to the sides and soffits of beams.
7 Apply acrylic waterproof coating to the upper surface of the beams.

8 Fix flashings, gutters, downpipes and channels to ensure complete drainage of any surface water.

13.17 High Pressure Water Concrete Repair Treatment

Restoring damaged concrete structures has normally meant using hydraulic breakers which often damaged the sound underlying concrete and reinforcement.

The water jet also removes rust and dust from any reinforcement which may have become exposed.

Water jetting at very high pressure can be used to remove concrete. This is an operation only to be undertaken by specialist contractors.

Vacuum suction is then used to remove any residual water.

Revision Questions

1 If the cover of the reinforcement in a bridge deck beam was less than 40 mm what would eventually happen?
2 There are three main reasons for waterproofing a bridge deck. What are they?
3 When using mastic asphalt as a waterproofing material, how many courses is it normally laid in and what are the recommended thickness?
4 What are the three main types of preservatives used in the preservative of timber used in bridgeworks construction maintenance?

Note: See page 305 for answers to these revision questions

14.1 Introduction

As soon as a road is constructed and brought into use, it begins to deteriorate, due to the effects of both traffic and weather.

A road is a major capital investment. Failure to maintain it properly is likely to lead to an increase in the number of accidents occurring on it and to the need for expensive reconstruction, to restore it to the original standard.

Maintenance, a complex procedure, requires the Highway Authority to adopt certain standards and policies with regard to maintenance of the roads for which it is responsible.

The standards for motorways and trunk roads have been laid down by the Highway Agency in their Trunk Road Maintenance Manual (TRMMS).

Standards for other roads are determined by the individual highway authority, with guidance provided under the various local associations, which have, together, produced a document entitled Delivering Best Value in Highway Maintenance.

These associations are:

1 County Councils (ACC)
2 District Councils (ADC)
3 Metropolitan Authorities (AMA)
4 Convention of Scottish Local Authorities (COSLA).

There is divergence of opinion as to what constitutes 'maintenance'. Some activities – e.g. sweeping, patching, reinstatement – are clearly maintenance; others, such as haunching, kerbing and overlays, can be classed either as maintenance or as improvement.

A common procedure is to group maintenance under three headings:

1 Amenity, or cyclic, or aids to movement
  e.g. Sweeping
       Gully emptying
       Grass cutting
       Weed killing
       Repair to signs
       Renewal of markings
2 Minor or routine
  e.g. Patching
       Reinstatement
       Surface dressing
       Haunching
3 Major or structural
  e.g. Resurfacing
       Overlays
       Widening
       Reconstruction

The key issue is standards of maintenance. Questions such as:

How often should a road be swept?
When does a footway 'trip' require attention?
How much kerb upstand must be maintained?
How frequently should gullies be emptied?

lead to policy decisions being made by elected members and senior officers of the Highway Authorities.

The *Highway Maintenance – Code of Good Practice* outlines a systematic approach as follows:

Assessment system
  Road classification
  Route inventory
  Traffic flows
  Accident records

Standards
  Values for safety (skid resistance, etc.)
  Warning levels, to identify the need for action
  Categories of deterioration
  Minimum values for kerb upstands, etc.

Resources
  Finances available
  Value for money

Programme
  Defects, their causes and remedies
  Specifications for work

Execution of work
  Carry out the work either by DLO or by contractor

Review performance
  Did the maintenance work achieve the objective?

A practical example of this cycle for a road surface on an 'A' class road, which is reported to be 'smooth and slippery', could be as follows:

(a) Determine the acceptable skid-resistance standard for the road.

(b) Test the surface for skid-resistance, to find the actual value. Is it low enough to justify action to improve it?

(c) Check whether other works are planned. If so, should they be called out before the action to improve skid-resistance, if undertaken?

(d) Determine the cause of the smooth surface. (Usually abrasion by traffic.)

(e) Consult any available accident statistics.

(f) Consider an economical solution such as surface dressing.

(g) Prepare a specification and estimate.

(h) Have the necessary work carried out.

(i) Monitor the quality of the repair.

The information given in the following sections cannot be regarded as comprehensive, but every effort has been made to give, in the space available, examples of activities that are common practice.

14.1.1 Maintenance Rating Systems

Recognition of a defect, determination of the cause and prescribing a suitable remedy are subjective: it is often possible for different people to look at the same problem and arrive at quite different solutions. This could lead to poor value for money in the remedial work specified or work being undertaken in one place where the need is far greater in another area. This is why systems of assessment have been developed.

Highways must be divided into categories, with an inventory made of each road's width, length, drainage, verges, footways, signs, markings, etc. This information, read in conjunction with predetermined standards and warning levels of deterioration, together with a form of assessment, enables a Highway Authority to summarize the scale and cost of the remedial works needed to return the road to an acceptable standard for its particular category.

The two major systems used are CHART and MARCH, although there are several others along similar lines used by highway authorities. (CHART has to be used for trunk roads to acquire funding by the Highway Agency.)

The essential difference between these two computer-based systems is that CHART locates defects within 10 m sections whereas MARCH works on maintenance lengths that are typically between 100 and 500 m in length. MARCH is thus the simpler system to use, and it provides simple costing facilities. CHART has the advantage of precision and this makes it more attractive for detailed planning of maintenance and for use in conjunction with other highway data banks containing, for example, traffic and accident data.

The key operation is the ability of the Inspector to identify defects and their severity and relate these to a common standard. Typical examples of carriageway defects are:

Cracking – transverse, longitudinal, random, reflective
Crazing
Fretting
Ravelling
Wheel track rutting
Adverse camber or unsatisfactory shape
Edge deterioration
Inadequate drainage
Footway deterioration
Kerb – upstand height deterioration

In addition, equipment, such as the SCRIM machine, which is used to find skid-resistance, and the *Deflectograph*, which measures road deflection under load, are used to assist in objectively checking the road.

SCRIM measures the skidding resistance of the wet carriageway surface by measuring the force on a special test wheel mounted on the side of the vehicle at 20° to the direction of travel. The ratio of the force developed at right angles to the plane of the wheel (the sideways force) to the load on the wheel is the sideways force coefficient (SFC). The mean SFC and speed are calculated for sub-lengths of 5, 10, or 20 m with the SCRIM machine travelling at 50 kph. The results obtained are compared with both maintenance and accident reaction standards to identify areas where remedial action is required.

Deflectograph. The deflection of road pavements under the passage of a heavy wheel-load is related to the long-term structural performance of pavements. The deflectograph is a machine which measures the maximum deflection of the pavement under a moving standard wheel load using a vehicle-mounted beam based on the Benckleman beam principle. The results from the machine identify areas of structural weakness. Subsequent analysis of the results can provide a series of recommended overlay thicknesses and residual lines for the road pavement. As the machine is very slow (2 km/hr), normally it is only used on roads where major works are likely, therefore some measurement of the structural problem is required.

CHART considers 10 m lengths of the highway and calculates a rating value from the measured type and quantity of the defect present. The relationships between the rating value and the percentage of defect present are contained within the computer system. For each defect, an average rating is produced for every 100 m of road. Treatment ratings are produced by averaging the ratings for continuous 100 m lengths of road which have been recommended for the same treatment. These ratings are used to assess the priority of the scheme.

There is no provision for the engineer to amend or 'weight' the recommendations within the system, nor can the system produce complete maintenance requirements along a particular length of road. CHART produces separate output lists for each treatment and

Treatment	Minor det. (% area)	Major det. (% area)	Wheel track rutting 14 mm 19 mm (% length)		Wheel track cracking (% length)	Skidding resistance SFC 0.1 below target (% length)	Edge det. (% length)	Adverse camber (% length)	Existing patching (% area)
Edge patching/ haunching	–	–	–	–	–	–	20–50*	–	–
Surface dress	20	7	–	–	10	10*	–	–	–
Resurface	–	20	20	2*	30	–	–	20–100*	–
Major strengthening	–	–	–	20*	–	–	–	–	–

* These warning levels vary with the severity of the defect

Table 14.1 CHART warning levels and treatments

Treatment	Minor det. (area %)	Surface course (area %)	Major det. (area %)	Wheel track rutting 25 mm (length %)	Edge det. (length %)
Patch	–	5	5	5	5–20
Haunch/Recon.	–	–	–	–	20–80*
Surface dress	25	–	–	–	–
Resurface	90	20	20	80	–
Strengthen	–	–	40	–	–

* These warning levels vary with the severity of the defect.

Table 14.2 MARCH defect warning levels

cross comparison between them must be produced manually outside the system. Engineers must assemble, and subsequently cost, maintenance works from the various printouts.

Typical CHART warning levels and treatments for a variety of carriageway defects are given in Table 14.1.

With MARCH, to arrive at the priority value for each maintenance length, the system compares the various amounts of deterioration, the importance of one type of defect with that of another, the traffic and footway flows, and finally employs a correction to allow for length variations of maintenance lengths. It is the most critical item within each maintenance length that gives the priority value used in the final works list.

The computer system has a separate facility for identifying the 'suggested treatments'. This treatment is based upon engineering experience. The system produces both an individual treatment for each defect and an overall treatment for each section so as to restore it to a given maintenance standard.

The engineer has the option to select the various values for calculating the priority ratings and may vary the treatment suggested to suit the policy for his authority. The suggested treatment may also be varied between different types of road so as to reflect the varying standards of maintenance which are applied to different highways. The final priority may be printed in several forms, of which the main one is a full printout of each job in the list, in priority order, giving full details of works in terms of quantities, rates and costs. The engineer may print out the lists in several orders and use financial cut-offs to limit the amount of work suggested by the system.

Typical MARCH warning levels and treatments for a variety of carriageway defects are given in Table 14.2.

14.1.2 Routine Maintenance Management Systems (RMMS)

The Department for Transport led the way with their system, which followed the Audit Commission's advocation, in 1988, of the adoption of Highway Maintenance Management Systems.

The DfT system consists of the following:

(a) a method of network referencing against which data can be held;

(b) a method of handling inventory data;

(*c*) a method of handling information (data) collected in accordance with the Code of Practice for Routine Maintenance regarding inspection and reporting, action to be taken, and standards to be met, including local variations needed to match local conditions;

(*d*) a works order interface.

The system of referencing is based on CHART and divides the network into *links* (a road with a unique classification) and *sections* (not normally exceeding 2 km in length). The ends of the sections are called *nodes*; therefore any item or defect can be identified by its distance from a node, linked to a system of cross-section locations.

Inventory

This information is usually collected by means of a walked survey, using a hand-held portable computer, and is then fed into the database.

Inspections and Reporting

Defects, categorized according to priority and response time, are formally reported to help management:

1 to produce lists of inventory items which can be used as check lists for inspectors;

2 to produce schedules of inspections to be carried out and their timing;

3 to produce schedules of cyclical maintenance tasks to be carried out at any time;

4 to produce programmes of work;

5 to generate works orders based on inspection records;

6 to monitor the effectiveness of maintenance processes;

7 to monitor the effectiveness of standards;

8 to allow checks on performance and provide a record of maintenance work done;

9 to analyse the results of adopting particular maintenance strategies;

10 to produce estimates of resource requirements and costs.

The development of this type of system is still in progress. Many Highway Authorities are designing their own systems, but, of course, have to use RMMS for the highways for which they are the agents of the Department for Transport.

14.2 Amenity and Aids to Movement

14.2.1 *Cleansing of Streets and Footways*

According to the Code of Good Practice, 'The objective, from the highways point of view, is mainly to remove debris from the side channels, to prevent surface water from ponding in them, and to prevent an excess of detritus (grit etc.) from being washed down into highway drains; to keep the carriageway surface generally clean, so that road markings are clearly visible and windscreens not obscured by dirty spray.'

The assumption is made that unkerbed roads need only be swept when it can be seen that it is necessary, rather than on a regular basis.

Footways must be swept to maintain safety of pedestrians using them.

14.2.2 *Sweeping and Cleansing Standards*

In the Code, these are dealt with under two headings and relate to the frequency of sweeping.

1 Highways in Rural Areas

(*a*) Motorways and trunk roads	twice/year
(*b*) Other roads	once/year

Standards will depend upon local conditions, with more frequent sweeping at lay-bys, frequently used stopping places, and in areas where dirty industries are located and channels quickly become blocked. In these and similar circumstances, unkerbed roads may need sweeping.

In addition, on motorways and other roads carrying high-speed heavy traffic, more frequent scavenging for objects lying on the carriageway and, hence, dangerous to road users may be necessary.

2 Highways in Urban Areas

(*a*) Town centres and shopping areas	weekly
(*b*) Trunk main distributory roads	monthly
(*c*) Other roads	three-monthly

In urban areas, higher standards will normally be required than is the case in rural areas, for both highway and cleansing needs. Footways in urban and built-up areas should be swept as often as the carriageways.

14.2.3 *Methods of Cleansing of Streets and Precincts*

Two main methods of using the labour force for street cleansing are: the beat system and the team system. These methods refer particularly to urban areas, mechanical sweepers being used in rural situations.

The Beat System

Generally manual sweeping is used with one operative having a beat of about 1.0–1.5 km of road length for which he is responsible. The actual length is dependent upon the amount of footway and pedestrian precinct adjacent to it.

The refuse collected by the operative may be removed either by:

(*a*) collection by a vehicle which visits a number of areas (or beats), subsequently carting the refuse to a suitable and convenient dump or tip; or

(*b*) collection in a centrally located bulk container or skip which is replaced when full by an empty container. These containers are carried on special vehicles.

The Team Method

A team of operatives is used to cleanse a series of areas, in rotation. The likely composition of such a team would be:

2–3 operatives (sweepers);
1 pick-up truck and driver;
1 mechanical road sweeper and driver.

Generally, the team will first sweep the footways, collecting rubbish into heaps, if there is a considerable quantity, or brushing small quantities into the road channel. The heaps are then collected into the truck and removed to a tip or incinerator plant. The mechanical sweeper follows the hand sweepers, collecting dirt and rubbish from the road and cleaning the channel. (*See* Chapter 11 for principle of operation of mechanical sweepers.)

In rural areas, often there is no footway, and it is usual for the mechanical sweeper to operate over a considerable area of roads, cleaning only the surface, and in particular, the channel.

For a shopping precinct, the sweeping and cleansing should preferably take place outside shop opening hours. The evening, soon after shops have closed is preferred, but early morning is a possible alternative time. Sweeping of the precinct may be done by either hand brooms or mechanical vacuum sweepers, the method used depending upon the size and layout of the precinct.

For a very busy area of precinct, it may be necessary to collect rubbish from the waste bins at midday in addition to the main cleansing time. This requires a mechanical trolley into which the bin contents can be emptied and carted away.

14.2.4 *Maintenance of Weed-covered Paved Areas*

Elimination of weed growth is ideally tackled in the spring before the weeds are sufficiently mature to have run to seed. The weather for spraying weedkiller should be fine and calm. Spraying of liquid chemical weedkiller is most effective when applied to the whole plant, as the chemical is absorbed through the leaves. Therefore, unless the growth is very high, plants should not be cutback before spraying.

Chemicals used should be those which are non-poisonous (i.e. not harmful to children and animals).

If the area is extensive, mechanical spraying methods would be used, with a hand lance with a hooded spray head to direct the spray at up to 3.5 g/m^2 at the weeds in each joint.

The operator should be trained to comply with the Pesticide Regulations and wear appropriate protective clothing, e.g. gloves, face shield or mask, etc.

A granular form of chemical weedkiller may later be scattered over the area and brushed into the joints to provide a longer lasting weedkiller.

14.2.5 *Maintenance of Embankments, Sides of Cuttings, Grips and Ditches*

This is done with the objective of preserving the stability of slopes or cuttings and to prevent damage by erosion, particularly during periods of heavy rainfall. Sides of deep ditches should be included under the general definition of slopes.

Vegetation must be cut, to encourage root growth, thus binding the soil on the slopes. Sides of deep ditches should be included under the general definition of slopes.

Vegetation must be cut, to encourage root growth, thus binding the soil on the slopes. Ditches must be cleared of vegetation to allow full unimpeded flow of storm water.

14.2.6 *Siding and Verge Maintenance*

The objective of siding is to prevent the encroachment of verge soil and vegetation growth on to the carriageway and to maintain verges in a condition that allows the cutting of grass without damage to the mowing machines.

Verges in urban areas tend to require more attention than in rural areas (due mainly to the parking of vehicles on verges in urban residential areas).

Clearing and cutting of grips (open drainage channel cut across a verge from the road channel) to drain surface water from the carriageway must be carried out as part of the maintenance of the verge.

14.2.7 *Gullies and Drainage Systems*

In addition to the sweeping of the surface, there is an accompanying need to empty the road gullies and to maintain the road drainage systems in good order.

The frequency of emptying the road gullies depends upon local conditions, such as the presence of dirty industries. The normal minimum suggested frequency of emptying is twice a year for all roads.

Drainage systems should be checked every two years to make sure that they are functioning properly. Additional inspections may be required in the autumn when there tend to be accumulations of fallen leaves.

Also, all trunk and principal roads should be inspected during periods of heavy rain, to check whether the drains are doing their job properly. If possible, other less important roads should be included in this procedure.

Clearing of drains should take place whenever necessary.
Dangerous problems arise with the theft of gully grids
and grates. Because of the wide variety of sizes, there
is often a considerable delay before a replacement can
be fitted. Now available are single size 'cut and fit'
polyester composite replacement grates which can be
cut down to size with a disc cutter as a temporary
measure.

14.2.8 Clearing of Blocked Drains and Sewers

Properly designed and constructed systems do not easily
become blocked, but unless regular checking of the sewers
is carried out, even the best designed and constructed can
become obstructed in extreme conditions.

It is more usual for blockages to occur in systems where
due to circumstances (such as level ground) it is not possible to lay the pipes to good gradients or where external
factors may cause subsequent damage to the pipes. One
such example is tree roots which can grow round a pipe
and crack it, thus providing a sharp edge against which
solid matter will tend to accumulate.

Where the pipeline itself becomes damaged, the only
remedy is to replace the pipes.

Blockages within the pipes can be either partial or
complete. Most blockages are cleared using a jetting
unit. This is a power-driven water jet, delivered through
a strong, flexible rubber hose to a nozzle. The issuing
high-pressure jet flushes out the blockage debris. (*See*
Chapter 11, Section 11.10.)

It is preferable and safer, when rodding in these circumstances, to work from the up-stream side of the blockage,
but this is not always possible, due to surcharge of the upstream manhole. In that case the rodding must be done
from downstream, taking care to work from the surface
wherever possible.

Drain clearing rods may be of:

(*a*) polypropylene;
(*b*) rattan cane;
(*c*) steel cores;
(*d*) helical steel spiral on a steel core.

A selection of the various drain cleaning and clearing
equipment is shown in Figure 14.1.

Due to jetting units and other mechanically driven
rodding systems, these fittings are now rarely used.

14.2.9 Manual Clearance of Blocked Drains

This is normally undertaken by opening access covers
at both ends of the blocked length and working from
above ground. If, however, this is not possible and the
chamber/manhole has to be entered, great care must
be taken before doing so, to ensure that it is safe to
enter. Details of safe practice are given in Chapter 9,
Section 9.7.

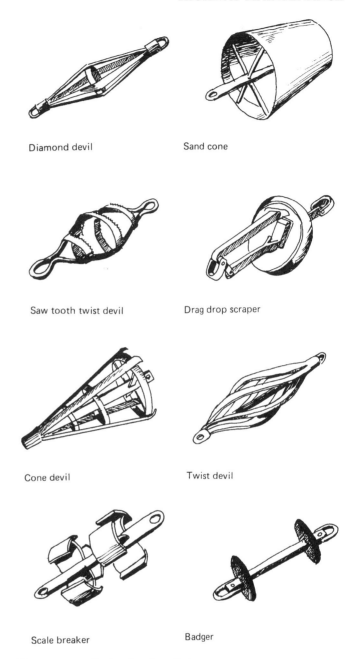

Diamond devil   Sand cone

Saw tooth twist devil   Drag drop scraper

Cone devil   Twist devil

Scale breaker   Badger

Figure 14.1 Drain-clearing fittings

14.3 Weather Emergencies

The Code states that the objective is to provide a service
which, as far as is reasonably possible, will permit the safe
movement of vehicular traffic whilst minimizing delays
and accidents.

It may be useful to know the overall organization for
dealing with these matters. These procedures are generally as follows:

1 Prepare lists of available personnel, vehicles and equipment.

2 Check stocks of salt, grit, etc.

3 Arrange for overhaul of all equipment prior to onset of winter.

4 Prepare gritting routes as follows:

(a) try to start and finish at depot;

(b) cover trunk roads, bus routes, shopping centres, roundabouts, known danger spots, etc.;

(c) put particularly dangerous places on more than one route;

(d) relate the length of the routes to the size of the gritters if possible to reduce empty running to a minimum;

(e) number routes in order of priority and note length of each route;

(f) prepare route cards for issuing to drivers at commencement of operations.

The Meteorological Office provides a weather forecasting service which is used in conjunction with ice-alert systems and local knowledge to decide whether or not to apply any treatment.

In spite of all the warning systems, it is still necessary to rely on regular inspection of weather conditions by personnel on duty. The rate of change of weather conditions make this operation a very 'inexact science'.

14.4 Brief Notes on Materials and Equipment

Dealing with snow and ice is not strictly highway maintenance, but is included as it can be said to be an aid to mobility.

14.4.1 Calcium Chloride

May be used to melt ice and will do so at temperatures above $-40°C$. As the temperature rarely falls below $-12°C$ in the UK this material is rarely used, particularly as it is more expensive than salt.

14.4.2 Salt

Salt melts ice at temperatures below $0°C$ but loses its ability to do so as temperature drops until it reaches $-10°C$, when it will not melt the snow at all.

Salt is available in the following forms:

(a) pure undried vacuum salt;

(b) soiled salt;

(c) crystal salt;

(d) ground rock salt (Grades 4 and 5).

Crystal salt and ground rock salt show less tendency to cake during storage and for this reason are preferred to the others.

14.4.3 Abrasives

(a) Many different materials such as, crushed stone, sand, grit, cinders, grog, granite chippings, etc. may be used.

(b) Material should be 10 or 6 mm single-sized aggregate or a 5 mm sand with a low percentage of fines.

(c) Particles should be angular.

(d) Abrasives can be stored out of doors.

14.4.4 Snow Fences

These are used in rural areas to prevent drifting across roads; they should be placed on the windward side at a distance from highways varying according to height and type of fence (e.g. 1.5 m high solid fence should be 21 m from carriageway on windward side). See Chapter 8 for more detailed information about snow fencing.

14.4.5 Salt Storage

It is important that salt is stored correctly. In so doing there is less risk to the environment and the salt is likely to be more dry. Dry salt improves the degree of control of spread rates and therefore economy in salt usage.

When siting salt piles it is important to avoid trees and hedges. Under no circumstances should piles be located within the rooting area of trees or set against a deciduous hedge. Ideally salt piles should be covered.

14.4.6 Spreading Salt

To be effective, salt must be spread evenly and at rates to suit prevailing weather conditions. Spreading, therefore, is best done from automatic machines, BS 1622 gives details of such equipment. Calibration procedures should be carried out to check the rates of spread. On some types of spreader body the rate of spread varies with road speed. Some purpose-built bulk spreader vehicles and some trailer spreaders, are speed-sensing and spread at the same rate at all speeds. Care should be taken to ensure that spread widths are neither too wide nor too narrow. It is desirable to calibrate and mark the controls of spreading machines for, say three spread rates:

$10 g/m^2$ for precautionary spreading:
$20 g/m^2$ and
$40 g/m^2$.

14.5 Treatment of Snow and Ice

14.5.1 Precautionary Salting

The recommended spread rate for precautionary salting is $10 g/m^2$. Every effort should be made to spread rock salt at the recommended precautionary rate prior to the onset of freezing conditions. This not only prevents the formation of ice, but, by preventing snow from adhering to the road surface, the passage of snow ploughs is eased.

14.5.2 Treatment of Snow

The salt spreading rate recommended for melting up to 40 mm thickness of fresh snow at 0°C under traffic, is up to 40 g/m². Although repeated applications of salt can remove a heavy snowfall and can be a useful method of operation in urban areas where congested conditions make the use of snow ploughs difficult and snow removal impracticable, this approach is not recommended unless there is no other way.

14.5.3 Ploughing

(*a*) Preparations for ploughing should be made as soon as the depth of snow exceeds 50 mm.

(*b*) Wet snow is heavier than dry snow; because of this it may not be possible to use the whole of the blade.

(*c*) When snow is 150 mm deep or more, only ploughing will be successful.

(*d*) Snow of 300 mm depth or more may have to be ploughed in layers.

(*e*) Ploughing should be followed by salting, usually at 10 g/m², to remove the thin layer of snow which remains.

(*f*) When ploughing in town, to avoid floods when thawing starts, cut diagonal channels through banks of snow and clear road channels and gullies.

(*g*) At intersections, where ploughing causes snow to become banked up, snow may have to be removed by hand.

(*h*) If snow has to be carted away, dispose of it in rivers, streams, sea, sewers, or dump on waste ground.

14.5.4 Treatment of Ice

Where ice has already formed on the road surface it is recommended that rock salt be applied at rates up to 40 g/m² depending on the amount of ice to be removed and the temperature. This ensures a rapid melt and an improvement in road conditions in a reasonable time.

Salt will not penetrate ice more than 12 mm thick: the salt solution will make the surface more slippery so an abrasive should be put down.

14.5.5 Treatment of Hard Packed Snow and Ice

If the foregoing recommendations are followed, hard-packed snow and ice should be rare. However, if hard-packed snow or ice does form down to −5°C, provided it is not more than 20 mm thick, removal is possible by using successive salt spreads at 20 g/m². At a temperature below −5°C, where the hard-packed snow or ice is more than 20 mm thick, the use of salt alone will result in an uneven and slippery surface. In these circumstances abrasives added to the salt spread will assist in making the surface less slippery. A single-sized abrasive aggregate is recommended of particle size 6–10 mm, or a 5 mm sand

having a low percentage of fines. Particles over 10 mm in size are not recommended because of the risks to pedestrians and vehicles. The particles should be angular in shape, but not sharp enough to cut vehicles' tyres. Reversion to salt only should be made as soon as possible since abrasives contribute little to the removal of the snow and ice and may block drains and gulleys in addition to impairing the skid resistance of the surface once the snow and ice have melted.

14.5.6 Sustained Low Temperatures

Sustained low temperatures are rare in Britain. For each degree drop in temperature below −5°C the amount of salt to maintain a given melting effect increases by about 14 g/m². Where the road is subject to traffic, however, little or no increase is required until sustained temperatures below −10°C are reached. At this stage the amount of salt required increases markedly and the addition of some calcium chloride to the salt might usefully be considered. However, the cost of calcium chloride is high and, because it absorbs moisture more rapidly from the air, its storage is difficult. Calcium chloride is usually supplied in flake form contained in plastic bags. A mixture of four parts salt to one of calcium chloride should suffice on the very rare occasions where a need may arise on a few isolated sites.

14.6 Identification of Road Maintenance Problems

14.6.1 Maintenance of Flexible Pavements

The routine maintenance of surfaces has essentially two interrelated objectives:

(*a*) Sealing against ingress of water and weathering due to oxidation, ultraviolet light and alternating freeze/thaw conditions, and directing surface and subsoil water away from the pavement foundation.

(*b*) Dealing with the effects of the action of traffic tyres and loads that cause abrasion and stress leading to fatigue failure of the road structure.

Causes of Pavement Deterioration

1 *Weathering/surface, water/fuel spillage*, leads to embrittlement and failure of the binder, with progressive loss of fine and coarse aggregate. Ingress of water weakens the sub-grade and renders it vulnerable to frost heave.

2 *Traffic loading*: flexing of pavements under repeated loading, leads to fatigue, crazing and structural failure, particularly over weak sub-grades and where the depth of carriageway construction is inadequate to provide sufficient spread of loads from commercial vehicles;

3 *Thermal movement* due to temperature changes between summer/winter and day/night. This is particularly significant for macadam overlying concrete foundations. Fatigue failure occurs at reflective cracks in surfacing.

4 *Moisture movement*: sub-grade materials, primarily cohesive soils (clays and silts), are prone to shrinkage or swelling movement induced by variation in moisture content. The extent of consequent reflective cracking or heaving in the pavement structure is dependent upon the type and thickness of the construction.

5 *Differential movement flexure* caused by stress at the boundaries of different constructions, e.g. between shallow/deep haunches or old/new foundations; in backfilled trenches or kerb/channel foundations; with varying quality/load capacity of sub-grade; at embankment edges or the interface of rigid/flexible foundations.

6 *Slippage cracking*: slippage cracks are crescent or half-moon shaped cracks, having the two ends pointing away from the direction of traffic flow. They are produced when braking or turning wheels cause the pavement surface to slide or deform. This defect usually occurs when there is a soft, binder-rich surface mix, or a poor bond between the surface and underlying layer (e.g. basecourse).

7 *Reflective joint cracking* occurs in bituminous surfacings laid on a concrete foundation. Thermal or moisture induced movement of a rigid slab foundation causes cracking to develop in the more flexible surfacing over joints and shrinkage cracks. This defect is not load-related. However, traffic loading may cause breakdown of the flexible surfacing, through fatigue, either side of the crack. If the surfacing is fragmented along a crack, the crack is said to be *spalled*. This latter term applies also to expansion and contraction joints in exposed concrete pavements.

8 *Pushing* is permanent longitudinal displacement of a localized area of the flexible pavement surface and is caused by traffic. It normally occurs only in soft or binder-rich surfacing materials. Pushing also occurs where bituminous macadam butts up to a concrete carriageway and is subjected to differential thermal movement.

9 *Potholes* are small (usually less than 0.9 m in diameter) bowl-shaped depressions in the pavement surfacing. Generally, they have sharp edges and vertical sides near the top of the hole. Their growth is accelerated by water collecting inside the hole. Potholes are produced by traffic abrading and dislodging small pieces of the surfacing, usually when the binder has become embrittled or is subject to stripping in the continued presence of water. The condition is progressive in both area and depth, ultimately affecting the road foundation if no corrective action is taken.

14.6.2 Structural Defects

1 *Reinstatement of public utility openings.* Undoubtedly the greatest cause of surface and structural irregularity is due to public utility openings. This is covered in detail in Chapter 15.

2 *Alligator crazing.* Alligator or fatigue cracking in a bituminous surface shows as a series of interconnected cracks. These are caused by fatigue failure under repeated traffic loads. Cracking begins at the bottom of the surfacing, or stabilized base, where tensile stress and strain are the highest under a wheel load. The cracks spread to the surface initially as a series of parallel longitudinal cracks. After repeated traffic loading they join up, forming many-sided sharp-angled pieces which develop into a pattern resembling chicken wire or the skin of an alligator. The pieces are generally less than 0.6 m long on the longest side. This defect occurs primarily in the wheel tracks and is often accompanied by rutting.

3 *Block crazing.* Block cracks are interconnected and divide the pavement into approximately rectangular pieces, varying from 0.3 × 0.3 m to 3.0 × 3.0 m in size. This defect is caused mainly by shrinkage of the pavement sub-grade reflecting through a thin road construction.

4 *Bumps and hollows.* Bumps are small, localized, upward displacements of the pavement surface and can be caused by a number of factors, which include:

(*a*) buckling of an underlying foundation
(*b*) frost heave
(*c*) infiltration of detritus into a crack, in combination with traffic loading.

5 *Corrugation.* This consists of a series of closely spaced ridges and valleys, or ripples, occurring at fairly regular intervals, usually less than 3 m apart along the pavement. This type of fault is usually caused by traffic action, particularly braking and accelerating on a soft surface, or by an unstable pavement foundation.

6 *Depressions.* These tend to be localized pavement defects, highlighting weakness in the underlying pavement structure, such as poorly compacted reinstatement. Shallow depressions are often not noticeable until after rain, when 'ponding' is seen. A succession of depressions can create serious unevenness in the riding quality and, when filled with water, can lead to aquaplaning as well as causing much 'splash and spray'.

7 *Rutting.* A rut is a surface depression in the wheel tracks. Pavement uplift may occur along the sides of a rut. Ruts are usually most noticeable after rainfall, when the wheel tracks are filled with water. Rutting stems from a permanent deformation in any of the pavement layers or the sub-grade, and results from traffic loading concentrated on a narrow width of the carriageway. Significant rutting can lead to major structural failure of the pavement. Binder-rich surfacing materials subject to repeated trafficking (channelization), such as bus lanes and HGV slow lanes, are cases in point.

14.6.3 Surfacing Defects

1 *Polished aggregate.* This is caused by repeated trafficking. When the aggregate in the surface becomes smooth to the touch, adhesion with vehicle tyres is considerably reduced, particularly when the road surface is wet. This is further aggravated by the consequent reduction in texture

depth, which reduces the surface drainage efficiency of the aggregate matrix. This defect is often highlighted by a significant increase in the number of wet skid accidents.

2 *Bleeding/sweating/fatting-up.* This is the formation of a film of bitumen on the pavement surface, creating a shiny, glass-like, reflecting surface which becomes sticky in hot weather. Bleeding is caused by excessive bitumen and/or low air voids content in a macadam mix, or by over-application of surface dressing binder. It occurs when binder fills the voids of the mix during hot weather and, under traffic, exudes on to the pavement surface. Fatting-up is essentially the same, being caused by loss of surface dressing chippings through over-embedment into the underlying surface or substrate, leaving behind a surface film of bitumen. These defects remain during subsequent cold weather. The bitumen accumulated on the surface will remain until corrective action is taken, such as carbonization. As with polished aggregate, wet weather skidding is an ever-present hazard.

3 *Weathering/ravelling/fretting.* This is the wearing away by erosion of the pavement surface, caused by the embrittlement of the binder and de-bonding of the fine aggregate. Ravelling may also be caused by metal-tracked vehicles, studded tyres or snow chains, which abrade the road surface. Softening of the surface and dislodgement of the aggregate due to fuel spillage is included in this category of defect.

4 *Stripping.* This is the loss of stone due to breakdown of the bond between aggregate and binder (bitumen or tar), generally caused by intrusion of water, or by the aggregate not being thoroughly dried before coating.

5 *Scuffing/scabbing off.* This is the loss of surface stone from a bituminous macadam or surface dressing due to traffic abrasion. Inherent weaknesses in bond between the aggregate and binder, whether due to stripping, binder embrittlement, or deficiency, will show up under the shear stresses induced by turning, or braking, vehicle tyres.

6 *De-lamination.* The separation of one layer of pavement surfacing from another is called 'de-lamination'. The most usual fault is the separation of a thin 'veneer' surface course from the underlying binder course or concrete foundation. The failure process is progressive, with increasing embrittlement of the bituminous binder, consequent growing loss of elasticity and the crazing of the veneer and separation of this from the undersurface, due to ingress of water and detritus.

7 *Adverse camber.* This is described as an irregular transverse profile of carriageway, where camber or crossfall is steeper than current design parameters. It may be due to successive resurfacings in an urban area, where it has been necessary to maintain kerb upstand while overlaying the centre of the carriageway. Alternatively, this condition may be due to subsidence, or to a road having been widened or realigned and not yet re-profiled by overlaying.

8 *Heave.* Heave is characterized by an upward bulge in the road surface, typically a long gradual wave of more than 3 m in length. It is frequently accompanied by surface cracking, and is usually caused by frost action in the sub-grade, or by movement and swelling of a saturated, moisture-susceptible soil.

9 *Edge cracking.* These cracks are parallel to and usually within 0.3 to 0.6 m of the outer edge of the pavement. This defect is accelerated by heavy traffic loading and can be caused by an inadequate depth of construction and/or a weak sub-grade. The area between the crack and the pavement edge is classified as 'ravelled' if it breaks up, frequently to the extent that pieces are removed, forming potholes.

10 *Haunch erosion/deformation.* In the absence of edge support, such as kerb or hard shoulder, pavement haunches are vulnerable to structural damage, i.e. cracking, crazing, erosion, rutting and general deformation from over-running heavy vehicles. This is particularly noticeable on the inside of bends, in the vicinity of ditches, and where no positive drainage exists. Persistent overrunning in narrow, unkerbed and undrained country lanes leads to structural failure and to rutting of the adjacent water-softened verge.

11 *Whole carriageway failure/collapse.* This can include potholing, rutting, crazing, heaving and general deformation of the whole carriageway profile. In the worst instances 'pumping' (exuding of liquid mud) of sub-grade material indicates the need not only for full-depth reconstruction, including subsoil replacement/stabilization, but also for an effective system of highway drainage, of sub- and surface water, to be provided before reconstruction.

14.6.4 Stages of Deterioration

From the premise that 'prevention is better than cure' (and also considerably more cost effective), there is clearly a need for maintenance personnel to be able to recognize the defects, stages and causes of deterioration of the road structure.

Initial stages of deterioration are:

(*a*) 'Fretting' – due to embrittlement and/or deficiency of the binder and loss of fine aggregate. This could also be due to segregation of the bituminous mix and/or inadequate compaction following laying. Fretted surfaces have an open rugous appearance and remain damp longer than sealed surfaces after rain.

(*b*) 'Formation of minor potholes' – a progression of fretting, but at this stage depressions are limited to the depth of the wearing course.

(*c*) 'Development of excessive potholing and surface irregularity' – a further stage of deterioration leading to the ultimate disintegration of both wearing and base courses, if no remedial action is taken.

To identify these stages the tell-tale signs are:

(*a*) 'Hungry', irregular, open-textured appearance – emphasized by a shadow effect when viewed towards the sun.

(*b*) Loose aggregate on the surface, or in carriageway channels. This is not to be confused with an initial surplus of chippings following surface dressing.

(*c*) Damp patches in an otherwise dry surface following rain.

Separation of Surfacing Courses

This is caused by the embrittlement of surfacing material, often overlying a concrete foundation, and separation through loss of bond with the underlying structure.

The tell-tale signs are crazing, breaking-out of random areas to reveal the substrate.

Surface Erosion

This is caused by the action of water and frost, often aggravated by lack of positive drainage. Shaded damp locations are particularly vulnerable.

The tell-tale signs are extensively pitted appearance and irregularity of surface, often accompanied by deeper potholing deep into the road foundation; also standing water, or an area shaded by trees and a persistently damp surface.

Edge Failure

Includes crazing, cracking of and deformation of the carriageway edge, usually in the absence of kerbs and positive surface water drainage, where vehicle over-run is common.

The tell-tale signs are:

(*a*) an irregular edge to the carriageway, usually accompanied by extensive rutting of the adjoining verge;
(*b*) surface water standing in potholes and ruts;
(*c*) deformation and crazing of the carriageway haunch, where water has penetrated the foundations.

Structural Deformation

This is readily seen and can include wheel tracking, rutting, crazing and potholing, accompanied by deformation of the carriageway profile. It is indicative of inadequate depth of construction and/or a weak sub-grade. The extent of the crazing can best be seen as a 'tracery' of wet cracks in an otherwise drying surface following rain.

14.7   Repairs

Some examples of the various types of carriageway repair are given below.

14.7.1   Patching

Patching is not clearly defined in highway terms. However, in this chapter it is considered to be the replacement of defective flexible pavement with new flexible material, hand laid, to any depth not less than the thickness of the surface course or more than 150 mm. The intention of patching is to provide a permanent restoration of the stability and riding quality of the pavement.

Hence, patching *excludes* the:

(*a*) filling in of depressions and cracks not involving removal of existing surfacing;

(*b*) sealing of pavement surfacing by the application of dry or pre-coated aggregate and/or binder;
(*c*) repair or replacement of the base, sub-base or sub-grade;
(*d*) use of concrete.

In general, patching covers the repair of separate random-sized areas of defective pavement, larger than potholes, but not the replacement of continuous lengths of the pavement of the whole width or part width of the carriageway.

There is a great variety of bituminously bound material suitable for patching pavement, covering cold, warm and hot laid materials (*see* Chapters 5 and 10).

It is apparent that most highway authorities have their own preferred methods of patching and materials. However, there is a reasonably defined procedure for good practice in patching and the emphasis of the following information is based on these procedures.

14.7.2   Equipment

In addition to the basic hand tools, which are:

pick, shovel, broom/brush, asphalt rake, tarmac fork;

with possibly,

spreader can, tar brush, barrow and hand rammer,

typical items of plant are:

1 power breaker with asphalt cutter;
2 plate vibrator or auto tamper, of the pedestrian controlled, hand operated type;
3 roller, self-propelled, vibratory or dead weight, up to 2.3 tonne weight (vibratory) or 6 tonne (dead weight);
4 emulsion sprayer and ancillary equipment, either hand or power operated.

Measuring equipment required by each gang is as follows:

1 tape;
2 straight edge;
3 thermometer: 70°–175°C pocket probe type;
4 line: 15 m minimum;
5 spirit level: 1 m size;
6 boning rods;
7 chalk, crayon, pencil and notebook.

14.7.3   Sequence of Operations

1   Preparation

(*a*) Break up defective areas of surfacing to the limits marked out, using breaker equipment. The area should be marked out with straight lines as indicated in Figure 14.2 to form rectangular patches.

(*b*) Excavate and remove, cutting back edges to a sound vertical face.

(*c*) Cleanse the cavity excavated.

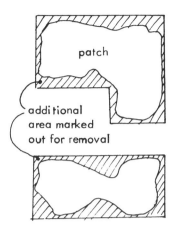

Figure 14.2   Preparation of patches

(*d*) Dispose of all excavated material.

(*e*) Cleanse the adjacent area of pavement.

2   Cavity Treatment

The cavity should be tack coated on the base and sides:

(*a*)   With cold emulsion if cold or warm laid materials are to be used.

(*b*)   With filled bitumen to the sides and cold emulsion to the base if hot laid materials are to be used.

3   Binder Course

(*a*) Application – fill the cavity to the level of the top of the binder course (when compacted) with a suitable binder course material.

(*b*) Compaction – compact, using compaction plant where possible and completing with hand tools.

(*c*) Check the level of the surface of the compacted binder course. If the level is such that the depth of the surface course will be less than twice the maximum aggregate size for that material, the whole of the binder course in the patch should be removed and replaced with new material, compacted and re-checked.

4   Surface Course

(*a*) Application – fill the cavity above binder course level so that, on compaction it will, as nearly as possible, be level with the surrounding pavement.

(*b*) Compact with powered compaction plant if possible, hand compaction being used for completion of the compaction where necessary.

(*c*) Check the finished level. The surface course should be removed and replaced with new material if:

any part of the patch is below the level of the surrounding pavement;

any part is more than 6 mm above the surrounding pavement.

5   Initial Sealing

Coated sealing grit should be applied to all open-textured surfacing materials. The sealing grit may be applied immediately the surface course is laid and compacted with it, or may be applied separately, for instance, to a series of patches being sealed on the same day, as a separate operation. Sealing grit is not capable of sealing the surface against water, but it will prevent detritus (mainly gritty and heavy solid waste matter) from penetrating the surface of the patch.

6   Pavement Cleansing

Any loose and/or waste materials should be brushed aside and collected for disposal.

14.8   Planing

Where a surface has become defective, or where resurfacing is required and it is not possible to raise the level of the surface because of, say, a low kerb upstand in an urban area, or a low bridge over the road, it may be necessary to remove the existing surfacing. This is done by using a planer.

Heaters are often used with planers, to soften the surface prior to the application of the cutters. Improvements in machinery have now made it possible to use a cold planer, which is safer.

The machine is set to the required depth of cut. Up to 50 mm of surfacing can be taken off in one pass. The surface to be rectified must previously have been probed. The loosened material is normally collected by a scoop and conveyor (part of the machine) and loaded directly on to lorries.

Resurface by either new surface dressing or new surface course. Prior to resurfacing, as well as at the start of the work, the surface needs to be swept clean (using a mechanical sweeper) and the footways should also be cleaned.

Time has to be allowed after heating and planing, before the surface is relaid. This time is required to allow carbon deposits from the burning to be worn away. These deposits have an adverse effect on the new surfacing materials.

14.9   Surface Dressing

Surface dressing has been the most widely used method of road maintenance since the beginning of the century. Today about 200 million m^2 of roads of all types, from country lanes to motorways, are treated annually.

The unit cost of surface dressing is low and the process of application is simple and direct, but with the ever-increasing volume and speed of traffic, close control of the operation is necessary to ensure successful results.

14.9.1 General Principles

The purposes of surface dressing are:

(a) to make the road surface waterproof;
(b) to arrest disintegration;
(c) to provide a non-skid surface.

The binder, when applied, must be fluid, so as to 'wet' both the chippings and the surface of the road; it must then set in order to hold the chippings firmly in place. In the early life of a surface dressing, the binder acts as an adhesive between the stone and the road surface. Once the stone is held firmly, it tends to be pushed into the original surface by traffic. As a result, the binder is forced up the sides of the chippings so that they are secured in a mortar. If this does not develop, premature failure through loss of stone (scabbing) may occur because the thin bond between road and stone may fracture due to the binder becoming brittle.

The opposite condition may arise when the binder is forced up to the top of the chippings, usually by embedment. The effect of the chippings is then lost and the surface appears to have an excessive amount of binder (fatting-up). The main factors which influence the avoidance of fatting-up are the size of chippings and the rate of spread of the binder, taken in conjunction with the hardness of the road surface to be dressed and the amount of traffic which uses the road.

14.9.2 Types of Dressing

Single dressing consists of a layer of binder with the chippings spread shoulder to shoulder.

Racked in dressing consists of a layer of chippings spread to provide about 90% cover. A second layer of smaller chippings is applied to fill the spaces, thus helping to interlock the chippings.

Double dressing is as the name implies, two dressings, the first layer having about 95% cover, the second about 100–105%.

Inverted double dressing requires two layers spread at about 95–100%, with any surplus being removed before the second application.

Sandwich dressing is a process to remedy a 'fatty' surface by applying chippings of two sizes with no binder being used for the first layer of chippings.

14.10 Materials

14.10.1 Chippings

The essential properties of chippings concern particle shape, flakiness, grading, resistance to crushing, abrasion, polishing, cleanliness and adhesion to binder.

Coated chippings are often used with a cut back bitumen binder.

14.10.2 Binders

Historically the binder was tar, thus the term 'tar spraying'. Tar and tar bitumen blends are still used occasionally, with Flux bitumen and bitumen emulsions being used for the majority of sites. Due to increased traffic and tyre damage to the dressing, modified binders are used where these conditions are likely to occur.

14.10.3 Modified Binders

Normal binders do not have sufficient strength to hold chippings in position, under heavy traffic, until embedment takes place. This results in chipping loss, and is thus an unsatisfactory dressing. Various materials, including polymers, have been developed to provide the strength necessary to cope with these conditions. Typical examples include:

(a) rubber latex on powder blended with bitumen or tar;
(b) styrene-butadiene blended to form a bitumen emulsion (polymer);
(c) ethylene-vinyl-acetate blended with 200 pen bitumen.

They are usually proprietary materials, and those that are increasingly in use include Surmac (Premium or Surfix 80), Shelphalt SX, Crodagrip Mobilfoam and Styrelf. They are considerably more expensive than traditional binders, but with carefully selected use and proper application they are very cost-effective.

14.11 Design of the Specification

Road Note 39 produced by the Transport Research Laboratory and The Road Surface Dressing Association Code of Good Practice, provided detailed information on the design and application of surface dressing.

This summary is based upon these documents.

14.11.1 Design Criteria

Road Note 39 flow chart lists the following factors.

Traffic Category, Volume and Speed

The number of commercial vehicles needs to be determined as it is this type of traffic which causes the embedment of chippings.

The speed of traffic is also a major factor in the design. Table 14.3 shows the nine traffic categories.

Commercial vehicles per lane per day	0 to 20	21 to 100	101 to 250	251 to 500	501 to 1,250	1,251 to 2,000	2,001 to 2,500	2,501 to 3,250	Over 3,250
Free flowing traffic	H	G	F	E	D	C	B	B	A
Restricted flow of traffic	G	F	E	D	C	B	A	A	A
New Roads & Street Works Act road type*	4	4†	3†	3	2	1	1	S	S

* Converted from m.s.a. to cv/l/d based on a growth rate of 2 per cent.
† Division between type 3 and type 4 roads at 125 cv/l/d not 100 cv/l/d

Table 14.3   Traffic categories

Road Hardness

Due to latitude and altitude the hardness of the road surface varies throughout the country.

The amount of embedment to be achieved is related to this factor.

Figure 14.3 shows examples of these, with probe depth and road surface temperature for the five hardness categories.

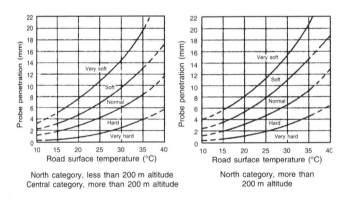

North category, less than 200 m altitude
Central category, more than 200 m altitude

North category, more than 200 m altitude

Figure 14.3   Hardness categories from depth of penetration and road surface temperature for different combinations of latitude and altitude categories

Road Condition and Radius of Bends

The condition of the road surface needs examination, not only to determine any preparatory works but to identify potential stress areas.

14.11.2   Design Selection

Figure 14.4 shows a flow chart to assist in the choice of dressing.

Once a choice has been made the next stage is to determine the chipping type and size with the binder type and rate of spread.

The chipping size varies due to the type of dressing chosen and the hardness of the road surface.

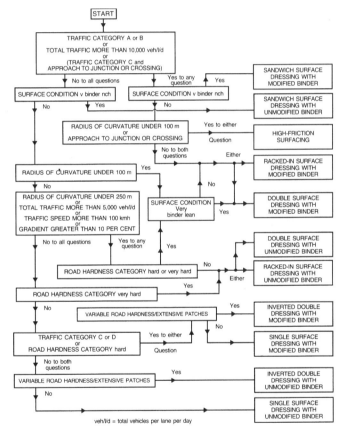

Figure 14.4   Selection of type of surface dressing

Figure 14.5 shows the recommended sizes for single surface dressings, and sizes for racked in and double surface dressings.

Binder rates are determined by consulting tables relating to: size of chipping, type of binder, road surface condition, shade, etc.

The rates vary from 1 litre per square metre to 1.8, and have to be carefully selected to ensure a successful dressing.

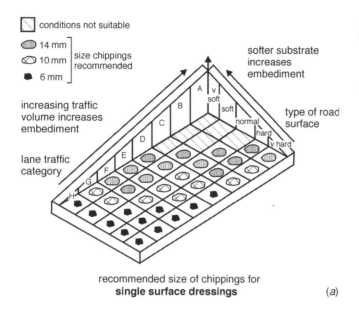

recommended size of chippings for
single surface dressings (a)

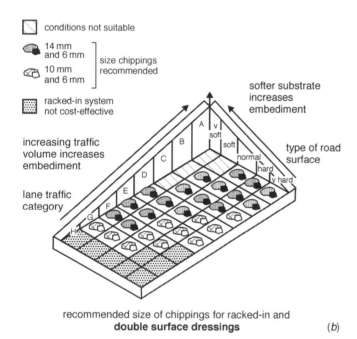

recommended size of chippings for racked-in and
double surface dressings (b)

Figure 14.5(a) and (b)

Tables in Road Note 39 fourth edition, cater for all situations. Rates given in the previous edition of this book are shown in Table 14.4.

14.12 Outline of Procedure for Surface Dressing

Surface dressing contracts vary in relation to who designs the specification, the client or the contractor. The following list consists of operations which take place, the type of contract determining which party carries them out.

14.12.1 Pre-planning – Six Months or More Before Work Starts

1 *Inspection* of roads to be surface dressed.
2 *Specification*
 of materials
 of type and size of chippings
 of type and viscosity of binder
3 *Order equipment*, such as
 lorries/chippers (number, condition)
 sweeper
 rollers
 loader
4 *Plan labour allocation to the work*
 chargehand
 gang size
5 *Consider signs and traffic control*
 number and types of signs
 arrangements for traffic control
6 *Programme*
 grouping of work for continuity
 arrangements for flexibility (for adverse weather, etc.)

14.12.2 Preparation – Three Months Before Work is to Start

1 *Road surface check*
 is the haunch in need of strengthening?
 is patching required?
 does it need sealing?
 does it need reshaping?
2 *Chippings*
 order chippings to storage areas
 control chippings on delivery
 checks on grading, size, shape, cleanliness, resistance to polishing, free flow (coated only)
3 *Binder*
 ordering and arrangements for delivery and application testing (or certification) of spraying equipment
4 *Adhesive agents*
 type and percentage of adhesive to be added to the binder
5 *Workers*
 select and train gang
 brief gang supervisor

14.12.3 Treatment Day

1 *Before spraying*:
 is protective clothing worn?
 erect signs
 sweep
 cover ironwork
 check air/road temperature
 consider weather conditions

Types of surface	LANE TRAFFIC CATEGORY (c.v.d.)							
	2 1000–2000		3 200–1000		4 20–200		5 Less than 20	
	Chipping size (mm)	Binder rate (litre/m²)	Chipping size (mm)	Binder rate (litre/m²)	Chipping size (mm)	Binder rate (litre/m²)	Chipping size (mm)	Binder rate (litre/m²)
Very hard	Not recommended		6	1.3	6	1.5	6	1.6
Hard	14	1.5*	10	1.3	6	1.3	6	1.4
Normal	14	1.4*	10	1.3	10	1.3	6	1.3
Soft	20†	1.3*	14	1.2	14	1.3	10	1.3
Very soft	Not recommended		20†	1.2	14	1.2	10	1.2

* Further evidence is being sought for successful use of these binders in these categories.

† 20 mm chippings may be used at the discretion of the Engineer. Because of the high risk of windscreen breakage, no loose chippings should remain on the surface when the road is opened to unrestricted traffic.

Note Uncrushed gravel may be used at the discretion of the Engineer for Traffic Categories 4 and 5 only. When using gravel, target rates of spread of all binders should be *increased* by 0.1 litre/m².

Table 14.4 Crushed rock or slag chippings – cationic bitumen emulsion class K1-70 binder: target rates of spread of binder at spraying temperatures (litre/m²). Rates of spread of binder actually applied should not vary by more than ±10 per cent of the target figure

tanker ready?
check temperature/pressure
perform jet test
plant and equipment ready?
workers briefed?
2 *During work* – check:
binder temperature
spray bar pressure
cover on road surface joints
chipping size (visual)
chipping cleanliness (visual)
rate of spread of binder
rate of spread of chippings
rollers 'well up', i.e. following closely behind
rollers not crushing aggregate
weather conditions
traffic control

14.12.4 Post Treatment – One Week

1 Sweep
2 Remedial work
3 Sanding
4 Completion of work returns/records

14.12.5 Slurry Seal

This is a process by which a bituminous surfacing which is extensively crazed can be improved. It provides a fine textured, dense, even riding surface which can be used as a surface course or subsequently surface dressed. The slurry seal consists of:

1 modified bitumen emulsion
2 fine aggregate
3 filler
4 water

to produce a thick workable slurry.

The surface to be treated must be cleaned thoroughly (i.e. loose material and oil spillage removed, surface cleaned by power broom, cracks cleaned out by compressed air).

Material is pre-mixed, spread from truck/tanker into spreader box. The thickness to be laid (in a single coat) is about 2–3 mm. The thickness is determined by a squeegee set to the required height across the spreader box. If a greater thickness is required, this should be laid in two coats. A time lag of up to three hours should be allowed between the laying of the two coats. With two-coat surfacing, it may be desirable to use a pneumatic-tyred roller on each coat to provide some additional compaction.

As with surface dressing, traffic speed should be restricted to about 15 mph for the first hour after laying the slurry, to allow the material to achieve stability.

14.13 Resurfacing

Resurfacing of a road pavement or other paved area is undertaken for one or more of the following reasons:

to add strength to the construction to prolong its life;
to correct the surface profile and thus improve riding quality and surface water drainage;

to restore skid-resistance to an old surfacing polished by traffic;

to restore the aesthetic appearance of a worn aged surface.

14.13.1 General Principles

1 *The strength of the existing construction* – is it adequate to carry future anticipated loadings and if not, what degree of strengthening is required?

2 *The shape (regularity) of the existing construction* – is it adequate to ensure satisfactory riding quality, surface water drainage, and uniform compaction of any overlay material or does it need regulating?

3 *Thickness* – will existing fixed levels such as kerbs, accesses, bridge headrooms, permit an overlay to be used or will a particular thickness of the existing construction need to be removed before a new surfacing is laid?

4 *Mechanical key/adhesion* – will the existing surface provide adequate key/adhesion to any overlay applied or is it too smooth?

5 *Skid-resistance* – in restoring the surface what level of resistance to skidding is required?

14.13.2 Detailed Considerations

1 Strength

The strength of the existing construction is all-important when considering resurfacing, as laying of new material on an unsound structure is simply wasting time and money. Cracks and crazing on an existing surfacing are an indication of inadequate strength in the road structure below and such areas should be taken out and reconstructed.

Potholes may simply be localized failures of a surface course or they may be an indication of more deep-seated weakness. A very poor shape in a road, even without cracks, is often also an indication of insufficient strength and sooner or later more definite signs of weakness in the form of cracks are likely to appear. Such areas therefore may also need reconstruction.

2 Shape

Frequently, particularly in urban areas, resurfacing is required to restore the uneven surface of a pavement which is otherwise reasonably structurally sound. This often happens on roads which have been excavated during the repair or installation of services. Weak places such as potholes may well require re-instatement of the pavement to some depth.

Excessive variations in thickness of bituminous surfacing layers due to the materials being laid on a poorly shaped substrate should be avoided as they can cause variations in initial compaction with subsequent variations in durability of the surfacing and reduction in the riding quality.

Levels permitting (such as kerbs, accesses, headroom) and any obviously weak areas being reconstructed, it is advisable to strengthen and reshape an existing road by superimposing a regulating binder course followed by a surface course.

Where an existing road is structurally sound but limited improvement in shape is required (e.g. where only the surface course is deformed or showing signs of wear) or where existing levels/headroom do not permit any appreciable thickness of overlay, an appropriate thickness of existing surfacing can be removed by hot or cold planing prior to application of a new surface course. Accumulations of surface dressings and fine-textured carpets which might have given rise to an excessively soft or 'fatty' surfacing may also need to be removed and this can be achieved by planing.

3 Mechanical Key Adhesion

A smooth or highly polished surface will not give any significant mechanical key to an overlay. In such circumstances planing or scabbling of the surface is advisable, particularly if on a gradient.

Whilst full bond between old and new surfacings will often take time to develop under traffic, some measure of initial bond is desirable and to provide this an application of bitumen emulsion tack coat should be made to the existing or planed surface prior to laying the new material. Special grades of bitumen emulsion are available for this purpose and in applying them the following points should be borne in mind:

(a) A rate of spread of 0.35–0.55 litres/m² is generally sufficient.

(b) The application should preferably be by pressure sprayer and the rate of spread should be uniform with no puddles allowed to form.

(c) The emulsion must be allowed to break (that is, to change from brown to black) before the surface course is laid, otherwise it will form a slip coat.

(d) The emulsion should not be spread so far in advance of the surfacing work that it is removed by traffic or otherwise rendered ineffective.

4 Selection of Surfacing

Traditional materials

(a) The surfacing material must be appropriate for the expected traffic conditions. For example, when resurfacing a main road, the surface course would normally be hot rolled asphalt to BS 594 (with pre-coated chippings applied for skid resistance). The appropriate binder course/regulating course beneath this would be either hot rolled asphalt binder course to BS 594 or dense macadam binder course to BS 4987.

On a relatively lightly trafficked road, dense macadam surface course to BS 4987 should be adequate, although hot rolled asphalt would be more durable. In either case a dense macadam binder course/regulating layer would be appropriate.

For lightly trafficked or pedestrian areas, the open or fine-textured macadams of BS 4987 are normally adequate, except in areas subjected to high stress such as tight turning and manoeuvring, particularly by heavier vehicles, when the dense macadam/rolled asphalt materials would be advisable.

In general, the open and fine-textured surface courses are less durable than the dense macadams which in turn are less durable than hot rolled asphalts. However, as a general guide the cost of the three categories of material increases with the durability. In the case of coated macadam, which relies for its strength on interlock of the aggregate particles, it is essential that the material is laid sufficiently thick to ensure adequate interlock. For hot rolled asphalt, the minimum thickness is related to the coarse aggregate content in order to ensure that when compacted the material has the least possible void content.

The thickness of the various layers of material mentioned above will generally be in accordance with the requirements of the relevant British Standards.

(b) *Developments in materials* – as stated previously, the traditional materials cater for the majority of situations. Due to several factors. new or modified materials have been developed to satisfy problems such as:

rutting due to heavy traffic,
spray from tyres,
noise,
skidding,
durability to increase life and reduce the frequency of disruption to traffic. (See Chapters 5 and 10.)

A major impetus to change is the introduction of DBFOs (Design, Build, Finance and Operate). This encourages the agent to use a wider range of materials and techniques. Examples are:

stone mastic asphalt to reduce deformation, noise and spray;
porous asphalt providing significant reduction in spray and noise;
thin surface courses reducing spray, cost and improving skid resistance.

5 Overlays on Rigid Roads

A common requirement is for an existing concrete road or paved area to be overlaid with bituminous surfacing. Apart from the considerations given above, there is a specific problem associated with surfacing over concrete and this arises from the joints in the concrete.

Concrete slabs are subjected to regular thermal movement and any thin bituminous overlay will fairly quickly crack over joints and cracks in the concrete. Experience has shown that even with relatively thick overlays this 'reflective cracking' will in time appear, although the thicker the surfacing the slower will be the rate of appearance. Research has indicated that a surfacing thickness of 100 mm or more may be needed in some instances to delay significantly the onset of reflective cracking. In view of this, it is normally recommended that a two-layer construction (binder course plus surface course) is used when surfacing over concrete, wherever existing levels permit. If only a single surface course layer can be accommodated, reflective cracking must be anticipated early in the life of the surfacing and when it occurs, appropriate maintenance to prevent ingress of water may need to be taken, such as sealing the crack with a flexible bituminous compound.

Many proprietary metal and synthetic fibre mesh materials are also used to prevent reflective cracking.

The procedures for laying, compaction and quality control are described in Chapter 5.

14.14 Strengthening or Widening of Carriageway by Haunching

Many roads are now 'improved' by widening at each side. A haunch is constructed to tie into the existing road structure, where there is sufficient verge width available. This is considerably cheaper than a full reconstruction of the road to the greater width.

In this case, the procedure is similar to that of repair, but initial excavation of the verge to road formation level is first required; the edge of the existing construction is cut back to a sound and solid structure. This additional width is then constructed in the same manner as the repair.

The best practice is to provide a kerb, with full bed and backing, to contain the new structure that is to be constructed on the widened sub-base. This not only provides an adequate edge marking, but also is strong enough to withstand lateral thrusts due to traffic on the road.

However, in rural areas, on roads bearing light traffic, it may be adequate to abut the verge directly on to the edge of the widened haunch. Care must be exercised to ensure that drainage systems are not damaged or disturbed by these works and that adequate drainage is provided for the widened carriageway.

14.15 Repair of Road Haunch

Care has to be taken, when planning, to use a surface 'repair' method, that the damage is not deeper into the road structure.

Consider a situation where the haunch of a flexible constructed carriageway has crazed and is moving under load. This has to be investigated to determine the cause of such a condition and the possible remedial work.

Possible causes
1 Poor drainage or lack of drainage.
2 Inadequate road structure.

3 Insufficient width.
4 Inadequate lateral support.

Methods of investigation
1 Visual check.
2 Examination of drains, if any.
3 Trial holes in road structure and verge.

It is most important to check the volume and type of traffic, particularly commercial vehicles which may cause extensive damage to road edges.

Remedy
1 Remove a damaged haunch (i.e. remove all material from the road edge until a sound and solid structure is reached).
2 Excavate below the damaged structure until a firm formation is obtained.
3 Ensure that the excavation width is sufficient to allow the use of machinery, particularly for compaction of the replacement material.
4 Assuming that the drainage extends below road formation, the repair up to formation level may be by:

(*a*) filling with dry stone, well consolidated (e.g. Type 2).
(*b*) filling with lean concrete.

5 If necessary, provision of new drains should be made to prevent possible build-up of the water table and to take surface water from the road; additional gullies being provided if necessary.
6 Provide kerbing, adequately bedded and haunched to act as a retaining wall for the road haunch, to form a suitable channel for removal of surface water, and to define an easily visible limit to the road structure.
7 Build-up the new road structure to match the existing, layer by layer. Where laying the bituminous surfacing, the joint with the existing surface should be painted with a tack coat.

14.16 Mix-in-Place Recycling

In recent years, there has been a gradual increase in the reuse of existing materials. The methods of re-using 'recovered' material fall into four broad categories:

1 Retread
2 Repave and re-mix
3 Off-site reprocessing
4 Mix-in-place

Retread and re-pave processes are dealt with in more detail in this section. *See* Section 5.25 for details of off-site reprocessing and mix-in-place.

Mix-in-place recycling is often used for road haunch repair, as an alternative to the excavation and replacement method which was described earlier in Section 14.15.

14.17 The Retread Process

This is an old process which, due to the trend towards the re-use of existing material, may increase in use.

The retread process is carried out in the following sequence:

(*a*) scarification of the old road surface;
(*b*) harrowing and re-shaping. New aggregate is added if required;
(*c*) application of binder, mixing-in-place, and preliminary rolling;
(*d*) filling in of surface voids with fine aggregate;
(*e*) surface treatment.

Some deviations from the above sequence, depending on the conditions on the site, are permitted. On larger works the process needs to be mechanized to a large extent but some very good work has been done using only hand tools. A suitable selection of plant is as follows:

8–10 tonne road roller fitted with a scarifying attachment;
light tractor;
set of spike harrows or cultivators;
5 tonne lorry fitted with a gritting attachment;
5000 litre (approximately) pressure-tank distributor.

14.17.1 Scarification

The existing worn road surface is scarified to a depth of 75 mm using an 8–10 tonne roller fitted with a scarifying attachment. Usually, in order to avoid unnecessary disturbance to traffic flow, only half the road width is treated at one time but on very narrow country roads the entire width of the road must be scarified. The maximum depth to be broken is 75 mm and every effort should be made to avoid disturbing any material below this depth.

This part of the process is generally the slowest and it is often economical to use two rollers for this initial work. Back-rolling should be used to help break down larger lumps. The objective is to produce a fairly evenly graded broken-stone surface suitable for the application of the binder. At this stage all large lumps of surfacing bound together with 'live' binder should be removed by hand, together with any stone larger than 75 mm. 'Live' binder left in the surface is liable to produce fat patches in the finished surface.

14.17.2 Application of Binders

For rapid economical work, binders are best applied by pressure tank distributors: greater penetration of the binders into the loosened aggregate, together with better control of the rate of application, is obtained by pressure methods. Small hand sprayers can also be used, one advantage being that there is less marking of the road surface than with the heavy tankers. The procedure, thereafter, varies depending on whether a hot binder or cold emulsion binder is to be used.

Type of Grading	Type of road material being treated	Rate of spread of binder required litre/m²	
		Cold emulsion	Hot binder
Coarse type	Water-bound macadam	4½ – 5½	2 – 3½
	Coated macadam	3½ – 4½	2
Dense type	Water-bound macadam	5½ – 9	3½ – 4½
	Coated macadam	Not usually treated	

Table 14.5 Recommended rates of application of the binder in the retread process

Tar or Bitumen Binders

Binder is sprayed on to the prepared surface and penetrates into the surfacing to an amount dependent on the grading of the aggregate, the road temperature and the temperature and viscosity of the binder used. Recommended rates of application of the binder are given in Table 14.5. This spray is followed immediately by a light dressing of 12 mm uncoated chippings to fill in surface voids and to prevent 'pick up' of the tarred surface under traffic. Mix-in-place methods are not recommended as the cooling action of the stone leads to rapid setting of the binder and good mixing cannot be obtained. The surface is then lightly rolled and is opened to traffic.

Cold Bitumen Emulsion

Bitumen emulsions are commonly used for retread work. The binder is applied in two or more runs, the harrow being passed over the work between runs in order to turn over the coated stone. No more than 3 litre/m² should be applied at any one run and therefore, at times, three or more passages of the spray tanker are required. Recommended rates of application of the binder are given in Table 14.5. Harrow close behind the spraying unit and make successive applications of binder in opposite directions. The road is then lightly rolled. Emulsions which have considerable resistance to breakdown on mixing with aggregate should be selected for use with aggregates containing higher proportions of fine material.

14.17.3 Compaction

Compaction of the coated road material must be carried out with care. Excessive rolling, especially with a steel-tyred roller on a newly treated surface, may seal the surface unduly and lead to water being trapped within the mixture which then remains soft and may deform seriously under traffic. This is particularly important where cold emulsions are used as about 50% of the emulsion composition is water and provision must be made for the evaporation of this water before the surface is sealed. A single pass of an 8–10 tonne roller is sufficient at this stage and the roller should be taken off if any signs of 'pushing' in front of the rolls are observed. In normal weather rolling may be carried out within an hour of the time of application of the binder.

14.17.4 Filling in Surface Voids

If the aggregate used in the road surfacing contains a fairly high proportion of fines, the road should present a firm, tightly packed, dense surface. The road can then be opened to traffic which should be allowed to run on it for some days followed by further compaction by the roller and before the surface dressing is laid.

With a coarse-type aggregate the road surface will more probably appear open in texture and it is necessary to fill the surface voids by the application of a suitable blinding material. The objective is to produce a dense surface to receive the final surface dressing. When hot binders have been used on a coarse-graded aggregate, 12 mm stone will be required, and occasionally two applications of stone are necessary, the second application being of smaller-sized stone. The road surface should be lightly brushed and then further compacted by rolling on the following day.

When an emulsion has been used for the retreading, it is normal practice to apply a very light coat of emulsion to hold the blinding material before opening the road to traffic. The rate of application of this coat depends on the size of chippings used; with 12 or 9 mm stone approximately 0.7 litre/m² is required, while for smaller sizes up to 0.33 litre/m² may be used.

14.17.5 Surface Treatment

The period before application of the final surface dressing is normally not less than three to four days, and preferably longer. During the period when the road is open to traffic, any irregularities become apparent and it is possible to remedy these before the final surface treatment.

This can be either surface dressing or a new binder course and surface course.

14.18 Repave and Re-Mix

The repave process consists of:

1 Planing the top 20 mm from the existing surface. This material is then discarded.

2 Heating the undersurface and laying a new 20 mm thick surface course of hot rolled asphalt and coated chippings.

Re-mix is a similar process, but carried out in two planing operations, the lower 20 mm of surfacing material being reused by being re-mixed with hot rolled asphalt to form the new 40 mm thick wearing course.

Both these methods are particularly useful for upgrading a carriageway by providing a new surface course without raising the surface level of the road.

14.19 Retexturing Road Surfaces

Often roads require surface treatment due to bitumen bleeding through, causing a severe reduction in the skid resistance value. One method of treating such dressing failures is by the hot compressed air (HCA) system.

A crew of two with vehicle mounted equipment comprising a compressor, propane gas to heat the compressed air to some 1650°C, flexible hose and hand-held lance with flared head, can treat some 500 m² of road surface a day. When treating fatted surfaces with hot compressed air, the excess bitumen is vaporized and blown away without dislodging the aggregate.

It is a more expensive method than surface dressing, but, whereas surface dressing usually has to extend over the whole carriageway, the HCA treatment is applied only to those areas suffering from fatting (surplus binder working its way on to the surface). In many cases it is only the wheel tracks which have deteriorated. A further advantage is that work can continue under almost any weather conditions when using this process.

14.20 Pressure Grouting: Rigid Pavements

This is to restore slab levels. Settlement of slabs destroys the alignment and the riding quality and results in unacceptable steps at joints and cracks, particularly if load transfer devices are not included in the design. Slabs can be restored to their original level by raising them mechanically and by injecting cement grout or a bituminous mixture under pressure, so raising and undersealing in the one operation. However, concrete slabs are normally raised by pumping in grout under pressure.

14.20.1 Principle of Pressure Grouting

Grout is pumped under pressure into a hole under the pavement. As this happens, the grout exerts an upward pressure on the bottom of the slab around the hole (Figure 14.6).

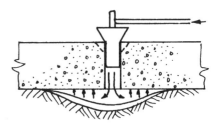

Figure 14.6 Pressure grouting principle

This upward pressure lessens as the distance from the hole increases due to the viscosity of the grout and skin friction created by the flow of the grout (i.e. the resistance which the grout meets as it flows away from the hole). It is due to this frictional resistance that it is possible to raise one corner of a concrete slab without raising the entire slab.

A stiff grout will, of course, not flow as freely as a more fluid grout. Hence, the stiffness of the grout mix can be used to control the area affected by a single pressure grout injection.

The depth of the cavity under the slab also affects the upward pressure over a given area. The more shallow the cavity, the greater the upward pressure around the hole.

If the grout mix is too stiff, it may form a 'pyramid' under the hole, leaving much of the cavity unfilled and therefore, unsupported (Figure 14.7).

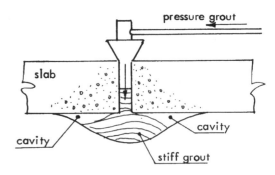

Figure 14.7 Problems encountered with too stiff grout

14.20.2 When to Pressure Grout

Although concrete slabs do not require strong sub-base support, they do require reasonably uniform support, in order to prevent eventual tilting or cracking of the slab, under traffic loading. If this uniform support is lacking, pressure grouting is a possible solution to the problem.

Frequent heavy loads passing over the pavement cause deflection of the slabs at joints. If the subgrade is of fine-grain texture and, especially when also saturated with subsoil water, this 'flexing' at the joints can cause material to be forced upwards through the joints and at the slab edges (road channel). This action is a form of 'pumping' and can cause voids to be formed, with subsequent settling of the slab.

14.20.3 Process of Pressure Grouting

(a) Generally, holes should be 300–450 mm from a transverse joint and at 1.8–2.0 m centres, so that not more than 2.0–2.5 m² of slab is raised by pumping into any one hole. If the slab is cracked, additional holes may be required.

(b) Where 'pumping' has started, but the adjacent slabs are still in alignment, a minimum of two holes can be used. If one corner has faulted, the hole at the corner should be set back sufficiently to avoid lifting the adjacent slab. The precise location of holes is a matter for careful determination in each particular case and site.

(c) Holes of 50–60 mm diameter are drilled, either by core drill or pneumatic drill.

(d) If the slab is tight against the sub-base an air line or blow pipe may be necessary to form a cavity which will expose enough slab area for the subsequent pressure grouting to be effective (see Figure 14.8).

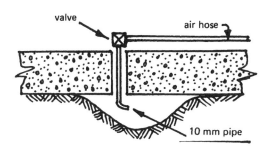

Figure 14.8 Cavity formation prior to grouting

(e) No pressure grouting should be carried out when the ground is frozen, or the air temperature is below 5°C.

(f) Equipment:

concrete/mortar mixer
pump of positive displacement type
small dumper (for transporting grout)
water tank
air compressor
pneumatic hammer
wooden taper plugs

It is essential that all the equipment be kept clean and be cleaned immediately after use every time.

14.20.4 Grout Mix

Usually a 4:1 or 5:1 mix of fine sand:Portland cement is found to be suitable. Alternatively, finely ground limestone or hydrated lime may be used instead of the sand.

One of the most important characteristics of the grout is its flowability: generally, the more uniform the material is in size and the finer grain size, the greater will be its flowability.

Wetting agents may be used in the mix. The use of a wetting agent lubricates the grout and makes runs of up to 2 m possible. It also tends to prevent pyramiding and increases sub-base permeation.

When pumping a long run, it may be found necessary to plug adjacent holes with hardwood plugs to prevent loss of grout.

A mix of a stiff consistency is normally used for raising slabs and a more fluid mix is used to fill voids.

Temperature, wind and humidity can all have a noticeable effect on the consistency of a grout mix.

Specified strengths of grout range from 1.0–20 N/mm² at seven days. A commonly specified strength is 2 N/mm².

14.20.5 Levelling of Concrete Slabs

(a) Determine the amount by which the slab has to be lifted. Also the realignment must be carefully set out and checked (probably a chalk line or straight edge). For larger slabs, it may be necessary to check levels with dumpy level and staff.

(b) Cool weather is desirable – the joint or the joints are then more open than in hot weather.

(c) Lifting should begin at the low point of the sag and be carefully planned to reduce gradually the sag over the whole area, not to lift up completely at one point before moving to the next.

(d) Slabs should not be lifted by more than 6 mm by pumping into any one hole at any one time.

(e) If pumping into two holes simultaneously, the holes should not be adjacent to each other.

(f) To avoid pyramiding, the grout should be sufficiently fluid to flow to adjacent holes – i.e. it should be possible to observe the movement of the grout.

14.20.6 Leakage or Blowouts

This is a common problem in raising concrete slabs. It can usually be controlled by varying the degree of consistency of the grout. A small reduction in the grout water content will result in an appreciable increase in stiffness. By stiffening, the grout stops many of its own leaks.

Leakage may be avoided by pumping a few holes along the outer edge of a slab and allowing the material to set before the slab-raising or pressure grouting is begun.

If a blowout occurs on earth or aggregate sub-base a shovel of dry aggregate tamped into the blow opening may correct the fault.

14.20.7 Completing and Checking Grouting Operations

When the pumping nozzle is removed from a hole, the grout should be cleaned from the hole – additional pumping may be required later at the same hole and the hardened grout would have to be drilled out before pumping could be resumed.

After grouting operations are completed, the holes should be filled with a stiff 1:3 cement grout, well tamped and floated to a smooth finish.

Pressure grouting operations may be checked as follows:

(*a*) Has the cause of damage been determined and, if possible, corrected before the start of pressure grouting?

(*b*) Are sufficient materials on hand to permit a continuous pressure grouting operation?

(*c*) Is the mixing and grouting equipment clean and in good working order?

(*d*) Are grout holes spaced closely enough and properly located to raise the slab without cracking or to fill without leaving voids.

(*e*) Has the total amount that each slab is to be raised been accurately determined by means of a chalk line or straight edge before the start of grouting.

(*f*) Is the grout stiff enough to raise the slab without blowing or leaking and yet fluid enough to prevent pyramiding?

(*g*) Are adjacent holes being checked that the grout is showing during pressure grouting to make sure that pyramiding is not occurring?

(*h*) Are means available for accurately measuring the amount a slab is raised during pumping in any one hole? Is care being exercised to limit this amount to 6 mm?

(*i*) Are grout holes being cleaned immediately after each pumping operation to permit additional pumping, if necessary, without the need for re-drilling?

(*j*) Has a final check been made to ensure that the slabs are at proper grade and level?

14.21   Replacement of Failed Slabs

When a slab is badly cracked it is cheaper and more satisfactory to replace the whole slab. The cause of the damage should be ascertained and dealt with before the repairs are carried out if it is likely to affect the new work. After breaking out the defective concrete, the sub-grade and base should be examined. Any soft areas should be removed and replaced with suitable hard material.

When the slab has been removed, any dowel bars and tie bars revealed in the adjoining concrete should be straightened, cleaned, the dowel bars coated with a thin layer of bitumen and provided with a cap as for new work. If dowel bars have not been provided originally they can be inserted in holes in the old slab made by a rock drill or in a slot made by sawing and breaking out. After cleaning, the holes should be washed out with water and filled with 3:1 cement mortar or with epoxy resin mortar. Then push in the dowel bar, the free end being coated with bitumen and capped.

Revision Questions

1   What is the definitive document related to highways maintenance in the UK as from 2001?
2   At what maximum depth of compacted snow, will road salt be effective?
3   What is 'fatting-up'?
4   What is haunching?

Note: See page 305 for answers to these revision questions

15 New Roads and Street Works Act 1991 Second Edition 2002

15.1 Background

Historically highways (roads) were created or built and became rights of passage (*see* Chapter 1). As society developed, the provision of utility services such as water, sewage disposal, gas, electricity and telephone became the standard expected by everyone, in order to maintain present day lifestyle.

As these services were introduced, the government provided statutory or licence rights to the undertakers which gave them the right to operate and place apparatus in the street. (Note: The word street is used in the legislation to cover all roads, footways, etc.)

15.2 The Public Utilities Street Works Act (PUSWA) 1950

In 1950 the Public Utilities Street Works Act regulated these Provisions into one Act in order to prescribe the rights and obligations between the Street Authority and Statutory Undertakers when working in streets. This Act operated until the early 1970s when it was clear that it required a review, due to

(*a*) growth in traffic,
(*b*) increase in undertakers,
(*c*) increase in customer usage,
(*d*) public complaints on its operation, particularly in the poor quality of reinstatements.

This review lead to the Model Agreement and Specification.

15.3 Model Agreement and Specification

This 'agreement' was produced to concentrate, in particular, on the issue of reinstatement.

It was purely voluntary and required the agreement of both parties (i.e. the Undertakers and the Street Authority) before it could be entered into.

Understandably, it was only of marginal success in endeavouring to resolve the problems, but it did provide some help in providing the background to the full review of the Act which was undertaken by Professor Michael Horne, on behalf of the Secretary of State, in 1985.

15.4 Horne Report

This comprehensive report of over 250 pages contained 73 recommendations, the main ones relating to the execution of works are summarized as follows:

* Undertakers should be fully responsible for the reinstatement.
* There should be a new national specification, with performance standards.
* There should be a guarantee period on the reinstatement.
* Research should be carried out on the long-term damage of excavation and reinstatements.
* In order to improve the quality of work, a training and certification scheme for operatives and supervisors should be introduced.
* The notice procedure needed to be more simple and up-to-date with modern communication systems.
* A street works register and improved coordination procedures was required.
* A code of practice on signing and guarding street works was required.
* Mapping and recording of apparatus needed improvement.

The report contains an interesting extract from:

'*An Act for better paving, improving and regulating the Streets of the Metropolis, and removing and Preventing Nuisances and Obstructions therein, dated 1817.*'

This clearly shows that the problems have been around for a very long time and that they may not be easy to resolve.

15.5 New Roads and Street Works Act 1991

The Horne Report led to the decision to introduce new legislation.

As the legislation was being drafted the government chose to combine two issues into one, i.e. street works and the provision of toll roads using private capital.

This resulted in the Act being in five parts, i.e.

Parts I and II covering toll roads
Part III Street works in England and Wales
Part IV Road works in Scotland
Part V being general.

The reason for there being separate parts of the Act which relate to Scotland is that different terminology is used in that country. The parts concerned are II and IV, though in the practical application of the Act, there is no difference between them and those relating to England and Wales.

As the Act is very comprehensive, other than summarizing Parts III and IV, this chapter will deal only with the roadworks aspects of the legislation.

15.5.1 Regulations and Codes

During the development of the Act it was recognized that the detailed requirements were very complex. Consequently the Act became an enabling Act, which allows the Secretary of State to prescribe Regulations and Codes of practice.

A summary of Parts III and IV of the Act is shown in Table 15.1, and the current Codes of Practice in Table 15.2.

These Codes were prepared by an organization known as the Highway Authority and Utilities Committee (HAUC), whose role is to advise the Secretary of State on the operation of the Act and to prepare the revision of Codes as requested.

15.5.2 Street Works Procedure

In order to have an understanding of the process, Table 15.3 provides an outline of the key issues. As stated earlier, this chapter will concentrate on the roadworks aspects, these being excavation and reinstatement.

15.5.3 Working Near Trees

The 2002 Specification included a new section aimed at protecting trees and more importantly, their roots.

1 Precautionary area
This is determined by measuring the girth of the tree at chest height and multiplying it by 4. This figure will be the radius of a circle having the tree at the centre. This area is known as the 'precautionary area' and in which when working, it is essential to take precautions.

2 Precautions during excavations and reinstatement are laid down by the Specification, with reference being made for liaison where necessary, with the local authority's trees officer.

SUMMARY OF THE ACT

Introductory provisions
Sections 48 to 52 (107 to 111)
Schedule 3

the definition of the basic terms 'street, 'street works', 'street authority' and 'emergency works' and (with Schedule 3) provisions for street works licences;

The street works register
Section 53 (112)

one of the new key provisions;

Notice and co-ordination of works
Sections 54 to 60 (113 to 119)

the principal structure for the new co-ordination arrangements, namely the revised notices regime and the new statutory duties of the street authority to co-ordinate works and the undertaker to co-operate in the arrangements;

Streets subject to special controls
Sections 61 to 64 (120 to 123)
Schedule 4 (Schedule 6)

the three types of street where the street works powers of undertakers may, in some way, be curtailed, namely protected streets, streets with special engineering difficulties, and traffic sensitive streets;

General requirements as to the execution of street works
Sections 65 to 69 (124 to 128)

the obligations of undertakers in respect of safety, avoidance of delays and obstructions, use of qualified persons and the protection of other undertakers' apparatus;

Reinstatement
Sections 70 to 73 (129 to 132)

the new reinstatement regime:

Charges, fees and contributions payable by undertakers
Sections 74 to 78 (133 to 137)

various financial obligations of undertakers in relation to their street works (including the two cases where the introduction of the obligation is dependent upon experience of the operation of the Act, namely charges for occupation of road space (sections 74 (133)) and long-term damage (section 78 (137));

Duties and liabilities of undertakers with respect to apparatus
Sections 79 to 82 (138 to 141)

various other obligations of undertakers in relation to apparatus, including the important revision in the law relating to undertakers' liability for loss or damage (section 82 (141));

Apparatus affected by highway, bridge or transport works
Sections 86 to 85 (142 to 144)

Provisions for the protection of undertakers' apparatus affected by the works of other bodies and for diversionary works
Sections 86 to 106 (145 to 165) and 166 to 171

the remaining sections and schedules contain provisions with respect to particular authorities and undertakings (sections 86 to 93 (145 to 152)), street authority and district council powers (section 94 (153)) and various supplementary and general provisions (including a very useful 'Index of defined expressions') (sections 95 to 106 (145 to 165) and 166 to 171).

Table 15.1 (N.B. Sections listed above refer to streetworks in England and Wales. Those in brackets refer to those in Scotland.)

CODES OF PRACTICE

Reinstatement Code of Practice
Code of Practice on 'Specification for the Reinstatement of Openings in Highways' (published in September 1992)

Diversionary Works Code of Practice
Code of Practice on 'Measures Necessary where Apparatus is Affected by Major Works (Diversionary Works)' (published on 22 October 1992)

Inspections Code of Practice
'Code of Practice for Inspections' (published on 22 October 1992)

Safety at Street Works and Road Works booklet
Code of Practice on 'Safety at Street Works and Road Works' (published on 11 February 1993)

Co-ordination Code of Practice
'Code of Practice for the Co-ordination of Street Works and Works for Road Purposes and Related Matters' (published on 22 February 1993)

Table 15.2

STREET WORKS – ACTION FLOW CHART (Key issues)		
Undertaker (U)	**Pre-works** Co-ordination meetings Agreement on sample inspection fees	**Street Authority (SA)** *Define:* Road/footway categories, Special surfaces, etc., Traffic sensitive streets, Sites with SED, i.e. special engineering difficulty, Protected streets, Sec. 58 restrictions.
Inspect site (if necessary) to identify problem areas.	U and SA meet (if necessary).	
Issue Notices (check for SEDs or Bridges).		Enter into street works register. Record units for inspection. Check for Sec. 56 or 58 restriction.
	During works	
Inform public of major works. Supervision. Safety. Qualifications of Supervisors and operatives.		Sample inspections. Complete checklist 1/2. Third party inspections. Retain on files.
Quality, i.e. compliance with Specification.		Record of defects for report (if applicable)
Notification to SA within 24 hrs.	**Interim completion**	Record notification on register. Third party inspections.
Send R1 to SA within 7 days. Maintain reinstatement.		Check permanent is completed within 6 months.
Notification to SA within 24 hrs. Send RI to SA within 7 days.	**Permanent completion**	Record date on register for Guarantee period to begin. Inspections; sample, checklists 3/4/5, third party, investigatory. Proceed with defect system. Record defects. If over 10% consider issue improvement notice.
	U and SA to meet (if necessary)	
	Accounting	Record number of sample and third party inspections, with check lists. Submit 3 month account and monthly account for defects.
Settle accounts.	**Performance monitor** Review with undertaker as necessary, or at HAUC meeting.	

Table 15.3

3 Precautions will – No machinery.
include   – Use of trenchless techniques.
      – Hand digging.
      – Roots over 25 mm not to be cut, unless by agreement.
      – Other roots to be clean cut.
      – Reinstatement of roots to include a mixture of soil and sand.

15.6 Excavation and Reinstatement of Trenches

Excavation of trenches made by undertakers have to conform to the Regulations and the HAUC code of practice Specification for the Reinstatement of Openings in Highways.

The Specification with its Appendices and Notes for Guidance is a very comprehensive document, which in the event of non compliance could result in prosecution. Therefore it is very important to recognize that the actual document should be consulted when necessary, as this chapter is only intended as a guide.

15.7 The Specification

This consists of 12 sections, appendices and notes for guidance.

S1 Introduction.
S2 Performance Requirements.
S3 Excavation.
S4 Surround to Apparatus.
S5 Backfill.
S6 Flexible and Composite Roads.
S7 Rigid and Modular Roads.
S8 Footways, Footpaths and Cycletracks.
S9 Verges and unmade ground.
S10 Compaction requirements.
S11 Ancillary Activities.
S12 Remedial Works.

15.7.1 Sections 1 and 2 Introduction and Performance Requirements

The basic concept is to prescribe 'operational principles', 'guarantee periods' and 'performance requirements' which are obligatory, regardless of which method is selected for the reinstatement.

Some of the 'operational principles' and 'performance requirements' are as follows:

(a) The as-laid profile of the reinstatement should be substantially flush with the adjacent surface and if, at any time within the guarantee period, it fails the performance requirements, then remedial work should be carried out.

(b) Permanent reinstatement must normally be made within 6 months.

(c) Roads and footways should be classified, by the Street Authority, according to their traffic loading.

(d) The permanent reinstatement must be guaranteed for 2 years and, for trenches over 1.5 m deep, the period of guarantee is 3 years.

(e) There are criteria for surface regularity, structural integrity, and skid resistance.

(f) If, at any time during the interim or guarantee periods, the surface profile of a reinstatement exceeds any of the intervention limits, remedial action should be carried out in order to return the surface profile of the reinstatement to the as-laid condition described. These intervention levels are shown in Table 15.4.

| Intervention | Reinstatement width | | | | |
	nom 250 mm	nom 500 mm	nom 750 mm	nom 1 m	over 1 m
Edge depression	10 mm	10 mm	10 mm	10 mm	10 mm
Surface depression	10 mm	13 mm	19 mm	25 mm	25 mm
Surface crowning	10 mm	13 mm	19 mm	25 mm	25 mm

Table 15.4 Reinstatement intervention levels

If, at any time during the interim reinstatement or guarantee periods, the surface profile of a reinstatement exceeds any of the intervention limits, remedial action shall be carried out in order to return the surface profile of the reinstatement to the as-laid condition described.

15.7.2 Section 3 Excavation

The excavation has to be;

stable,
wide enough for compaction plant,
vertical,
free from unnecessary water,
supported if necessary

and causing minimum damage to the road structure.

Excavated material that is to be re-used shall be stored in a manner to maintain its suitability.

15.7.3 Section 4 Surround to Apparatus

Due to the nature of the apparatus, the undertaker has the right to select suitable material for the surround and up to 250 mm above the crown of the pipe.

This is subject to its suitability, described in the next section.

15.7.4 Section 5 Backfill

In the past, this layer was often regarded to be of little importance, the emphasis being placed on the pavement layers.

If this layer fails, then all the reinstatement over it will be likely to fail.

The specification addresses this issue by using simple site tests to determine the suitability of materials for use.

(For a greater understanding of soils and tests, refer to Chapters 3 and 10.)

Backfill materials are classified as:

Class A, Graded granular, this implies a mixture of stone sizes, a typical example being granular sub-base.

Class B, Granular, this having less variation in the sizes of stone.

Class C, Cohesive granular, this is a mixture of granular with silt or clay.

Class D, Cohesive, as its name suggests this has self-binding properties, i.e. clay.

Class E, Unacceptable, these are materials which should be rejected, i.e. organic matter, wet clay, frozen, likely to burn, or chemically contaminated.

Granular materials interlock naturally, and are only affected by being very wet.

Cohesive is very susceptible to moisture, thus may be regarded as too risky to use.

There are additional requirements for frost susceptibility and particle size, i.e. 75 mm maximum and 37.5 for trenches less than 150 mm wide.

Guidance is provided in the appendices for simple site tests to classify backfill and its suitability for use.

The class of backfill will determine the thickness of the sub-base layer.

15.7.5 Section 6 Flexible and Composite Roads

The reinstatement of the pavement layers is based upon the form of its construction and the road category, as determined by the Street Authority. Table 15.5 shows these categories, types 0 to 4. If the road has an excess

Category	Traffic Capacity
Type 0	Roads carrying over 30–125 m.s.a.
Type 1	Roads carrying over 10–30 m.s.a.
Type 2	Roads carrying over 2.5–10 m.s.a.
Type 3	Roads carrying over 0.5–2.5 m.s.a.
Type 4	Roads carrying up to 0.5 m.s.a.

Table 15.5 Road categories

of over 125 million standard axles (m.s.a.) the reinstatement is agreed between the Street Authority and the Undertaker.

Table 15.6 shows the key to reinstatement materials, the Undertaker having the right to choose the material and layer thickness with reference the type of road.

There is a general requirement that the surface reinstatement shall match the existing as closely as possible.

Deferred set mixtures are not permitted in permanent surfacing.

Permanent cold binder and surface courses have to be approved, the procedure being outlined in an appendix to the specification.

Figure 15.1 shows an example for a type 4 flexible road, with the four reinstatement methods.

The thickness of the sub-base is determined by the type of backfill, varying in thickness from 150 mm for graded granular to 300 mm for cohesive.

15.7.6 Section 7 Rigid and Modular Roads

If a rigid road has an overlay of over 100 mm it is regarded as composite and is reinstated accordingly. Rigid roads only have two reinstatement methods, i.e.

Method A, all permanent at the first visit,
Method B all permanent up to sub-base, work being completed on the second visit.

Reinstatement of rigid roads is more complex and expensive than flexible or composite but fortunately the percentage of such roads is very low.

The Specification requires C40 concrete to be laid to match the thickness of the existing; reinforcement, if any, also to match the existing.

The 'essentially' straight edges is a general requirement, which is achieved by saw cutting.

There are two methods which can be used to support the new slab.

If the slab is cut to full depth it is necessary to tie the new slab into the existing using dowel bars, this is a complex and expensive operation.

The alternative is to saw cut about 30 mm, then break out the remainder at an angle of 27 degrees, thus forming a wedge shaped infill.

There are several general conditions, i.e.

(a) Edge preparation
(b) A slip membrane to be laid beneath the slab
(c) A curing membrane to be applied.

If reinforcement is present, a length of 150 mm in the slab has to be left exposed and tied into the new reinforcement. In the case of modular roads, the modules can be laid on an existing sub-base and road base of flexible, composite or rigid construction. The reinstatement of this has to match the existing, with the blocks being re-used or matching replacements provided.

▨ { HRASC / CGSC	▦ { HRASC / CGSC / DBSC / PCSC	▨ { HRABC / DBC	▤ { HRABC / DBC / PCBC

HRASC – Hot rolled asphalt surface course to BS 594: Part 1 1985. All roads – 30/14 Design Type F mix, 50 pen (2 to 8 stability) to Table 3, Column 9.
 – Types 2, 3 and 4 roads – 30/14 Recipe Type F mix, 50 pen to Table 5, Column 21. Footways – 15/10 Recipe Type F mix to Table 5, Column 19.

CGSC – Close graded surface course macadam to BS 4987: Part 1 1988. All roads – 10 mm size close graded, 100 pen to Clause 7.4.

DBSC – Dense surface course macadam to BS 4987: Part 1 1988. Types 3 and 4 roads – 6 mm size dense, 100 pen to Clause 7.5. Footways – 6 mm size dense, to Clause 7.5.

HRABC – Hot rolled asphalt binder course to BS 594: Part 1 1985. All roads – 50/20 mix, 50 pen to Table 2, Column 3. Footways – 50/20 mix to Table 2, Column 3.

DBC – Dense binder course macadam to BS 4987: Part 1 1988. All roads – 20 mm size dense, 100 pen to Clause 6.5. Footways – 20 mm size dense, to Clause 6.5.

PCSC
PCBC } – In accordance with Appendix A10.

DSM – Deferred set macadam 20 mm binder course or 10 mm of 6 mm surface course macadam to BS 4987: Part 1 1988. Minimum binder viscosity of 30 secs STV – approximately equivalent to 10 days deferred.   ▨ { DSM / PCSM

Concrete – to SHW Clause 1001. All roads – C40 mix. Footways – C30 mix.   ▨ Concrete

CMB 3 – Cement Bound Material Category 3 to SHW Clause 1038.   ▨ CBM 3

GSB 1 – Granular Sub-base Material Type 1 to SHW Clause 803 used in accordance with Appendix A1.   ▨ GSB 1

Notes on HRAWC

1 Natural gravels not permitted as coarse aggregate in HRAWC for use in Type 1 and 2 roads.
2 A design mix may give better performance where queuing of heavy traffic is likely to occur. Also, a design mix may be more economical and easier to lay, compact and provide with surface chippings.
3 Chippings shall be 20 mm or 14 mm nominal size, pre-coated.

Note on Appendices A3 to A7 – All layer thicknesses in millimetres.

BACKFILL MATERIALS

Class A Graded Granular   ▨ Class A

Class B Granular   ▨ Class B

Class C Cohesive Granular   ▨ Class C

Class D Cohesive   ▨ Class D

Table 15.6   Key to materials

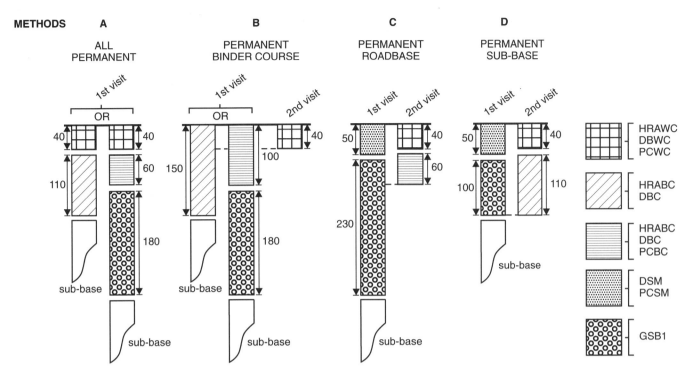

Figure 15.1   Type 4: flexible roads

15.7.7   *Section 8 Footways, Footpaths and Cycletracks*

Flexible reinstatements are listed as three methods, which are:

Method A, all permanent at the first visit.
Method B, all permanent on the first visit up to binder course, a permanent surface course to follow later.
Method C, similar to B with a binder course and surface course to follow later.

Generally, the material should match the existing, with a minimum thickness of 60 mm of surfacing and 100 mm of sub-base. As shown in Table 15.6, the approved materials are 20 mm dense binder course, 6 mm dense surface course with a 50/20 mix of hot rolled asphalt when necessary to match the existing material.

As in flexible roads, permanent cold-lay surfacing materials are being developed and, subject to them being approved, may be used.

Where the Street Authority has designated the footway as high duty or amenity the reinstatement must match the existing.

Rigid reinstatements require a minimum concrete strength of C30, the thickness and surface matching existing as closely as possible.

Modular reinstatements require the undamaged modules to be cleaned and reused, with matching replacements if necessary.

15.7.8   *Section 9 Verges and Unmade Ground*

Where the road construction layers are extended into the verge or unmade ground, the reinstatement must match the existing.

Grassed areas should have the original turf re-used or be re-seeded with 100 mm of soil.

15.7.9   *Section 10 Compaction*

Effective compaction is the key to a successful reinstatement.

It depends upon:

The type of material
Layer thickness
Weight and performance of the compacting equipment
Skill of the operative.

An appendix in the Specification shows a table covering cohesive, granular soils and bituminous material using a range of equipment.

The resultant layer thickness and number of passes, with several other general requirements, indicate good practice.

It is important to note that it is the reinstatement performance (on settlement) during the guarantee period, which determines compliance, whether or not these compaction requirements have been met.

15.7.10 Section 11 Ancillary Activities

These include such items as:

Test holes
Traffic sensors
Sewers drains and tunnels.

It outlines requirements for reinstatement and cooperation between the relevant authorities prior to the work commencing.

15.7.11 Section 12 Remedial Works

This section reinforces the responsibility of the Undertaker to satisfy the performance requirements. Contravention could result in legal action under the provisions of the Act. It outlines methods of repair for cracks, repair of settlement beyond the reinstatement limits and other significant defects. As stated earlier, this summary is only intended as a guide to Specification, with its Appendices and Notes for Guidance which must be consulted for detail where necessary.

15.8 Reinstatement Practice

In order to ensure that all aspects of the operation from initial excavation to the end of the guarantee period satisfy the Specification, it must be well organized. Because of the nature of the work, as many pre-site decisions as possible should be made before work commences. Variation of any method on site can be expensive in both time and resources.

Aspects of the job which should be considered prior to excavation include:

(a) skills and qualifications of the operatives
(b) method of excavation
(c) re-use of existing material
(d) conveyance of equipment and materials
(e) method of reinstatement.

Example 1: Assume an opening 1.0 m square and 1.0 m deep is to be made in a road of flexible construction. This will require four stages:

Stage 1 Prior to excavation
Stage 2 During excavation and backfill
Stage 3 Interim and permanent reinstatement
Stage 4 During guarantee period

Each operation/stage is considered and details are shown on the operation sheets shown in Figure 15.2. N.B.: It is assumed for the purpose of this exercise that all necessary signing and guarding have been provided at every stage.

Example 2 in Figure 15.3 shows a reinstatement in a type 2 road, in cohesive ground using method A, i.e. permanent at the first visit, compaction being carried out with a 1700 kg twin-drum roller. It is very important to realize that it is the performance of the reinstatement, not adherence to the specification which satisfies the guarantee period.

Obviously on inspection, the surface course and the joints are visible, the other work being hidden. As the Specification requires edges to be 'essentially' straight it is easy to see why poor joints are one of the commonest cause of defect notices.

Example 3 in Figure 15.4 shows a rigid reinstatement in a type 4 road on granular ground.

Example 4 in Figure 15.5 shows a method A reinstatement 600 mm deep in cohesive ground.

It should be noted that the Specification shows 60 mm of surfacing, it is not clear whether this has to be laid in one or two layers. Good practice following BS 4987 states two layers.

Example 5 in Figure 15.6 shows an interim and permanent reinstatement in a modular footway 600 mm deep in cohesive ground.

15.9 Inspection

The Street Authority carry out inspections under four distinct procedures, i.e.

Sample
Substantial investigatory works
Routine investigatory works
Defects

There is a fee structure for inspections and rectification of defects which is set by the Secretary of State.

15.9.1 Sample

Table 15.7 shows the five types of inspection. Inspection categories 1 and 2 take place when the works are taking place. 3, 4, 5 cover the guarantee period.

On each inspection a check list is used, these are shown in Tables 15.8 and 15.9.

15.9.2 Substantial Investigatory Works

The Street Authority have the power to carry out investigations to ascertain whether an Undertaker has complied with his duties. If defects are found the Undertaker has to pay for the investigation and rectification.

OPERATION	FACTORS TO CONSIDER	CHOICE	SPECIFICATION REFERENCE
STAGE 1 PRIOR TO EXCAVATION			
1) Location of Services	Contact with other 'stats' and Street Authority On site method (Local knowledge)		Mark as necessary Use cable location equipment
2) Method of Excavation	Size of hole Method of conveyance of equipment	3	Air/hydraulic compressor and breaker, then hand dig
3) Edge Support	Size of hole Duration of exposure Ground conditions Other services Conveyance	3.3.2 3.5	Have hydraulic frame available with a few struts and poling boards
4) Excavated Material (a) Surround to apparatus	Own service requirements Type of material from excavation	4	Use fine excavated material (to satisfy spec.)
(b) Backfill	Type of material Suitability (Local knowledge) Reusable or disposable	5.1... Appendix 2	Use excavated material Dispose of unsuitable Use Granular Sub Base as replacement
5) Compaction	Size of hole Conveyance of equipment Nature of material (say Granular)	5.1... 10.1... Appendix 2 Appendix 9	Hand compact around apparatus 100 mm Layers each having 4 passes (50 kg Vibro tamper)
6) Reinstatement of Pavement (a) 1st Visit	Type of road construction (say flexible) Type of road category (say 4) Conveyance of materials Availability of materials	6 Appendix 3 1.3.1.1	*(see diagram below)*
(b) 2nd Visit	Compressor Disc cutter Cold planer to remove 40 mm layer Cut edge and paint with emulsion Separate layers Match existing surface Availability of materials Temperature (use of 'hot box')	Appendix 3.4	*(see diagram below)* Cut out 40 mm Layer by hand cut vertical edges with disc cutter Paint edges with bitumen emulsion
STAGE 2 DURING EXCAVATION AND BACKFILL			
	Location of Services Suitability of excavated material Separate reusable from disposable Maintain moisture content Break reusable down to less than 75 mm particle size (40 mm for around apparatus)	5 Appendix 2	Road marked as necessary (services and work area) Check for compliance with Stage 1 Check layer thicknesses Check number of passes
STAGE 3 PAVEMENT REINSTATEMENT			
1st Visit Sub Base and Road Base	Material quality		Use as much reusable as possible Check compliance with Stage 1
Binder course Surface course	Material quality Surface profile	1 2.2	Check compliance Stage 1 Check surface profile Check compliance with specification
2nd Visit Surface course	Material quality Temperature Matching existing surface Surface profile	1 2.2	Check compliance Stage 1 Check surface profile compliance with specification
STAGE 4 PERFORMANCE IN GUARANTEE PERIOD			
Inspection	Specification/Guidance Notes Performance Criteria	2.2 Appendix 1 12	Check Profile, Stability and Intervention Levels Repair as necessary

1st Visit — separate top layers with sand

100 mm	P.C.B.C. in 60 mm & 40 mm Layers 6 passes 50 kgVT	Surface course / Binder course
180 mm	G.S.B. in 4 Layers 4 passes each Layer using	Road Base
200 mm	50 kg Vibro Tamper	Sub Base
	Granular Backfill	

2nd Visit
DBWC 40 mm layer
8 passes with 1400 kg Vibro Plate
..W.C...
..B.C...

Figure 15.2   Reinstatement practice

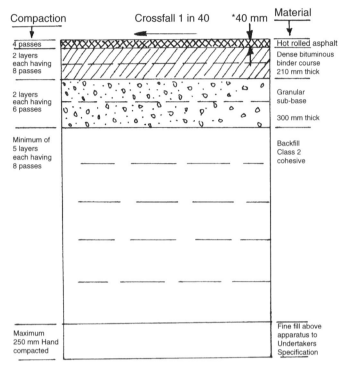

* Surface course options are: Hot rolled asphalt 30/14 mix (Type F)
to BS 594 or
Dense Bituminous Surface Course
to BS 4987 clause 7.4

Figure 15.3   Example 2 Type 2 road, cohesive ground Method A, permanent

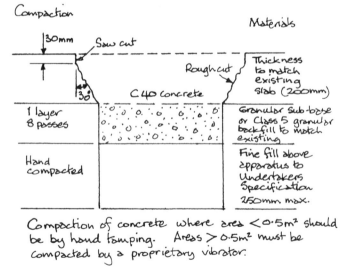

Compaction of concrete where area < 0.5m² should be by hand tamping. Areas > 0.5m² must be compacted by a proprietary vibrator.

Figure 15.4   Example 3 Rigid reinstatement Type 4 road, granular ground

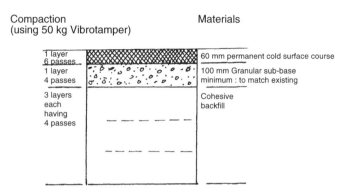

Figure 15.5   Example 4 Method A reinstatement 600 mm deep, cohesive ground

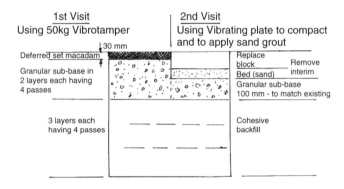

Figure 15.6   Example 5 Interim and permanent reinstatement, modular footway, 600 mm deep, cohesive ground

Type of inspection	Inspection category
During excavation	1
Whilst reinstating (including any interim phase)	2
Immediately after permanent reinstatement (within one month)	3
Between 6 and 9 months after permanent reinstatement	4
During the one month preceding the end of the guarantee period	5

Table 15.7   Types of inspection

UNDERTAKERS' WORKS
INSPECTIONS CHECK LIST

PAGE 1   FOR USE ON INSPECTIONS 1 AND 2   DATE

Authority reference   Undertaker's reference

Location o/s of or from _____ to _____

_____

_____

_____

Time taken for inspection _____

1 Signing & Guarding and Excavation

2 Signinq & Guarding and Backfill &   Insert inspection number ☐
  Reinstatement

Signing & Guarding	**Acceptable**	**Unacceptable[1]**	**Unseen/ inapplicable**
Correct type	☐	☐	☐
Distances	☐	☐	☐
Safely zones	☐	☐	☐
Barriers	☐	☐	☐
Traffic control	☐	☐	☐
Pedestrian control	☐	☐	☐
Other	☐	☐	☐
Excavation	☐	☐	☐
Backfill & Reinstatement			
Surround to apparatus	☐	☐	☐
Backfill material	☐	☐	☐
Compaction	☐	☐	☐
Sub-base	☐	☐	☐
Roadbase	☐	☐	☐
Binder course	☐	☐	☐
Surface course	☐	☐	☐

[1] State reason for unacceptability

Name   Signature   Date

Table 15.8   Inspections check list

UNDERTAKERS' WORKS
INSPECTIONS CHECK LIST

PAGE 2   FOR USE ON INSPECTIONS 3, 4 AND 5   DATE

Authority reference _____ Undertaker's reference _____

Location o/s of or from _____ to _____

_____

_____

_____

_____

3 Immediately following permanent reinstatement   Insert inspection number

4 Intermediate (6 months)

5 End of guarantee period   ☐

	Acceptable	**Unacceptable[1]**
Edge depressions/trips	☐	☐
Surface depressions	☐	☐
Surface crowning	☐	☐
Edge cracking	☐	☐
Texture depth	☐	☐
Surface regularity	☐	☐
Other	☐	☐

[1] State reason for unacceptability

Name   Signature   Date

Table 15.9   Inspections check list

15.9.3 Routine Investigatory Works

These are instigated in response to a report from a third party.

15.9.4 Defects

If in any of the previous inspections a defect is found, there is a procedure for rectification. In these cases the guarantee period starts again. The guarantee does not begin again if rectification takes place within the intervention levels.

15.10 General Summary

In spite of all the detail in the Specification and the Inspection Codes there are many issues of where clarification is required. These are usually where an item is not included or is not clearly defined.

This is inevitable with the complexity of all the detail.

The procedure requires these to be agreed between the two parties or raised at a HAUC meeting.

Ultimately these could lead to being included in a revised Specification or Inspections Code in the process of time.

Experience has already shown an improvement in reinstatement. Investigations carried out by Street Authorities indicate further improvement is still necessary.

Revision Questions

1 What is meant by the precautionary area when working near trees and how is it determined?
2 What is the traffic capacity of a type 2 road?
3 What is the minimum strength concrete used for reinstatements in rigid footways, footpaths and cycle tracks?
4 What are the four classes of suitable backfill material allowed under The Specification?

Note: See page 305 for answers to these revision questions

16

Site organization

16.1 Introduction

Prior to the Local Government and Land Act 1980, which is more commonly known as the DLO (Direct Labour Organization) Legislation, most highway maintenance and small construction projects were carried out by the Highway Authorities with a minimum of preparation of contract documents. It was only on major works that it was necessary to invite tenders from contractors.

The 1980 legislation, however, requires the Highway Authority to obtain a 'prior written statement' (in other words, an estimate) of the cost of the work before it commences.

For this estimate to be prepared, it is necessary for the Client (i.e. the Highway Authority Engineer) to specify the nature, description and details of the work; that is, to produce contract documents. Therefore, contract documents have come to be more widely used, even for very minor operations. This use has continued as best value has evolved with contractor/client joint teams.

16.2 Contract Documents

These consist of

drawings
specifications
bills of quantities.

16.2.1 Contract Drawings

These must show the scale and complexity of the work in sufficient detail for the work to be carried out. The bigger the job, the greater the need for special drawings relating to it. Small jobs can often be done, using only 'standard' drawings.

For drawing details, see Chapter 2.

16.2.2 Specifications

These are written documents, used in conjunction with drawings, to describe the work to be carried out, together with descriptions of materials, etc., and the quality of work, etc., required.

A specification serves three purposes, which are:

(*a*) to provide the estimator with the information which is required in order to prepare a realistic quotation;
(*b*) to provide the quantity surveyor with sufficient information for him to prepare a bill of quantities;
(*c*) to be used as instructions for carrying out the work. This should be done by supervisors.

The writer of a specification must be specific in his/her instructions and crystal-clear and complete in the detail. The specification, as part of the contract documents, must have all the preciseness of an agreement and convey exactly what is wanted. A specification is written to convey information which cannot be shown on drawings and covers three main areas:

Materials
Workmanship
Quality.

To write a successful specification, a sound knowledge of construction work is required, and valuable information can be found by reference to British Standards and Codes of Practice and to the Department of Transport Specification for Highway Works.

As an example, two clauses from the DoT Specification for Highway Works are outlined below, in order to give an appreciation of the precise nature of these specifications and illustrate the detail needed to avoid any ambiguity.

The two clauses are:

Cl.803 – this covers a particular material
Cl.2412 – this is more concerned with workmanship and quality.

Granular Sub-Base Material Type 1: from Cl.803

1 Type 1 granular material must be crushed rock, crushed slag, crushed concrete or well burnt non-plastic shale. The material is to be well-graded, and lie within the grading envelope in the table.

2 The material passing the 0.500 mm BS sieve must be non-plastic as defined by BS 1377, and tested in compliance therewith.

3 The material must be transported, laid and compacted without drying out or segregation.

4 The material must have a 10% fines value of 50 kN or more when tested in compliance with BS 812 except that samples shall be tested in a saturated and surface dried condition. Prior to testing the selected test portions have to be soaked in water at room temperature for 24 hours without previously having been oven dried. *See* Section 5.8.1, Table 5.4(*a*).

Brickwork and Blockwork: as Cl.2412

1 Brickwork and blockwork shall be laid on a full bed of mortar and bonded as described in the Contract. Single frogged bricks are to be laid with the frog uppermost. Perpends between bricks and blocks must be filled with mortar before the next mortar bed is laid. Whole bricks and blocks shall be used except where it is necessary to cut closers or where agreed by the engineer.

2 Brickwork and blockwork must be built uniformly. Corners and other advanced work shall be stepped back and not raised above the general level more than 900 mm. Courses are to be kept horizontal and matching perpends are to be in vertical alignment.

3 Unless agreed by the engineer, overhand work is not permitted.

4 Reinforcement in brickwork and blockwork must be completely embedded in the mortar joint.

5 Where pointing is required in the Contract the joint is to be raked out to a depth of 12 mm and after the completion of the entire facework, pointed in mortar as described in the Contract.

6 Where jointing is required in the Contract it must be done as the work proceeds and to the finish described in the Contract.

16.2.3 Bills of Quantities

A Bill of Quantities provides a brief description of the work, linked to the Specification, with the amounts of work to be done measured in a standard form as prescribed in the 'ICE Standard Methods of Measurement of Civil Engineering Quantities'.

Examples of the units in common use in roadwork are given in Table 16.1.

Bills of Quantities are always prepared to a particular format; an example of a typical extract from a Bill is shown in Figure 16.1, ready for the Contractor/Estimator to insert prices.

16.3 Tender Documents

These are beyond the scope of this book. They are fully covered in the General Conditions of Contract and Forms

Type of Work	Unit of Measurement
Removal of trees and tree stumps	Enumerated in stages (girth)
Demolition work	Lump sum
General excavation	Cubic metre
Excavation of pipe trenches	Linear metre in various stages
Excavation of top soil and stripping turf	Square metre (extra over)
Trimming slopes, and soiling and seeding or turfing	Square metre
Pitching to slopes	Square metre
Mass concrete	Cubic metre
Mass concrete in slabs, roofs and floors, not exceeding 12 in. thick (300 mm)	Square metre or cubic metre
Reinforced concrete	Cubic metre
Shuttering	Square metre
Bar or rod reinforcement	Tonne or Megagramme
Fabric reinforcement	Square metre
Heavy precast concrete blockwork	Cubic metre
General brickwork, up to 27 in. thick (685 mm)	Square metre
Facings and fair faced work	Square metre (extra over)
Chases, corbels, etc. in brickwork	Linear metre (extra over)
Rubble masonry	Cubic metre
Asphalt work	Square metre
Asphalt lining to small sumps	Number
Painting of steelwork by general contractor	Square metre
Road surfacings and footway pavings, including foundations	Square metre
Kerbs and channels	Linear metre
Concrete foundations to kerbs and channels	Linear metre
Channels formed in surface of concrete roads	Linear metre (extra over)
Expansion joints in concrete roads	Linear metre
Clayware and concrete pipe sewers and drains with cement joints	Linear metre
Clayware and concrete pipes with flexible joints	Linear metre (extra over)
Bends, junctions, etc. to ditto	Number (extra over)
Cuts to pipes	Number
Concrete beds, haunchings, etc. to pipes	Cubic metre
Gullies, penstocks, valves, etc.	Number
Manholes, valve chambers, etc.	Number (measured in detail)

Table 16.1 Civil Engineering Units of Measurement

Item no.	Description	Quantity	Unit	Rate	Amount
11	Remove existing fences of any height and description including wicket, field and gates and dispose or stack on site.	14	m		
12	Excavate for and construct PVC gulley with Class E concrete including fitting cover and frame Grade D on 2 courses Class B Engineering bricks	4	no.		

Figure 16.1 Bill of Quantities (unpriced)

of Tender Agreement and Bond for Use in Connection with Works of Civil Engineering Construction, published by the Institution of Civil Engineers.

16.4 Work Planning

When all the contract documents have been prepared and the work put out to tender, the successful contractor has to carry out the work. Their success, or failure, depends upon the ability of the organization to ensure that the work is done:

(*a*) at the right price – within that estimated, or the job will be done at a loss;

(*b*) to the required standard – as defined by the specification and drawings;

(*c*) within the time specified – if applicable. Some jobs have a contract period, with penalty clauses to be operative if the work is not finished on time.

To achieve all this involves work planning, programming, organization and control. These items are outlined in the following sections.

Before any roadwork operation begins, some form of work planning should take place, if the job is to be done efficiently, economically and to the right standard of quality. Obviously the scale of the planning operation will vary enormously from that required for, say, a £20 million motorway contract to a small trench reinstatement job.

Regardless of the size of the job, there are basic principles which always apply. The functions of *planning* and *controlling* a job are inseparable. No one can effectively control a job unless it has been planned.

The job must involve the following activities before work starts:

Planning
Programming
Organizing

and these activities continue whilst the work is being done, together with

Controlling.

16.4.1 *Planning*

Planning is defined as 'a scheme or arrangement to achieve a desired purpose'. Without planning, the job may eventually get done, some way and at some cost. Good planning will enable it to be done in the shortest time and/or in the most economical way.

It is worthwhile to consider some of the results of bad planning:

(*a*) A poor attitude to the work – 'Why bother, it will be changed anyhow.'

(*b*) Uneven workload – 'One day we are quiet, the next day we are rushed off our feet.'

(*c*) The issuing of conflicting instructions/information.

(*d*) Operatives and machines standing idle.

(*e*) Work done at a loss or uneconomically.

(*f*) The job running the workforce, rather than the workforce running the job.

The aim of planning is to ensure, as far as possible, that such defects are avoided.

16.4.2 *Programming*

A programme is 'a scheme or plan of things to be done with relevant details'. Hence, planning has been taken a stage further to become a programme of operations so that the work can be organized and controlled.

16.4.3 Organization

To organize is 'to arrange, coordinate and prepare for activity'. This is, perhaps, the most critical precondition for getting a job completed. The ability to organize effectively, with as little fuss as possible, is a priceless asset.

16.4.4 Control

To control is defined as 'to superintend and keep in order'. An activity that has not been properly planned and programmed will be difficult to control.

Control involves provision for coping with emergencies, producing contingency plans, etc., to cover such problems as:

(*a*) absence from work;
(*b*) plant breakdown;
(*c*) late delivery of materials;
(*d*) sudden changes in weather conditions, e.g. site flooding,

and ensuring that supervisors and foremen know what steps to take to overcome such problems.

16.5 Preparations for Starting Work

Firstly, it is necessary to plan ahead and decide:

(*a*) what is to be done
(*b*) what must be done
(*c*) what may not get done
(*d*) in what order jobs should be done
(*e*) what items need special attention
(*f*) how to do each job item
(*g*) who will do it
(*h*) time-tabling so that items of work can start and finish without upsetting other jobs on the list.

This initial check list will help to identify problems and enable resources to be:

of the right type,
of the right quantity,
available at the right time,
in the right place.

16.6 Resources of Production

It is helpful to work to a regular pattern. One system that can be applied is known as 'the five Ms' – representing the resources of production as shown in Figure 16.2.

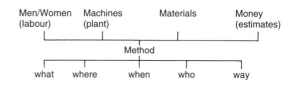

Figure 16.2 The five Ms

Men/Women

All jobs require labour to carry them out. This labour is called 'men' for convenience. Some of the factors to consider under this heading (the first 'M') are:

the skills required;
the number of men/women needed;
the output rate.

Machines

Most roadwork jobs require the use of some plant. With roadworks, operatives have to go to the site of the work, the job does not come to them, as is the case in factory production work.

Factors to consider are:

(*a*) types of plant and equipment needed;
(*b*) suitability of particular items of plant, or of the operator, for a particular job;
(*c*) output – the rate at which the plant can work.

In most roadwork jobs, it is the cost, suitability and availability of plant which are the key factors, for it is rare for plant to be utilized to maximum output (i.e. to be operating at full power all the time it is on site).

Materials

Almost every job has materials specified; these must be acquired in the right quantity and be in the right place at the right time.

Factors to be considered are:

(*a*) type
(*b*) quality
(*c*) ordering procedure – constraints, delays in delivery, etc.
(*d*) delivery, unloading, loading, storage, etc.

Money

This is a resource only in the sense that it influences and, in some cases, governs all the decisions on the three main resources.

Method

It is the fifth 'M', method, which produces an answer. It is not possible to determine effectively the labour, plant and materials needed until the method is established.

A methodical approach to the problem determines the allocation of the three main resources to the job, namely

men/women (labour)
machines (plant)
materials.

Many factors influence the choice of method of operation, which is determined on the basis of skill and experience and of knowledge of the work processes involved. In deciding upon method, the 'five Ws' may. be useful:

What	What has to be done?
	What clear objectives must be set?
Where	Where is the job located?
	Is there heavy traffic at the site?
	Is there far to travel from the depot?
When	When is it to be done?
	Is it a seasonal job (e.g. surface dressing)?
	Priority – how important is it in relation to other jobs.
Who	Who is to do it?
	What skills are available?
	Number and location of men?
Way	How is it to be done?

16.6.1 Allocation of Resources

Identification of the method helps to determine the number of men/women required, the skills that will be needed, and the time for which this number of operatives will be required, so that a target can be set for completion.

Basically, there are three methods of determining how to make the best use of available resources.

Evaluation by Experience

This requires considerable inherent skill and cannot be taught. It is simple and can be effective, but is only as good as the experience of the evaluator. Its major defect is the inability to monitor and adjust, should targets not be met.

Evaluation by Estimating or Calculation for Each Job

This can be very accurate, but it takes time and requires data (bonus values, outputs, rates of spread, etc.) on which to base the calculations. The disadvantage of this method is that, for small works, the time taken to work out a 'target time' may be out of proportion to the work involved. On larger jobs the method can be used effectively.

Use of a Computer to Extract Base Data

This method is very useful because it requires the planner only to input amounts of work to be done, based on an identified list of operations. As many roadwork activities are repetitive, and with the increasing availability of computers and suitable software, this method is coming into ever greater use.

Its main advantages include:

simplicity;
case of updating/adjusting data;
time saved, compared with other methods.

The major disadvantages are the effort, time and resources required to set up the system.

Essential information includes labour outputs, plant resources, materials and rates of spread. Tables 16.2 and 16.3 give examples of a resource estimate, and some resource data.

16.7   Work Programming

In programming, plans are set out graphically so that the work can be effectively organized and controlled by all those involved.

Methods used to programme work vary according to the scale and complexity of the job, from a simple check list to a network analysis. The ideal programme is one that contains all the necessary information and is easy to prepare and use. For small maintenance jobs a simple list may be adequate, showing headings controlling these jobs where resources are fairly constant (Figure 16.3(a)). Where it is necessary to build up a more complex work programme, a method statement, as shown in Figure 16.3(b), can be a very useful approach to analysis of a particular item.

Two examples of work programmes are shown in Figures 16.4 and 16.5. In these, Week 1 is a firm programme, Week 2 being provisional, and the programme is carried forward week by week.

16.7.1   Bar Charts

On larger works, it may be necessary to produce a bar (or Gantt) chart. This is a simple method of laying out the work schedule in block form, in relation to units of work, finance, labour or plant. A typical bar chart is shown in Figure 16.6. This allows planned, and actual, start and finish times/dates to be determined, and makes clear the dates when resources such as materials, plant, etc., must be on site – together with the quantities required – so that work can proceed according to plan.

The actual performance (i.e. measured work done) in each week can then be compared with the prediction and the work rescheduled, if necessary.

Item	Output per day	Labour cost/unit	Plant cost/unit	Material cost/unit	Total cost/unit
Site clearance					
Take up kerb and remove from site	160.00	0.70	0.60	—	1.30/lin m
Cut down and remove small tree (up to 300 mm girth)	18.00	6.00	4.50	—	10.50/no
Fencing					
Provide and erect chain link fence 1.2 m high	40.00	2.60	0.90	4.40	7.90/lin m
Drainage					
Excavate trench for 300 mm pipe 1.0 to 1.5 m deep using wheeled excavator	26.00	4.00	3.00	—	7.00/lin m
Provide and lay 300 mm ogee pipe	80.00	1.50	0.50	6.00	8.00/lin m
Backfill trench and dispose of surplus for 300 mm pipe trench 1.0 to 1.5 m deep	70.00	1.50	2.00	—	3.50/lin m
Excavate for manhole ring 1000 mm diameter 1.5 to 2.0 m deep	4.00	26.00	47.00	—	73.00/no
Construct manhole in 1000 mm rings (concrete) on concrete base and surround and metal cover	1.40	72.00	104.00	270.00	446.00/no
Construct *in situ* gully (concrete) including excavation, frame and reinstatement 450 × 450 × 840 mm deep	1.70	60.00	27.00	68.00	155.00/no
Sub-base and Roadbase					
Provide, lay and compact Type 1 sub-base by hand	28.00	3.60	2.40	20,00	26.00/m³
Provide, lay and compact DBM 100 mm thick using paver	2000	0.20	0.20	4.70	5.10/m²
Surfacing					
Provide lay and compact DBM surface course 30 mm thick using paver	2500	0.18	0.18	1.60	1.96/m²
Kerbs and Footways					
Excavate kerb trench by machine	135.00	0.40	0.60	—	1.00/lin m
Provide and lay PCC fig7 kerb* including concrete bed and surround	60.00	1.70	0.60	6.60	8.90/lin m
Provide and lay PCC paving slab 50 mm thick on sand bed	25.00	4.00	1.40	5.20	10.60/m²
Provide lay and compact DBM surface course by hand 15 mm thick	170.00	0.60	0.40	0.90	1.90/m²

Table 16.2 Resource estimate from a typical library of operations
N.B. Fig 7 kerb is now HB2.

	Rate of spread (m²/tonne)
Coated macadama*	
compacted thickness (mm)	
15	29–33
20	20–25
25	14–18
40	10–12
50	8–9
75 (open-graded)	7–8
75 (dense)	5–6
100	4–4.5
Dry stone, wet mix, Types 1 and 2	
compacted thickness (mm)	
75	7–8
100	5–6
Lean concrete	
compacted thickness (mm)	(m²/m³)
100	10
150	7
Kerbing concrete (bed and haunch)	(linear m/m³)
125 × 250 mm unit	14
ditto including 250 × 125 mm concrete channel	9
Concrete	(tonne/m³)
1:2:4 mix	2.3
	(kg/m³)
weight of cement	460
weight of sand	920
weight of stone	1840
	Loose fill value†
Aggregate	(tonne/m²)
Sand, hoggin Type 1	1.6
Chippings, 12–19 mm	1.8
Hardcore, concrete	1.23
Hardcore, brick	1.15
Clinker	0.8
Ash	0.96
	Rate of spread
Surface dressing	(m²/tonne)
14 mm nominal size	65–90
10 mm nominal size	90–110
6 mm nominal size	126–165
	Quantity
Brickwork	(bricks/m²)
1 brick thick	120
½ brick thick	60
Mortar	(kg/m²)
1 brick with frog	66
1 brick, no frog	55
½ brick with frog	29
½ brick, no frog	22

* BS 4987 gives a comprehensive list. The values given here represent approximate covering capacities for small areas. The higher value should be used for open-graded materials, the lower value for dense materials and HRA.
† Values assume bulking of up to 30%, depending upon material.

Table 16.3   Resource data for some roadwork materials

(a)

Order no. & job type	Date received	Date completed	Target man days	Gang members	Variation	Man days actual	Credit/debit

(b)

Item number	Description of work	Man days		Lorries and plant			Materials		Duration of work
		No.	D	Type	No.	D	Description	Qty	Days

Figure 16.3 (a) Job list for a simple job (b) Method statement

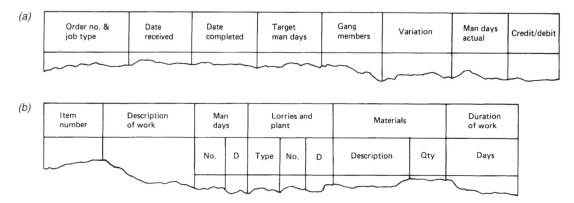

DIRECT LABOUR PROGRAMME — SMALL WORKS

.................................Area   14 days beginning................................

Area Supervisor ....

To be completed each week by Area Supervisor to cover activities of all gangs under their control.

1st week 'firm'   2nd week 'provisional'

To be prepared by   Thursday p.m. for Area Surveryor's agreement the following day.

When taking leave the Area Supervisor must ensure that a programme has been agreed to cover the period of his leave.

Gang	No. of men	Location and job description	Priority	Approx. duration	Materials and plant

....   Approved A.S. ....

Figure 16.4 Work programme

Date ....
(week commencing)

Works

Work planning sheet   Area .... Foreman .... Supt ....

Job no.	Location and work	Man days	Gang days	Days of week	Gang	Resources				Date finished	Time taken	Credit hours	Debit hours
						Materials	Qty	Plant	Time				
Week No. 1 Programme Copy to Client				→									
Week No. 2													

Figure 16.5 Work programme

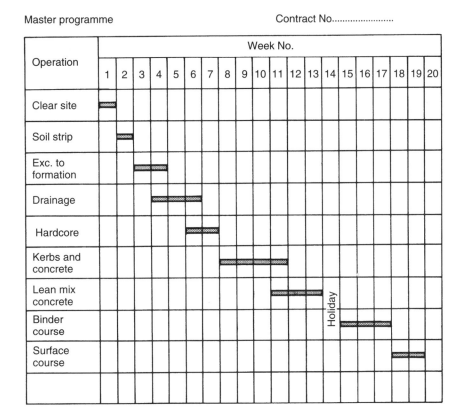

Figure 16.6 Bar, or Gantt, chart

16.7.2 *Programming Example*

Table 16.4 shows a Bill of Quantities for the minor improvement given in outline sketch form in Figures 16.7 and 16.8.

The Bill of Quantities has been priced, to give the reader an idea of how this document is completed. It has also been broken down into a method statement according to labour, plant and materials (Table 16.5). The labour values are based on a two-man gang, with lorry and digger operators in addition for the periods when their plant was on site.

The programme for a small job such as this could be a check list or, perhaps a bar chart. As the work is very sequential, a check list would suffice, as follows:

Operations	*General description*
1, 2, 3, 4, 5	Site clearance
6, 7, 8*, 9, 10	Drainage
12, 13, 14, 15	Kerb and edging
16, 17, 18	Footway construction
8*	Complete gullies
19, 20	Carriageway surfacing
11	Verges

N.B.: This check list should be read in conjunction with the method statement (Table 16.5).

16.8 Examples of Diagrams

For those who are not directly concerned with the planning and progressing of construction work, it may be useful to follow a simple example which shows how bar charts and other diagrammatic representations can be used.

Consider the job of excavating a trench in a grass verge, in which a 225 mm drainage pipe is to be laid for a distance of 800 m, with connections to ten new gullies.

Firstly, the turf has to be stripped and laid aside; then the trench must be excavated, granular pipe support placed, and the pipes laid. This is followed by backfilling, compaction and, finally, replacement of the turf.

This sequence can be shown as a bar chart, as in Figure 16.9. Alternatively, many of these items can be plotted on a graph which indicates whether or not the work is progressing satisfactorily, i.e. is being done within the planned time. For example, take the excavation of the trench: Figure 16.10 shows that this operation is consistently behind schedule, finishing one week later than planned.

From Figure 16.11, the graph of the pipe laying, it can be seen that, although the laying started later than planned, good progress was made, resulting in the work

Item No.	Description	Quant-ity	Unit	Rate	Amount
1	Clear all rubbish, uproot bushes, undergrowth or small trees not exceeding 0.6 m girth and tree stumps not exceeding 0.3 m diameter and dispose.	12	m²	2.10	25.20
2	Remove existing fences of any height and description including wicket, field and gates and dispose or stack on site.	14	m	1.10	15.40
3	Fell trees and dispose. Girth exceeding 0.6 m, not exceeding 1.00 m.	3	No.	7.20	21.60
4	E.O. for grubbing stumps and roots. Fill holes with suitable material.	3	No.	13.40	40.20
5	Trim hedge by hand, height 1.5 m to 2.2 m. Collect and burn trimmings.	40	m	1.10	44.00
6	Remove accumulated spoil and weed from a ditch by machine. Provide and lay 300 mm porous concrete pipes on ditch bottom and surround with rejects. Back fill with imported granular material. Ave. depth 1.4 m to invert.	65	m	9.20	598.00
7	E.O. for pipe junction.	4	No.	11.50	46.00
8	Excavate for and construct PVC gulley with Class E concrete inc. fitting cover and frame. Grade D on 2 courses Class B Eng. bricks.	4	No.	120.70	482.80
9	Excavate for and provide and lay 150 mm O.G. concrete pipes to gulley connections.	8	m	16.80	134.40
10	Excavate and build C/pit from 900 mm diameter concrete rings average depth 1.4 m. Fitted with Grade A cover and frame on 1 to 3 courses Class 8 Eng. bricks.	3	No.	310.00	930.00
11	Provide and lay topsoil average 100 mm thick and seed.	150	m²	1.30	195.00
12	Excavate in any material kerb trench alongside carriageway.	160	m	1.40	224.00
13	Lay, back and bed straight concrete kerbs in new concrete.	120	m	6.70	804.00
14	Ditto to radius.	40	m	7.80	312.00
15	Lay footway edging 50 mm × 150 mm including bed and backing in new concrete.	100	m	3.30	330.00

Item No.	Description	Quant-ity	Unit	Rate	Amount
16	Lay and compact 100 mm Type 2 granular material to footway sub-base. Apply weedkiller.	200	m²	3.20	640.00
17	Lay and compact 20 mm binder course to a depth of 40 mm.	200	m²	2.90	580.00
18	Lay and compact 10 mm surface course to a depth of 20 mm.	200	m²	1.90	380.00
19	Lay and compact single course bit. mac.	50	tonnes	30.50	1525.00
20	Apply sealing grit to carriageway surfacing.	450	m²	0.20	90.00
				TOTAL	£7417.60

Table 16.4   Bill of quantities (priced)

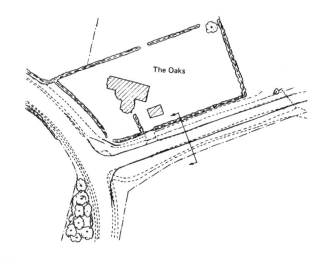

Figure 16.7   Plan of minor road improvement adjacent to 'The Oaks' – not to scale

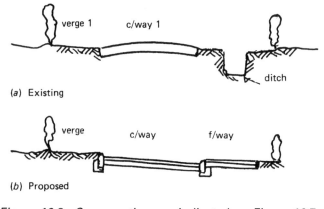

Figure 16.8   Cross-sections as indicated on Figure 16.7

Item No.	Labour		Plant		Materials
	No.	Duration (days)	No. and type	Duration (days)	
1	2	0.15	1 Lorry 1 Digger	0.15	
2	2	0.07	1 Digger	0.07	
3	2	0.14	1 Digger	0.14	
4	2	0.28	1 Digger	0.28	
5	2	0.34	1 Lorry	0.34	
6	2	0.8	1 Lorry 1 Digger 1 Roller	0.8	12 tonnes gravel rejects 85 tonnes hoggin 67 m 300 mm porous concrete pipes
7	2	0.07	1 Lorry	0.07	4 no. junctions
8	2	1.7	1 Lorry	1.7	4 no. PVC gullies 150 bricks 4 grids & frames 2.4 m³ concrete
9	2	0.3	1 Lorry 1 Digger	0.3	8 m 150 mm pipes
10	2	2.0	1 Digger	2.0	3 grids & frames 2.1 m³ concrete 3 cover slates 9 no. 300 × 900 mm rings 180 bricks

Item No.	Labour		Plant		Materials
	No.	Duration (days)	No. and type	Duration (days)	
11	2	0.94	1 Digger	0.94	(sweepings, etc.) 19 kg sand
12	2	0.17	1 Digger	0.17	
13	2	2.6	1 Lorry	2.6	124 m kerbs 9 m³ concrete
14	2	1.0	1 Lorry	1.0	42 m kerbs (radius) 3 m³ concrete
15	2	1.11	1 Lorry	1.11	105 m path edging 3.5 m³ concrete
16	2	0.5	1 Lorry 1 Digger 1 Roller	0.5	30 tonne Type 2 5 litres weed killer
17	2	0.63	1 Lorry 1 Roller 1 Digger	0.63	19 tonne binder course
18	2	0.37	1 Lorry 1 Roller	0.37	9 tonne surface course
19	2	1.87	1 Roller	1.87	50 tonne single course bit. mac.
20	2	0.01	1 Lorry	0.01	4.5 tonne grit
Summary:		15.05	Lorry Digger Roller	15.05 5.98 4.17	

Table 16.5   Method Statement relating to the Bill of Quantities shown as Table 16.4

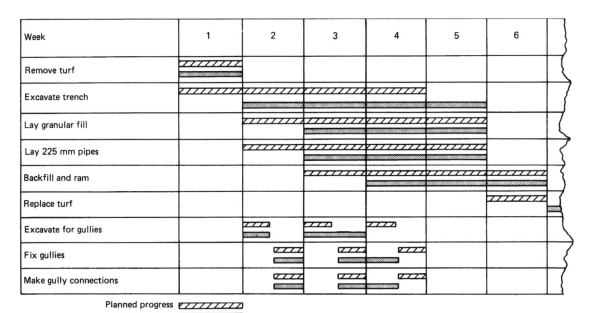

Figure 16.9   Bar chart example

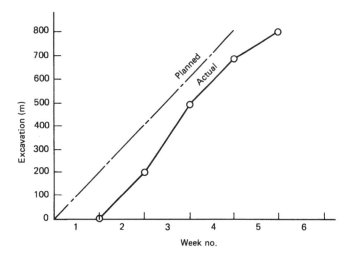

Figure 16.10   Trench length excavated per week

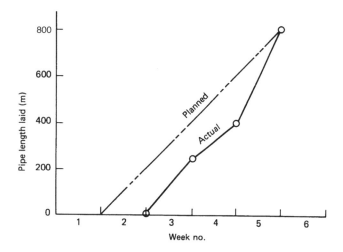

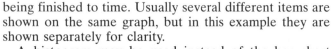

Figure 16.11   Lengths of pipe laid per week

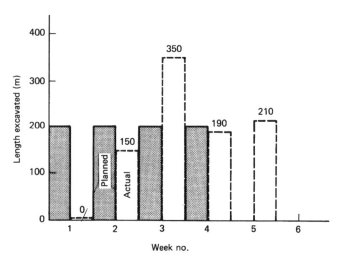

Figure 16.12   Histogram of trench excavation per week

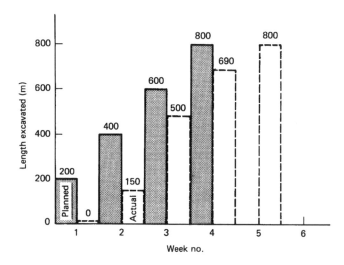

Figure 16.13   Trench excavation per week (cumulative)

being finished to time. Usually several different items are shown on the same graph, but in this example they are shown separately for clarity.

A histogram may be used instead of the bar chart of the graph. Again, the pipe laying is taken as the item to be shown; the information given in the graph is presented in a different form (Figure 16.12). Here each week's output is shown independently, but output could be shown on a cumulative basis, as in Figure 16.13.

It should be noted, however, that these forms of job control are unlikely to be employed on a job as small as that which is considered here.

Revision Questions

1   What are the three purposes of a specification in a contract document?
2   When planning a works programme what type of chart would facilitate planning or organization?
3   Detail out a simple method statement for a morning's operations on a simple site task?
4   Devise a simple master programme for the completion of a whole project within a 3 minute brainstorming session?

Note: See page 305 for answers to these revision questions

Answers to revision questions

Chapter 1
1 Refer to 1.1 (h) Para 4 Line 1
2 Refer to 1.4.3 Para 3 Line 1
3 Refer to 1.10 Para 3 Line 1
4 Refer to 1.11.1 Para 2 Line 1

Chapter 2
1 Refer to 2.1.6 Para 1
2 Refer to 2.2.2 Para 1
3 Refer to 2.4.5 Para 2 Line 1
4 Refer to 2.4.8 Para 2

Chapter 3
1 Refer to 3.2.5 (Springs) Line 1
2 Refer to 3.6 Para 1 Line 3
3 Refer to 3.11 Para 2 Line 1
4 Refer to 3.16.1 Para 1

Chapter 4
1 Refer to 4.8.3 Para 2 Line 4
2 Refer to 4.10.3
3 Refer to 4.17.2 Para 2
4 Refer to 4.20.1

Chapter 5
1 Refer to 5.2 Para 3(1), (2), (3) & (4)
2 Refer to 5.9 Table 5.5
3 Refer to 5.11.2 Para 3 Line 1
4 Refer to 5.21.3 Table 5.9

Chapter 6
1 Refer to 6.4 Para 4
2 Refer to 6.16.2 Para 1 Line 2
3 Refer to 6.18
4 Refer to 6.24

Chapter 7
1 Refer to 7.3.3 Para 1 Line 7
2 Refer to 7.6.3 Para 3 Line 6
3 Refer to 7.7.1 (whole bed method)
　　　　　Para 1 Line 1
4 Refer to 7.8.6 Para 3 Line 6

Chapter 8
1 Refer to 8.3 Para 6
2 Refer to 8.5.1 Line 2
3 Refer to 8.10.2 Figure 8.35
4 Refer to 8.11 Para 2 Line 3

Chapter 9
1 Refer to 9.3.1 ((d) Noise) para 2
2 Refer to 9.4.3 Para 2 Line 1
3 Refer to 9.5.4 Table 9.2
4 Refer to 9.5.9

Chapter 10
1 Refer to 10.8.2
2 Refer to 10.23.2
3 Refer to 10.32
4 Refer to 10.35

Chapter 11
1 Refer to 11.1 Para 2 Line 3
2 Refer to 11.7 Para 2 Line 1
3 Refer to 11.10.3 Para 1 Line 1
4 Refer to 11.15 Para 3 Line 1

Chapter 12
1 Refer to 12.1.2 (Placement)
2 Refer to 12.1.2 (Mounting Height)
3 Refer to 12.2.5 Para 2
4 Refer to 12.6.1 Para 6 (c)

Chapter 13
1 Refer to 13.5 Para 2 Line 2
2 Refer to 13.9 Para 2 Line 1
3 Refer to 13.9.1 (Mastic Asphalt)
4 Refer to 13.15 Para 2 Line 1

Chapter 14
1 Refer to 14.1 Line 18
2 Refer to 14.5.5 Line 4
3 Refer to 14.6.3 (2 Bleeding/sweating/
　　　　　fatting – up)
4 Refer to 14.14 Para 1

Chapter 15
1 Refer to 15.5.3 Para 2 Line 1
2 Refer to 15.7.5 Table 15.5
3 Refer to 15.7.7
4 Refer to Table 15.6 (Backfill materials)

Chapter 16
1 Refer to 16.2.2 Para 2
2 Refer to 16.7.1
3 Refer to Figure 16.3(a) & 16.3(b)
4 Refer to Figure 16.6

- Highways Agency (1990) 'Design Manual for Roads and Bridges'
- DETR (1998) 'Places, Streets and Movement'
- The Traffic Signs Manual – Chapter 8 'Traffic Safety Measures and Signs for Road Works and Temporary Situations. (ISBN 0-11-550937 2)
- DB32 1977 & 1992
- Safety at Street Works and Road Works. A Code Of Practice. (ISBN 0-11-551144-X)
- Construction (Design and Management) (CDM) Regulations, 1994
- Approved Code of Practice (ACoP) 2001 and guidance on the CDM Regulations 1994 'Managing Health and Safety in Construction' came into force on 1 February 2002
- Construction (Health, Safety and Welfare) (CHSW) Regulations, 1996
- Delivering Best Value in Highway Maintenance – July 2001 – LAA
- RoSPA Road Safety Engineering Manual – 3rd Edition 2002
- Transport In The Urban Environment – IHT
- CDM Regulations – work sector guidance for designers (CIRIA publications)
- CDM Regulations – practical guidance for clients and clients' agents
- CDM Regulations – practical guidance for planning supervisors

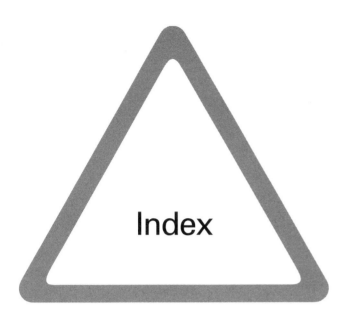

Index